Springer
Berlin
Heidelberg
New York
Barcelona
Hong Kong
London
Milan
Paris
Tokyo

Physics and Astronomy   ONLINE LIBRARY

http://www.springer.de/phys/

ADVANCES IN MATERIALS RESEARCH

Series Editor-in-Chief: Y. Kawazoe

Series Editors: M. Hasegawa   A. Inoue   N. Kobayashi   T. Sakurai   L. Wille

The series Advances in Materials Research reports in a systematic and comprehensive way on the latest progress in basic materials sciences. It contains both theoretically and experimentally oriented texts written by leading experts in the field. Advances in Materials Research is a continuation of the series Research Institute of Tohoku University (RITU).

1   **Mesoscopic Dynamics of Fracture**
    Computational Materials Design
    Editors: H. Kitagawa, T. Aihara, Jr., and Y. Kawazoe

2   **Advances in Scanning Probe Microscopy**
    Editors: T. Sakurai and Y. Watanabe

3   **Amorphous and Nanocrystalline Materials**
    Preparation, Properties, and Applications
    Editors: A. Inoue and K. Hashimoto

4   **Materials Science in Static High Magnetic Fields**
    Editors: K. Watanabe and M. Motokawa

Series homepage – http://www.springer.de/phys/books/amr/

K. Watanabe   M. Motokawa   (Eds.)

Materials Science in Static High Magnetic Fields

With 206 Figures

 Springer

Professor Kazuo Watanabe
Professor Mitsuhiro Motokawa

High Field Laboratory for Superconducting Materials
Institute for Materials Research, Tohoku University
2-1-1 Katahira, Aoba-ku, Sendai 980-8577, Japan
e–mail: kwata@imr.tohoku.ac.jp, motokawa@imr.tohoku.ac.jp

Series Editor-in-Chief:

Professor Yoshiyuki Kawazoe

Institute for Materials Research, Tohoku University
2-1-1 Katahira, Aoba-ku, Sendai 980-8577, Japan

Series Editors:

Professor Masayuki Hasegawa
Professor Akihisa Inoue
Professor Norio Kobayashi
Professor Toshio Sakurai

Institute for Materials Research, Tohoku University
2-1-1 Katahira, Aoba-ku, Sendai 980-8577, Japan

Professor Luc Wille

Department of Physics, Florida Atlantic University
777 Glades Road, Boca Raton, FL 33431, USA

Library of Congress Cataloging-in-Publication Data: Materials science in static high magnetic fields/
K. Watanabe, M. Motokawa (eds.) p. cm. – (Advances in materials research, ISSN 1435-1889) Includes
bibliographical references and index. ISBN 3540419950 (alk. paper) 1. Superconductors. 2. Materials–Magnetic
properties. 3. Magnetic fields. I. Watanabe, K. (Kazuo) II. Motokawa, M. (Mitsuhiro) III. Series. QC611.95.M37
2001 620.1'12973–dc21 2001049589

ISSN 1435-1889
ISBN 3-540-41995-0 Springer-Verlag Berlin Heidelberg New York

Springer-Verlag Berlin Heidelberg New York
a member of BertelsmannSpringer Science+Business Media GmbH

http://www.springer.de

© Springer-Verlag Berlin Heidelberg 2002
Printed in Germany

The use of general descriptive names, registered names, trademarks, etc. in this publication does not imply,
even in the absence of a specific statement, that such names are exempt from the relevant protective laws and
regulations and therefore free for general use.

Typesetting: Data conversion by LE-TeX, Leipzig
Cover concept: eStudio Calamar Steinen
Cover design: *design & production*, Heidelberg

Printed on acid-free paper   SPIN: 10836788   57/3141/ba - 5 4 3 2 1 0

Preface

There is no strict definition of the term "High Magnetic Field". It has been proposed to use this term for magnetic fields that are technically difficult to generate and therefore need special equipment or large resources. Static fields above 20 T are apparently high magnetic fields in this sense, but in the case of pulsed field 40 T is easy to obtain and any field lower than this approximate limit should not be considered as "high". When a static field is used for materials processing, even 10 T is considered as "high" because the long-term use of a conventional superconducting magnet is difficult.

Recently, there has been much technical progress in producing high magnetic fields, both pulsed and static; in large part this is due to the development of new materials. Complicated poly-helix coils are now replaced by simple Bitter coils made with plates of CuAg alloy with high strength and high conductivity; these are used in both water-cooled and hybrid magnets (now up to 45 T at NHMFL, the US National High Magnetic Field Laboratory at Tallahassee, Florida). By using CuAg wire, a nondestructive pulsed field record of 80 T has been achieved at Osaka University. For daily use in experiments, 70–75 T should soon become available. Major facilities for static high fields worldwide are planning to generate fields over 40 T by increasing the electric power.

On the other hand, the use of static high magnetic fields is expanding. Formerly these were used mainly in physics, i.e. for measuring physical phenomena, and for the characterization of materials. Application to so-called materials processing like crystal growth or heat treatment is more recent. Development of cryogen-free superconducting magnets surely contributes much to the expansion of high field applications.

The aim of this book is to survey the activities of high field applications, mainly those that are pursued at the High Field Laboratory for Superconducting Materials (HFLSM) that is attached to the Institute for Materials Research (IMR) at Tohoku University. Each chapter is composed of a general introduction to the subject and of detailed experimental results with explanations. Many books have been published concerning high magnetic fields and related subjects, but it is still considered worthwhile to publish this book as a volume of the series "Advances in Materials Research", in order to survey the current status of materials science in static high magnetic fields.

The editors are grateful to Dr. Satoshi Awaji for his help in compiling this book in its present form. They very much appreciate Prof. Yoshiyuki Kawazoe, the editor of this book series, for the original suggestion and encouragement to publish this book.

Sendai   *Mitsuhiro Motokawa*
August 2001   *Kazuo Watanabe*

Contents

Part V   Other High Field Physical Properties

List of Contributors

E. Abe
Tohoku University/Research
Institute of Electrical
Communication
2–1–1 Katahira, Aoba–ku
Sendai 980–8577, Japan

N. Akiba
Tohoku University/Research
Institute of Electrical
Communication
2–1–1 Katahira, Aoba–ku
Sendai 980–8577, Japan

J. R. Anderson
University of Maryland/Dept.
Physics
College Park
MD 20742-4111, USA

S. Awaji
Tohoku University/Institute for
Materials Research/High Field
Laboratory for Superconducting
Materials
2–1–1 Katahira, Aoba–ku
Sendai 980–8577, Japan

W.P. Chen
Tohoku University/Institute for
Materials Research
2–1–1 Katahira, Aoba–ku
Sendai 980–8577, Japan
and
CREST/Japan Science and
Technology Corporation
Sengen, Tsukuba, 305–0047, Japan

D. Chiba
Tohoku University/Research
Institute of Electrical
Communication
2–1–1 Katahira, Aoba–ku
Sendai 980–8577, Japan

T. Dietl
Tohoku University/Research
Institute of Electrical
Communication
2–1–1 Katahira, Aoba–ku
Sendai 980–8577, Japan

S.D. Durbin
Tohoku University/Institute for
Materials Research
2–1–1 Katahira, Aoba–ku
Sendai 980–8577, Japan

T. Fukase
Tohoku University/Institute for
Materials Research
2–1–1 Katahira, Aoba–ku
Sendai 980–8577, Japan

W. Giriat
Centro de Fisica/IVIC
Apartado 21827
Caracas, Venezuela

T. Goto
Sophia University/Faculty of Science
and Technology
7–1 Kioi–cho, Chiyoda–ku
Tokyo 102–8554, Japan

H. Hashidume
Tohoku University/Research
Institute of Electrical
Communication
2–1–1 Katahira, Aoba–ku
Sendai 980–8577, Japan

F. Herlach
Katholieke Universiteit Leuven/
Dept. Natuurkunde–LVSM Celestij-
nenlaan 200 D
B–3001 Heverlee, Belgium

T. Hirai
Tohoku University/Institute for
Materials Research
2–1–1 Katahira, Aoba–ku
Sendai 980–8577, Japan

Y. Iijima
National Institute for Materials
Science
1–2–1 Sengen
Tsukuba 305–0047, Japan

K. Inoue
National Institute for Materials
Science
1–2–1 Sengen
Tsukuba 305–0047, Japan

G. Iwaki
Hitachi Cable Ltd.
3550 Kidamari, Tsuchiura
Ibaraki 300–0026, Japan

R. Jahana
Gakushuin University/Faculty of
Science
1–5–1 Mejiro, Toshima–ku
Tokyo 171–8588, Japan

K. Katagiri
Iwate University/Faculty of
Engineering
4–3–5 Ueda, Morioka 020–8551,
Japan

H. Kato
Tohoku University/School of
Engineering
Aramaki Aoba, Aoba–ku
Sendai 980–8579, Japan

T. Kato
Tokai University/School
of Engineering
1117 Kita Kaname, Hiratsuka
259–1292, Japan

S. Kawaji
Gakushuin University/Faculty of
Science
1–5–1 Mejiro, Toshima–ku
Tokyo 171–8588, Japan

G. Kido
National Research Institute for
Metals
Tsukuba 593, Japan

A. Kikuchi
National Institute for Materials
Science
1–2–1 Sengen
Tsukuba 305–0047, Japan

N. Kobayashi
Tohoku University/Institute for
Materials Research
2–1–1 Katahira, Aoba–ku
Sendai 980–8577, Japan

H. Komatsu
Tohoku University/Institute for
Materials Research
2–1–1 Katahira, Aoba–ku
Sendai 980–8577, Japan

T. Komatsubara
Tohoku University/Center for Low
Temperature Science
Aramaki Aoba, Aoba–ku
Sendai 980–8578, Japan

K. Kumagai
Hokkaido University/Graduate
School of Science
Nishi 8 Kita 10 Jyo, Kita–ku
Sapporo 060–0810, Japan

N. Kuroda
Kumamoto University/Dept.
Mechanical Engineering and
Materials Science
2–39–1 Kurokami
Kumamoto 860–8555, Japan

H. Maeda
Kitami Institute of Technology
165 Koencho Kitami
090–8507, Japan

K. Maki
Hokkaido University/Graduate
School of Science
Nishi 8 Kita 10 Jyo, Kita–ku
Sapporo 060–0810, Japan

Y. H. Matsuda
University of Tokyo/Institute for
Solid State Physics
Kashiwanoha, kashiwa–shi
Chiba 277–8581, Japan

F. Matsukura
Tohoku University/Research
Institute of Electrical
Communication
2–1–1 Katahira, Aoba–ku
Sendai 980–8577, Japan

H. Matsumoto
Tokai University/School
of Engineering
1117 Kita Kaname, Hiratsuka
259–1292, Japan

S. Miyashita
Tohoku University/Institute for
Materials Research
2–1–1 Katahira, Aoba–ku
Sendai 980–8577, Japan

T. Miyazaki
Tohoku University/School of
Engineering
Aramaki Aoba, Aoba–ku
Sendai 980–8579, Japan

I. Mogi
Tohoku University/Institute for
Materials Research
2–1–1 Katahira, Aoba–ku
Sendai 980–8577, Japan

H. Moriai
Hitachi Cable Ltd.
3550 Kidamari, Tsuchiura
Ibaraki 300–0026, Japan

M. Motokawa
Tohoku University/Institute for
Materials Research/High Field
Laboratory for Superconducting
Materials
2–1–1 Katahira, Aoba–ku
Sendai 980–8577, Japan
email:motokawa@imr.edu

K. Murata
Osaka City University/Gradiate
School of Science
3–3–138 Sugimoto, Sumiyoshi-ku
Osaka 558–8585, Japan

T. Nakada
Tohoku University/Institute for
Materials Research
2–1–1 Katahira, Aoba–ku
Sendai 980–8577, Japan

N. Nakagawa
Hitachi Cable Ltd.
3550 Kidamari, Tsuchiura
Ibaraki 300–0026, Japan

K. Nakajima
Tohoku University/Institute for
Materials Research
2–1–1 Katahira, Aoba–ku
Sendai 980–8577, Japan

S. Nakajima
Tohoku University/Institute for
Materials Research
2–1–1 Katahira, Aoba–ku
Sendai 980–8577, Japan

T. Nishizaki
Tohoku University/Institute for
Materials Research
2–1–1 Katahira, Aoba–ku
Sendai 980–8577, Japan

K. Noto
Iwate University/Faculty of
Engineering
4–3–5 Ueda, Morioka 020–8551,
Japan

H. Ohno
Tohoku University/Research
Institute of Electrical
Communication
2–1–1 Katahira, Aoba–ku
Sendai 980–8577, Japan

Y. Ohno
Tohoku University/Research
Institute of Electrical
Communication
2–1–1 Katahira, Aoba–ku
Sendai 980–8577, Japan

T. Okamoto
Gakushuin University/Faculty of
Science
1–5–1 Mejiro, Toshima–ku
Tokyo 171–8588, Japan

T. Omiya
Tohoku University/Research
Institute of Electrical
Communication
2–1–1 Katahira, Aoba–ku
Sendai 980–8577, Japan

T. Sakon
Tohoku University/Institute for
Materials Research
2–1–1 Katahira, Aoba–ku
Sendai 980–8577, Japan

T. Sasaki
Tohoku University/Institute for
Materials Research
2–1–1 Katahira, Aoba–ku
Sendai 980–8577, Japan

G. Sazaki
Tohoku University/Institute for
Materials Research
2–1–1 Katahira, Aoba–ku
Sendai 980–8577, Japan
and
Tohoku University/Center for
Interdisciplinary Research
Aoba Aramaki, Aoba–ku
Sendai 980–8578, Japan

Y. Syono
Tohoku University/Institute for
Materials Research
2–1–1 Katahira, Aoba–ku
Sendai 980–8577, Japan

K. Tachikawa
Tokai University/School
of Engineering
1117 Kita Kaname, Hiratsuka
259–1292, Japan

K. Takamura
Tohoku University/Research
Institute of Electrical
Communication
2–1–1 Katahira, Aoba–ku
Sendai 980–8577, Japan

T. Takeuchi
National Institute for Materials
Science
1–2–1 Sengen
Tsukuba 305–0047, Japan

N. Toyota
Tohoku University/Faculty of
Science
Aoba Aramaki, Aoba–ku
Sendai 980–8578, Japan
and
Tohoku University/Interdisciplinary
Research Center
Aramaki, Aoba-ku
Sendai 980–8578, Japan

S. Tsuji
Hokkaido University/Graduate
School of Science
Nishi 8 Kita 10 Jyo, Kita–ku
Sapporo 060–0810, Japan

T. Ujihara
Tohoku University/Institute for
Materials Research
2–1–1 Katahira, Aoba–ku
Sendai 980–8577, Japan

S. Yanagiya
Tohoku University/Institute for
Materials Research
2–1–1 Katahira, Aoba–ku
Sendai 980–8577, Japan

H. Yoshino
Osaka City University/Graduate
School of Science
3–3–138 Sugimoto, Sumiyoshi-ku
Osaka 558–8585, Japan

K. Watanabe
Tohoku University/Institute for
Materials Research/High Field
Laboratory for Superconducting
Materials
2–1–1 Katahira, Aoba–ku
Sendai 980–8577, Japan
email:kwata@imr.edu

Part I

General Review
of Static High Magnetic Fields

1 Static High Magnetic Fields and Materials Science

M. Motokawa, K. Watanabe and F. Herlach

Like temperature or pressure, the magnetic field is one of the important thermodynamic parameters that are used to change the inner energies of materials. Materials are essentially composed of atomic nuclei and electrons, and the properties of a material are mainly characterized by the behavior of electrons. An electron has charge and spin, and any material responds to an applied magnetic field in the first place via the charge and spin of the electrons. Materials can be subdivided into classes according to their reaction to an external stimulus like a magnetic or electric field. For example, regarding the electrical properties, materials are classified into insulators, semiconductors, metals and superconductors; the materials containing free charge carriers respond to a magnetic field due to the Landau energy. From these responses, a wealth of important information on the electronic structure is obtained, among others by means of magneto-transport experiments (including the quantum Hall effect), cyclotron resonance and the de Haas-van Alphen effect. Magnetic materials contain magnetic moments due to electron orbits or spins that can be oriented by an applied magnetic field. In a paramagnet, the magnetic moments are disordered in zero magnetic field; in ferro-, ferri- and antiferromagnets the magnetic moments are aligned by quantum-mechanical interactions. In these materials, a variety of phase transitions can be induced by the applied magnetic field. Magnetization measurements and magnetic resonance experiments are the conventional techniques for studying the magnetic properties. In a diamagnetic material, magnetic moments in an atom or a molecule compensate each other such that there is no net magnetic moment. Even these materials respond to a magnetic field via Larmor diamagnetism, which is due to the influence of the magnetic field on the orbital motion of electrons. Of course, all these effects become stronger with increasing applied field. The interaction energy between a material and the magnetic field is quite small compared to the energy density of the field. For example, a field of $10\,\mathrm{T}$ corresponds to $4\times10^7\,\mathrm{J/m^3}$, and the force acting on the magnet is $4\,\mathrm{kg/mm^2}$. On the other hand, the interaction energy at $10\,\mathrm{T}$, i.e. the cyclotron energy of a free electron or the Zeeman energy of a free spin, is only about 2×10^{-22} J, corresponding to a wavenumber of $10\,\mathrm{cm^{-1}}$. This is the reason why very high magnetic fields are needed for this research.

Many efforts to generate higher magnetic fields have been made since the development of the electromagnet by Ampére, Sturgeon, Faraday and Ruhmkorff in the first half of the 19th century, and the discovery of superconductivity by Kamerlingh Onnes at the beginning of the 20th century. The field strength was limited to a few teslas at most by the saturation of the iron used in these magnets; the magnetization of the iron was needed because at the time no electrical current could be generated that was strong enough to generate a very high field. The early superconductors could not be used, because in these superconductivity is destroyed by a relatively weak magnetic field. The first to overcome these limitations was Kapitza at the Cavendish Laboratory, who generated a very high current in a pulsed mode; in 1924 he obtained almost 50 T in a 1 mm bore with a lead acid battery, and in 1927 35 T with 10 ms pulse duration with his famous flywheel-alternator [1]. He used these fields to do many pioneering experiments. Later on, the pulsed field method [2] was developed further to obtain fields up to 80 T, and with extreme methods that allow for the destruction of the coil, fields up to 1000 T were obtained that could be used in experiments; the absolute record is now at 2800 T. A static magnetic field is more difficult to produce than a moderate pulsed field of order 45 T; the pioneering work was done by Bitter at MIT, who succeeded in generating up to 10 T in 1939 [3]. His ingenious design with a stack of "Bitter plates" is still used today for generating the highest static fields up to 33 T. The idea of a "hybrid magnet" was proposed by Wood and Montgomery in 1966 [4]; the first of these magnets was realized at the Clarendon Laboratory. During the period from late 1970 to the beginning of the 1980s, 30 T class hybrid magnets were constructed at MIT, at the University of Nijmegen and at the Grenoble High Magnetic Laboratory, and in 2000 the 45 T hybrid became operational at the NHMFL at Tallahassee. The most recent survey on magnet technology can be found in the proceedings of the RHMF-2000 symposium [5], where all the major magnet laboratories presented their facilities in a dedicated session.

The Institute for Materials Research (IMR) at Tohoku University has a long history of the development of high magnetic field equipment. Professor Honda, who founded this institute in 1916, realized the importance of high magnetic fields for research on magnetic materials. In 1935, a Kapitza-type magnet was introduced first in Japan; a peak field of 30 T was obtained and this was used for magnetoresistance measurements on Bi. Unfortunately, however, it was destroyed by bombardment during World War II. After the war, a Bitter-type water-cooled magnet that generated fields up to 12.5 T was installed in 1955. One of the most interesting studies performed with this magnet was the discovery of two-step metamagnetic transitions in $CoCl_2 \cdot 2H_2O$ by Kobayashi and Haseda [6]. This led to a new concept proposed by Date and Motokawa [7] that the soliton-like local spin excitation called "spin-cluster excitation" was possible in an Ising spin system, besides the spin wave excitation that was known at that time as the unique elementary excitation of magnetic spin systems. This concept was confirmed later by Torrance and Tinkham using high-field far-infrared spectroscopy [8].

The new high field laboratory attached to IMR was inaugurated in 1981; Bitter magnets and a hybrid magnet that in 1985 produced 31.1 T were built. The original main purpose of this facility was the development of superconducting wires for use in a nuclear fusion furnace, but later the magnets were used for many other studies as they are pursued at other facilities. The most striking invention made in this laboratory was the development of cryogen-free superconducting magnets [9].

The use of high magnetic fields is expanding. High magnetic fields were used mainly for measurements in condensed matter physics and later on for the characterization of materials, for instance, measuring the critical field H_c of superconductors. The experiments using high magnetic fields, both pulsed and static, have been limited for a long time to the investigation of the electric and magnetic properties of materials. Quite recently, the research was extended to include diamagnetic materials. Since the effects of the magnetic field on diamagnetic materials are very small compared to those observed in electric conductors or magnetic materials, only a few experiments had been done. In the case of diamagnetic materials, high magnetic fields are used for the processing of materials rather than for the measurement of physical quantities. The orientation effect is most popular for this purpose and magnetic levitation is a new technology to control material synthesis. It is well known that the morphology of a crystal can be strongly influenced by growth in high magnetic fields. The magnetic force acting on a diamagnetic material is repulsive and if it is stronger than gravity, the material can be levitated. This effect provides the opportunity for processing materials under conditions similar to those in a space laboratory. Of course, these applications can be performed only in static magnetic fields, never in pulsed fields.

Recently, it has become much easier to generate high magnetic fields above 10 T due to the development of superconducting magnets and in particular cryogen-free superconducting magnets with fields up to 15 T; these can be used for a long term without the labor of filling with liquid helium. The cryogen-free superconducting magnet has opened a new field in science, especially in material-processing technology. This has inspired us to review materials science in static high fields with a focus on high-temperature superconductors, and magnetic, optical and other electric properties in high fields. An update of research subjects for advanced materials science in static high fields will be given, following this introductory chapter.

1.1 Static High Magnetic Field

The production of a high magnetic field is a technical challenge as much as the generation of ultra-low temperature and ultra-high pressure. The technical problems inherent in the generation of static and pulsed magnetic fields are of a different nature. A pulsed magnet works with a limited amount of energy (typically in the range 0.1–1 MJ) that has to be supplied during the short pulse duration; this requires high power that can easily be provided by an energy storage device such as a capacitor bank. It is relatively easy to generate a pulsed field of order 50 T for about 10 ms in a 20 mm bore [2]. The generation of higher fields is mainly limited by the mechanical strength of the coil; at present the highest field that can be obtained without damage to the coil is close to 80 T. Extreme techniques that allow for the destruction of the coil can now provide up to 300 T with the single-turn coil (without damage to the sample under investigation), up to 600 T with electromagnetic flux compression and up to 1000 T with explosive-driven flux compression. This involves current of the order of several mega-amperes, and the pulse duration decreases with increasing field from about 10–0.1 μs. Modern techniques of transient data recording have enabled the effective use of these extreme fields for research in condensed matter physics, and many important discoveries of new phenomena have been made. There is one major restriction for experiments in a pulsed field: at the surface of electrically conducting samples, very high current is induced by the skin effect; this results in heating and eventually in vaporization of the sample. The skin effect impedes the penetration of the magnetic field into the sample; consequently, magnetic pressure (Maxwell stress) is applied at the surface. At 100 T, this pressure amounts to 4 GPa. The skin effect can be minimized by reducing the physical size of the sample to dimensions of typically less than 0.1 mm, and thus a limited amount of measurements on metals in a pulsed magnetic field is feasible. However, for precise measurements on large samples, static fields are needed. Static magnetic fields are also necessary for material processing as mentioned above because the field must be maintained for a long time.

To generate a static field higher than that obtained by a superconducting magnet (at present, this is about 20 T), electric power of the order of 10 MW must be supplied to the magnet coil in order to overcome the loss of energy due to the resistance of the coil. This power is much less than in a pulsed magnet, but the consumed energy is huge: every second about 10 MJ are transferred to the coil. All this energy is converted into Joule heating of the magnet; therefore, a very large water-cooling system is required. The efficient removal of the heat in a water-cooled coil calls for a current that is not higher than about 10 kA. Accordingly, the number of turns for a static resistive magnet is of the order of 100. Figure 1.1 shows the relation between magnetic field strength and the corresponding time duration.

Table 1.1 shows the present status of static high magnetic field facilities in the world. It is evident that the high field technique is progressing rapidly and several facilities are aiming at static high fields over 40 T. One of the reasons for requiring 40 T is research on high T_c superconductors. With a view to the 21st century, there are many serious issues to be solved. In particular, these include energy, environment, advanced information, traffic systems, space development, ageing and important issues related to human welfare in general. Although all have of these have to be addressed individually, this will be

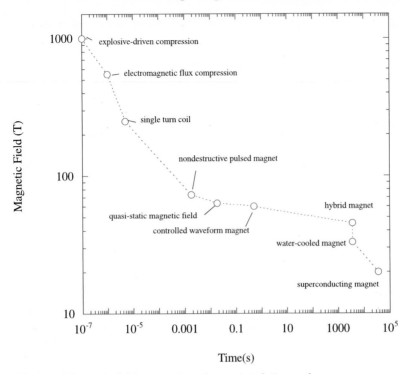

Fig. 1.1. Magnetic field strength and associated timescale

achieved on the basis of common technologies. One of these is the technology of superconductivity. High-T_c superconductors that were discovered in 1986 involve truly new phenomena and important characteristics are observed in the superconducting state as well as in the normal state. High-T_c superconductors are the most promising materials for superconducting technology. In order to develop high-T_c superconductors, the critical surface consisting of the critical current density, critical field and critical temperature has to be determined. Static high magnetic fields over 40 T are indispensable for this evaluation, because high T_c superconductors maintain their superconducting properties in extremely high fields exceeding 40 T at lowtemperature.

Furthermore, static high magnetic fields over 40 T will push the frontier of research forward and it is expected that new phenomena will be discovered by the ultimate condition of the highest fields combined with extreme values of high pressure and ultra-low temperature.

1.2   Materials Science in High Fields

Table 1.2 shows various types of experiments involving magnetic fields. High-T_c superconductors reveal unique behavior due to the low carrier density, the

Table 1.1. Hybrid magnet systems in the world generating static high fields over 30 T

	hybrid			superconducting		water-cooled		
	Max. Field [T]	Coil Temp. [K]	Warm Bore [mm ϕ]	Field [T]	Power [MW]	Warm Bore [mm ϕ]	Field [T]
NHMFL (USA)	45.2	1.8	616	14.3	24	32	30.9
Grenoble	31.4	1.8	420	10.5	10	50	20.9
(France)	(40)		(–)	(8)	(24)	(34)	(32)
Nijmegen	30.4	1.8	360	10.8	6	32	19.6
(The Netherlands)	(40)		(–)	(11)	(20)	(32)	(29)
NRIM (Japan)	37.3	4.2	400	14.1	15	30	23.2
IMR (Japan)	31.1	4.2	360	11.5	8	32	19.6

( ): under construction

strong anisotropy of the two-dimensional crystal structure and the very short coherence length that could not be observed in conventional low-T_c superconductors. For instance, the upper critical field for high-T_c superconductors does not have a clear phase boundary, since the superconducting fluctuation is extremely large due to a very short coherence length. Peculiar phenomena on vortex dynamics are investigated by measuring the magnetization and magnetoresistance in high fields. In particular, the detailed vortex motion in superconductors can be understood only by use of a constant magnetic field, although additional studies of the dynamics in pulsed fields are useful.

The important trends of research on magnetism in high fields can be classified into two major parts. One is basic research on low-dimensional magnetism and strongly correlated magnetic substances; here it is considered very important to investigate the field-induced phase transformations in high magnetic fields. The other is research for applications of magnetic substances and the development of strong permanent magnets and magnetic recording materials that are of importance to industry. The research on magnetism generally requires a very high field, where the Zeeman energy of the electron spin system is larger than the interaction energy, or the Landau energy of the electron is larger than the Fermi energy corresponding to the kinetic energy of the electron. Pulsed magnetic fields are useful in this case. While the measurement of magnetization, magnetoresistance and cyclotron resonance is frequently done in pulsed fields, static fields are needed for experiments on the specific heat; this is very important to determine the electron effective mass reflected in the specific heat.

Extremely high pressure and ultra-low temperature provide the ultimate experimental environment. Therefore, it is expected that the combination

Table 1.2. Objects of research in high magnetic fields

Experimental Method	Sample Substance
Susceptibility and Magnetization	Magnetic material
	· ferromagnetic
	· antiferromagnetic
	· paramagnetic
	Diamagnetic material
	· giant molecule
	· biopolymer
	· superconductor
de Haas-van Alphen effect	Superconductor, metal, semiconductor
Shubnikov-de Haas effect	Metal, semiconductor
Magnetoresistance	Superconductor, metal, semiconductor
Critical current density	Superconductor
Hall effect	Metal, semiconductor
Magneto-optic effects	Magnetic material, semiconductor
Faraday effect	Magnetic material
	Diamagnetic material
Cotton-Mouton effect	All materials
Magnetic Kerr effect	Magnetic material
Electron spin resonance (ESR)	Magnetic material
	· ferromagnetic resonance
	· antiferromagnetic resonance
	· paramagnetic resonance
Nuclear magnetic resonance (NMR)	All materials
Cyclotron resonance	Metal, semiconductor
Muon spin resonance (μ SR)	All materials
Mössbauer effect	All materials
Specific heat	All materials
Thermal conductivity	All materials
Nernst effect	Superconductor, metal, semiconductor
Ultrasonic absorption	All materials
Neutron diffraction	Magnetic material
X-ray diffraction	All materials

of multiple ultimate conditions with high magnetic fields will bring about the discovery of new phenomena. In particular, the combination of ultra-low temperature and a static high field will give much information in de Haas-van Alphen effect, Shubnikov-de Haas effect, ESR, NMR, cyclotron resonance, specific heat and ultrasonic absorption measurements.

By changing the wavelength of light from the microwave region to the far-infrared, infrared, visible light and ultraviolet regions, the photon energy can be easily controlled over the range 0.01–10 eV. The advantage of optical measurements is in the easy availability of different techniques such as absorption, reflection and scattering, and in techniques that involve polarized light (Faraday, Cotton-Mouton). This is the reason why the investigation of magneto-optic effects is a promising research area for the application of static as well as pulsed high magnetic fields.

Even if a static field is not combined with low temperature, it is an important tool for influencing chemical reactions and crystal growth. The recent development of a cryocooler has enabled the development of a "cryogen-free" superconducting magnet using high-temperature superconducting current leads. This new type of magnet does not have to be filled with liquid helium; it is quite easy to operate and can maintain a constant magnetic field for a very long time of the order of 10 000 h without maintenance. This has resulted in the rapid progress of research on chemical reactions and crystal growth.

Further, the cryogen-free superconducting magnets enable the development of new types of functional experimental equipment. For example, X-ray or neutron diffraction investigations are now possible in a static high field up to 10 T for a long-term experiment. In the following chapter, we report outstanding investigations in materials science using a static high field.

References

1. P. Kapitza: Proc. Roy. Soc. A **105**, 691 (1924) and **115**, 658 (1927)
2. F. Herlach: Rep. Prog. Phys. **62**, 859 (1999)
3. F. Bitter: Rev. Sci. Instrum. **10**, 373 (1939)
4. M.F. Wood and D.B. Montgomery: "Combined superconducting and conventional magnets", in *Proc. Int. Conf. "Les champs magnétiques intenses, leur production et leurs applications"*, Grenoble, 12-14 Sept. 1966, ed. R. Pauthenet, (CNRS, Paris, 1967) p. 91
5. *Proc. 6th Int. Symp. on Research in High Magnetic Fields*, eds. F. Herlach et al., Porto, 31 July to 4 Aug., 2000, Physica B **294&295** (2001)
6. H. Kobayashi and T. Haseda: J. Phys. Soc. Jpn **19**, 765 (1964)
7. M. Date and M. Motokawa: Phys. Rev. Letters **16**, 1111 (1966)
8. J.B. Torrance, Jr. and M. Tinkham: Phys. Rev. **187**, 505 (1969)
9. K. Watanabe, Y. Yamada, J. Sakuraba, F. Hata, C.K. Chong, T. Hasebe and M. Ishihara: Jpn. J. Appl. Phys **32**, L488 (1993)

Part II

High-T_c Oxide High Field Superconductors

2 Vortex Phase Diagram
of High-T_c Superconductor YBa$_2$Cu$_3$O$_y$
in High Magnetic Fields

N. Kobayashi and T. Nishizaki

In conventional type II superconductors, only an Abrikosov vortex-lattice is formed in the mixed state. On the other hand, in high-T_c superconductors a rich variety of vortex phases, such as a vortex-lattice, a vortex-glass, a vortex-liquid and so on, have been proposed by many theoretical and experimental works. These novel vortex phases are believed to result from characteristic features of high-T_c superconductors. Those are: the strong anisotropy which exceed 10^4 in mass ratio; the short coherence length (3 nm in the CuO$_2$ plane); the large thermal agitation; and a large number of weak disorder pinning sites. This means that vortex phases in high-T_c superconductors are determined by a delicate balance between the thermal energy, the vortex elastic energy and the vortex pinning energy.

In the presence of strong thermal fluctuations, the upper critical field $H_{c2}(T)$ is no longer a well-defined phase transition line and the thermodynamic phase transition in a magnetic field is replaced by a melting transition of the vortex array. In clean systems, Nelson and Seung [1] and Houghton, et al. [2], predicted that the vortex-lattice melts via a first-order phase transition when the mean-squared thermal displacement of vortices from the equilibrium positions exceeds a certain fraction of vortex distance. Experimentally, a sharp resistive kink [3–6], a discontinuous magnetization jump [7–15] and a calorimetric anomaly [16–18] have provided evidence of the first-order vortex-lattice melting transition in YBa$_2$Cu$_3$O$_y$ (YBCO) [3–5,7–10,16–18], Bi$_2$Sr$_2$CaCu$_2$O$_y$ (BSCCO) [6,11–13] and La$_{2-x}$Sr$_x$CuO$_y$ (LSCO) [14,15]. Computer simulations [19–23] have also been developed to clarify the first-order vortex-lattice melting transition.

On the other hand, vortex phases of high-T_c superconductors are strongly affected by disorder in the samples. In highly disordered systems, it was proposed that the vortex-glass phase is formed at low temperature, reducing the positional correlation between adjacent vortices and melts via a continuous second-order phase transition [24]. The second-order phase transition was experimentally confirmed in a YBCO epitaxial film [25], showing that the $I-V$ curves above and below the vortex-glass transition temperature T_g are scaled onto two universal curves.

When the magnetic field is applied to clean single crystals, the first-order vortex-lattice melting line $H_m(T)$ [or $T_m(H)$] terminates at a critical point H_{cp} [11–13,18,26] and changes to the continuous second-order vortex-glass transition line $H_g(T)$ [or $T_g(H)$] above H_{cp} [18,26]. The change of the order of the vortex melting transition is observed both in YBCO [18,26] and BSCCO [11]. The position of the critical point in the vortex phase diagram depends on the strength of the disorder and the material's parameters.

High-T_c superconductors show a characteristic enhancement of the magnetization hysteresis in an intermediate field region between the lower critical field H_{c1} and the vortex solid-to-liquid transition field, which is called the second peak. This phenomenon has been observed in single crystals of almost all of the high-T_c superconductors [27–31]. Many possibilities for the second peak have been discussed so far. For example, an enhanced vortex pinning resulting from weak-superconducting regions [27], a collective flux creep [32], a matching effect [33], a dimensional crossover [34,35], and so on. Recently, we have shown that the vortex phase diagram is more complicated and the vortex solid region is divided into at least two distinct phases by a characteristic field $H^*(T)$ at which the positional correlation of the vortex lattice is significantly reduced and the magnetization shows a steep increase toward the second peak in YBCO [36]. Similar results have also been observed in BSCCO [37] and $Nd_{1.85}Ce_{0.15}CuO_4$ [38].

In this paper, we review our recent study on the vortex phase diagram of high-quality untwinned YBCO single crystals by measuring the electrical resistivity and the magnetization in high magnetic fields up to 30 T. In particular, our attention is focused on the influence of oxygen deficiency on the melting transition and on the vortex phase change in high magnetic field. Samples used in this study are naturally untwinned or detwinned YBCO single crystals. The oxygen content of the untwinned YBCO samples was controlled by annealing at 1 or 8 bar of oxygen pressure at various temperatures. In Sects. 2 and 3, the vortex phase diagram of a slightly overdoped YBCO with $T_c \simeq 92$ K, which was annealed at 450°C for 7 days in flowing oxygen gas, where the oxygen content y is estimated to be 6.96, is described as a typical example. In Sect. 4, the effect of oxygen deficiency on the vortex phase diagram is presented for fully oxidized and slightly underdoped crystals.

2.1   Melting Transition of the Vortex System

2.1.1   First-Order Vortex-Lattice Melting Transition

Figure 2.1 shows the temperature dependence of the resistivity for the slightly overdoped YBCO single crystal at various magnetic fields H parallel to the c-axis [39,40]. Although the resistivity suddenly drops to zero in the absence of the field, it shows remarkable broadening in the magnetic field because the superconducting fluctuations are significant and the vortex-liquid phase exists over a wide range of the phase diagram. With decreasing temperature, the resistivity shows a clear discontinuous jump, which is attributed to the first-order vortex-lattice melting (freezing) transition. At the melting transition temperature $T_m(H)$, the resistivity shows a small hysteresis with a subloop.

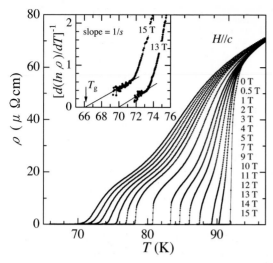

Fig. 2.1. The temperature dependence of the resistivity for slightly overdoped YBCO at various magnetic fields ($H \parallel c$). From right to left: 0, 0.5, 1, 2, 3, 4, 5, 7, 9, 10, 11, 12, 13, 14 and 15 T. Inset: the inverse logarithmic derivative of the resistivity near the vortex-glass transition temperature T_g

The nature of the hysteresis loop is consistent with the first-order transition [41]. This sharp drop in resistivity due to the first-order melting transition is observed in the low-field region below 10 T, as shown in Fig. 2.1.

Figure 2.2 shows the magnetization as a function of temperature in zero-field cooled (ZFC) and field-cooled on cooling (FCC) modes measured by a SQUID magnetometer for untwinned YBCO [36]. The inset of Fig. 2.2 is an expanded view of the magnetization $\mu_0(M - M_s)$ in the vicinity of T_m, where M_s is a linear extrapolation of the low-temperature magnetization expressed by the solid line. A discontinuous magnetization jump is clearly observed at the vortex-lattice melting temperature T_m, the value of which is consistent with that obtained from the resistivity measurement [9,10,42]. The magnetization jump at this temperature is about $\mu_0 \Delta M \simeq 0.4 \times 10^{-4}$ T, and the entropy change per vortex per CuO$_2$ double layer at the transition is estimated to be $\Delta S \simeq 0.7\,k_B$ using the Clausius-Clapeyron equation $dH_m/dT = -\Delta S/\Delta M$.

The characteristic values obtained above at the vortex-lattice melting transition are consistent with the results of calorimetric measurements [16–18]. In spite of the different anisotropy of the materials, BSCCO [11–13] and LSCO [14,15] have almost the same value of ΔS at low temperatures as YBCO [7,9,10,16–18,36]. These similarities of the thermodynamic property may suggest common mechanisms of the vortex-lattice melting transition in high-T_c superconductors.

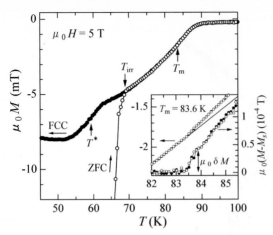

Fig. 2.2. Magnetization as a function of temperature in ZFC and FCC modes at 5 T for slightly overdoped YBCO. The inset is an expanded view of the magnetization and the jump in the magnetization $\mu_0(M - M_s)$ near the vortex-lattice melting temperature

2.1.2 Second-Order Vortex-Glass Melting Transition

In the high-field region above 10 T in Fig. 2.1, the resistivity jump disappears and the resistivity shows a continuous, broad transition. This result indicates that the first-order vortex-lattice melting transition terminates at the critical point $\mu_0 H_{cp} \simeq 10$ T and changes to a continuous second-order melting transition above H_{cp}.

The vortex-glass theory [24] predicts the temperature dependence of the linear resistivity,

$$\rho(T) \propto (T - T_g)^s , \tag{2.1}$$

near the glass transition temperature T_g, where s is the critical exponent. The resistivity above H_{cp} is well described by this equation, because a plot of $[\partial(\ln\rho)/\partial T]^{-1}$ versus T shows a straight line in the critical region of the vortex-glass phase, as shown in the inset of Fig. 2.1. The value of the critical exponent $s \simeq 6.9 \pm 1.2$ is obtained from the slope $1/s$; the obtained value is consistent with the previously reported value of $s \simeq 6$–8 [40,26,25]. These results indicate that the transition above the critical point is consistent with the vortex-glass transition [24] and vortices in the liquid phase are frozen into the vortex-glass phase with decreasing temperature in high magnetic fields.

2.2 Second-Peak Effect in Magnetization Hysteresis

The second-peak effect in the magnetization curve is one of the most remarkable features of the vortex-solid phase of high-T_c superconductors. In

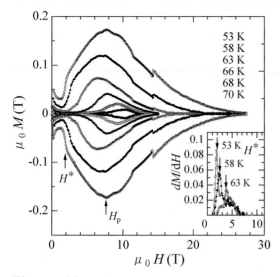

Fig. 2.3. Magnetization curves at various temperatures for slightly overdoped YBCO. Inset: the field derivative dM/dH shows a sharp peak at a characteristic field H^*

the early study of the second peak of YBCO [27,28,32,33], the observed second peak was difficult to associate with the phase transition of the vortex system because the second peak was broad and thermodynamic phase transition lines such as $T_{\mathrm{m}}(H)$ and $T_{\mathrm{g}}(H)$ were not well characterized. Recently, we have found a sharp onset of the second peak near the critical point of the first-order vortex-lattice melting line for clean single crystals of YBCO [36]. Figure 2.3 shows magnetization curves measured by using Hall probes at various temperatures. The magnetic field ($H \parallel$ the c-axis) up to 30 T was generated by a hybrid magnet system at the High Field Laboratory for Superconducting Materials, IMR, Tohoku University. The magnetization shows a steep increase at a characteristic field H^* defined by a sharp peak in the magnetic-field derivative dM/dH, as shown in the inset of Fig. 2.3. Above H^*, the magnetization increases gradually toward a second peak at H_{p}. The steep increase of the magnetization at H^* is remarkable in the local magnetization measurements [37,36,43] as compared with the bulk magnetization measurements. With increasing temperature, the minimum of the magnetization between the central peak and the second peak extends to the higher field region and the magnitude of the hysteresis becomes smaller. In the high temperature region above 70 K, the magnetization shows an anomalous re-entrant behavior. At $T = 70$ K, for example, the magnetization hysteresis decreases and disappears in the intermediate field region ($3.5\,\mathrm{T} \leq \mu_0 H \leq 6\,\mathrm{T}$) with increasing field. Above the reversible region, irreversibility appears again and the second peak is observed around 10.5 T. The re-entrant magnetization is observed only in high-quality samples with a low pinning force [36].

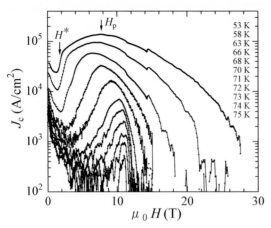

Fig. 2.4. Magnetic-field dependence of the critical current density J_c for slightly overdoped YBCO at several temperatures. From top to bottom: 53, 58, 63, 66, 68, 70, 71, 72, 73, 74 and 75 K

Using Bean's critical state model, the magnetic- field dependence of the critical current density J_c is calculated from the magnetization hysteresis curve as shown in Fig. 2.4. Contrary to the previous reports for YBCO [27,32,33,28], characteristic fields such as H_p and H^* increase toward the critical point ($\mu_0 H_{cp} \simeq 11\,\mathrm{T}$, see Fig. 2.1) of the first-order vortex-lattice melting line and disappear above 75 K with increasing temperature. This result indicates that a new line $H^*(H)$ defined by the magnetization anomaly may be related to the phase-transition line of the vortex system. The characteristic fields are not described by the simple power-law dependence of $H(T) \propto (1 - T/T_c)^n$ [28]. Therefore, it is difficult to interpret the second peak in Figs. 2.3 and 2.4 in terms of previously proposed mechanisms such as field-dependent pinning in the oxygen-deficient region [27], collective flux creep [32,44], a matching effect [33,45], or dimensional crossover [34,35].

The heavily-twinned YBCO shows a larger value of J_c with a broad second peak in the intermediate-field region below H_p. However, J_c of the heavily twinned crystal becomes comparable to that of the untwinned one above H_p. This result indicates that the pinning force of the untwinned sample is remarkably reduced in the intermediate-field region and some common pinning mechanism such as from an oxygen vacancy may work for both untwinned and heavily twinned samples in the high-field region.

2.3 Vortex Phase Diagram

Figure 2.5 shows the magnetic field versus temperature phase diagram including the vortex melting lines $H_m(T)$ and $H_g(T)$, the second peak $H_p(T)$ and the irreversibility field $H_{irr}(T)$ defined by a criterion of $J_c \simeq 90\,\mathrm{A/cm^2}$.

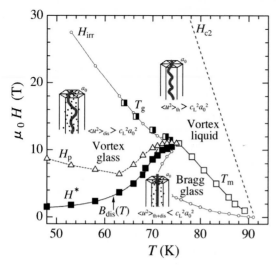

Fig. 2.5. Vortex-matter phase diagram of slightly overdoped YBCO. The transition lines $T_m(H)$, $T_g(H)$, and $H^*(T)$ terminate at the critical point and divide the diagram into three different phases: the vortex-liquid; the vortex-glass; and the Bragg-glass. The solid curve is a fit to the field-driven transition line $B_{dis}(T)$

Below H_{cp}, $H_{irr}(T)$ shows re-entrant behavior in the intermediate field region of $5 \leq \mu_0 H \leq 11 - 13$ T and meets the melting line at H_{cp}.

Above H_{cp}, $H_{irr}(T)$ almost coincides with the second-order phase transition line $T_g(H)$. The magnetic-field dependence of $T_m(H)$ is well expressed by an expression proposed by Houghton et al. [2],

$$B_m(T) = \frac{\phi_0^5 c_L^4}{16\pi^2 \lambda_{ab}^4(0)\gamma^2 k_B^2 T^2}\left(\frac{T_c}{T} - 1\right)^2, \tag{2.2}$$

using an anisotropy parameter $\gamma = 6$, the Lindemann criterion $c_L = 0.12$ and the penetration depth in the ab plane $\lambda_{ab} = 150$ nm. These values are reasonable.

The characteristic field H^* increases monotonically with a nearly T^2 dependence and also terminates at the multicritical point. $H^*(T)$ divides the vortex-solid phase into two regions, the high-field (low-temperature) region with strong pinning and the low-field (high-temperature) region with weak pinning. The vortex phase diagram in Fig. 2.5 suggests the existence of the Bragg-glass phase and the field-driven disordering transition [46–48]. The Bragg-glass phase forms an almost perfect vortex lattice with long-range translational order at low fields, although the vortex array is distorted by the weak pinning. Therefore, one can expect that the Bragg-glass phase melts into the vortex-liquid phase through a first-order transition when the temperature is increased. On the other hand, with increasing field, the random pinning induces dislocations of the vortex array and the vortex system should undergo a transition into the disordered vortex-glass phase. Then, the field-driven transition line from the Bragg-glass to the vortex-glass states is expected to terminate at the multicritical point for both $T_m(H)$ and $T_g(H)$. Our experimental results are consistent with the theoretical prediction. Thus, $H^*(T)$ is attributed to the field-driven disordering transition line [36].

Ertas and Nelson [47] estimated the disordering transition field B_{dis} on the basis of the Lindemann criterion, taking account of the effect of both thermal and disorder fluctuations. In the intermediate temperature region between about 40 K and 70 K for YBCO, the disordering transition line is given by

$$B_{dis}(T) \approx B_{dis}(0)(T_{dp}^s/T)^{10/3} \exp[(2c/3)(T/T_{dp}^s)^3] \,, \qquad (2.3)$$

for $T < T_{dp}^s$. Here, c is a constant of order unity [49] and $B_{dis}(0)$ is expressed by

$$B_{dis}(0) \approx \phi_0/\xi_{ab}^2(0)[\varepsilon_0\xi_{ab}(0)/\gamma T_{dp}^s]^{10/3}c_L^{16/3} \,, \qquad (2.4)$$

where T_{dp}^s is the single vortex depinning temperature, ξ_{ab} the in-plane coherence length, $\varepsilon_0 = \phi_0/4\pi\lambda_{ab}$, and λ_{ab} the in-plane penetration depth. For YBCO, the value of $B_{dis}(0)$ is estimated to be 1.4–5.7 T, which is reasonably consistent with the H^* value obtained by extrapolating to the low temperature region. As shown by the solid lines in Fig. 2.5, the temperature dependence of $H^*(T)$ is well described by this model using the reasonable fitting parameters $T_{dp}^s = 37.6$ K and $B_{dis}(0) = 0.7$. In the higher-temperature region, the disordering transition line $B_{dis}(T)$ coincides with the vortex-lattice melting line given by (2.2), which decreases with increasing temperature because the vortex displacement is dominated not by disordering fluctuations but, rather, by thermal fluctuations.

2.4   Effect of the Oxygen Deficiency

As mentioned in the previous section, disorder in the sample affects the vortex phase diagram of high-T_c superconductors. We have systematically examined the effect of disorder on the vortex phase by controlling the amount of oxygen deficiency [50,51]. The oxygen content was controlled by changing the annealing conditions.

Figure 2.6 shows the magnetic-field dependence of the resistivity ρ up to 30 T for the fully oxidized untwinned YBCO single crystal ($y \simeq 7$ and $T_c \simeq 87.5$ K). As shown in Fig. 2.6, the resistivity shows a discontinuous jump due to the first-order vortex-lattice melting transition for fields up to 30 T, indicating a high critical point H_{cp} beyond 30 T for the extremely clean YBCO single crystal. The magnetization of the fully oxidized YBCO sample shows a steeper increase at H^* than that of the slightly overdoped YBCO one, similar to the results for BSCCO and indicating the drastic change of the vortex state.

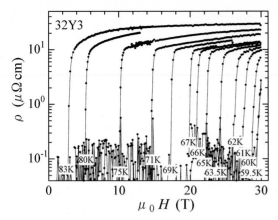

Fig. 2.6. Magnetic-field dependence of the resistivity up to 30 T for fully oxidized YBCO

Figure 2.7 shows the vortex phase diagram of the fully oxidized sample together with the transition lines of the slightly overdoped YBCO sample for comparison. Although the vortex-glass melting line $T_g(H)$ is not observed in the fully oxidized YBCO sample, H^* seems to approach the critical point $H_{cp}(> 30\,\mathrm{T})$, similar to the phase diagram of the slightly overdoped YBCO sample. Since the increase of the point disorder due to the oxygen deficiency stabilizes the vortex-glass phase, the decrease of H_{cp} and $H^*(T)$ is naturally expected from the disordering transition theories [46,47]. The value of H_{cp} for fully oxidized YBCO is at least three times larger than that for slightly overdoped YBCO. Thus, the vortex phase diagram is strongly affected by the oxygen deficiency.

As mentioned above, H_{cp} decreases with increasing oxygen deficiency for overdoped YBCO. However, the vortex phase diagram for slightly under-doped YBCO shows a new feature. Figure 2.8 shows the normalized resistivity versus temperature for slightly underdoped YBCO (thick lines), including the data for slightly overdoped YBCO (thin lines) for comparison. The resistivity for slightly underdoped YBCO shows first the sharp drop corresponding to the first-order vortex melting temperature T_m. However, the resistivity remains below T_m and decreases gradually with decreasing temperature, indicating the second-order transition. As shown in the figure, $T_m(H)$ for slightly underdoped YBCO agrees well with the well-defined $T_m(H)$ for slightly over-doped YBCO. The gradual change in the lower resistivity tail is described by the vortex-glass theory [24], showing the temperature dependence of the resistivity $\rho(T) \propto (T - T_g)^s$ near the glass transition temperature $T_g(H)$. Plotting $[\partial(\ln \rho)/\partial T]^{-1}$ versus T, T_g and the reasonable value of the critical exponent $s \simeq 5$–6 are obtained from a extrapolation of the line with a slope $1/s$.

Figure 2.9 shows the vortex phase diagram for slightly underdoped YBCO. The vortex-glass transition line $T_g(H)$ decreases with increasing temperature

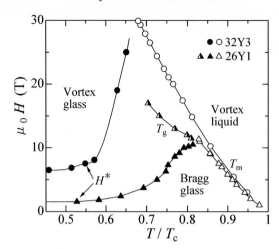

Fig. 2.7. Vortex phase diagram for fully oxidized YBCO. The result for the slightly overdoped sample is also shown for comparison

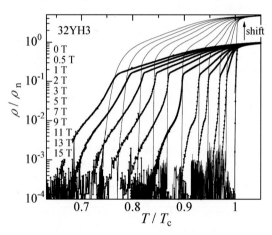

Fig. 2.8. Normalized resistivity as a function of temperature for slightly under-doped (thick lines) and slightly overdoped (thin lines) YBCO. From right to left: 0, 0.5, 1, 2, 3, 5, 7, 9, 11,13 and 15 T

and approaches the first-order melting line $T_m(H)$ well below the critical point of $T_m(H)$, in contrast to the case of overdoped YBCO (see Fig. 2.5). Although the magnetization shows the broad second peak H_{cp} in low fields, no remarkable feature is observed near H_{cp} [36,50]. Therefore, there is no clear evidence of the Bragg-glass phase and the field-driven disordering transition for slightly underdoped YBCO. Worthington et al. [52] discussed the two-step nature of the resistivity similar to Fig. 2.8 and introduced a concept of the vortex slush regime between $T_m(H)$ and $T_g(H)$ in proton-irradiated YBCO. The results in this study indicate that the Bragg-glass phase disappears with

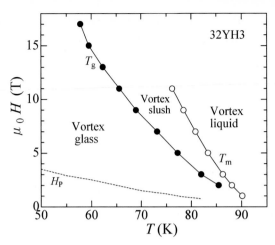

Fig. 2.9. Vortex phase diagram for slightly underdoped YBCO

increasing oxygen deficiency and that the vortex-liquid phase undergoes two successive transition lines of $T_m(H)$ with no symmetry change and of $T_g(H)$ with a symmetry change into the vortex-glass phase. For underdoped YBCO with further oxygen deficiency ($T_c \leq 92\,\mathrm{K}$), only the vortex-glass transition is visible, so the vortex-slush is observed in YBCO with a limited value of the oxygen deficiency.

2.5   Conclusion

The vortex phase diagram of YBa$_2$Cu$_3$O$_y$ has been investigated by measuring the electrical resistivity and the magnetization of high-quality single crystals with different oxygen content and disorder in high magnetic fields up to 30 T. It has been confirmed that the first-order vortex-lattice melting transition takes place at low magnetic fields and the first-order vortex phase transition is replaced by the second-order vortex-glass melting transition at high fields. The anomalous increase in magnetization hysteresis is observed in the vortex-solid state and is attributed to the field-driven disordering transition from the vortex-lattice (or Bragg-glass) to the disordered vortex-glass states. This result is consistent with the change from first-order to second-order in the melting transition. Furthermore, it has been clarified that the vortex phase diagram is strongly affected by disorder in the samples such as that from oxygen deficiencies.

Acknowledgement

The authors would like to thank K. Shibata, T. Naito and M. Maki for collaboration in this work and T. Sasaki, T. Nojima, S. Awaji and K. Watanabe for valuable discussions. Part of this work was carried out at the High Field Laboratory for Superconducting Materials, IMR, Tohoku University. We thank K. Sai, Y. Ishikawa, and Y. Sasaki of the HFLSM for operating the hybrid magnet. We also thank S. Otomo, H. Miura, S. Tanno and K. Hosokura of the Center for Low Temperature Science, Tohoku University for supplying helium.

References

1. D.R. Nelson and S.H. Seung: Phys. Rev. B **39**, 9153 (1989)
2. A. Houghton, R.A. Pelcovits and A. Sudbø: Phys. Rev. B **40**, 6763 (1989)
3. H. Safar, P.L. Gammel, D.A. Huse, D.J. Bishop, J.P. Rice and D.M. Ginsberg: Phys. Rev. Lett. **69**, 824 (1992)
4. W.K. Kwok, S. Flesher, U. Welp, V.M. Vinokur, J. Downey, G.W. Crabtree and M.M. Miller: Phys. Rev. Lett. **69**, 3370 (1992); W.K. Kwok, J. Fendrich, U. Welp, S. Flesher, J. Downey and G.W. Crabtree: Phys. Rev. Lett. **72**, 1088 (1994); W.K. Kwok, J. Fendrich, S. Flesher, U. Welp, J. Downey and G.W. Crabtree; Phys. Rev. Lett. **72**, 1092 (1994)
5. M. Charalambous, J. Chaussy and P. Lejay: Phys. Rev. B **45**, 5091 (1992); M. Charalambous, J. Chaussy, P. Lejay and V.M. Vinokur: Phys. Rev. Lett. **71**, 436 (1993)
6. S. Watauchi, H. Ikuta, J. Shimoyama and K. Kishio: Physica C **259**, 373 (1996)
7. R. Liang, D.A. Bonn and W.N. Hardy: Phys. Rev. Lett. **76**, 835 (1996)
8. T. Nishizaki, Y. Onodera, N. Kobayashi, H. Asaoka and H. Takei: Phys. Rev. B **53**, 82 (1996)
9. T. Nishizaki, Y. Onodera, T. Naito, N. Kobayashi, H. Asaoka and H. Takei: J. Low Temp. Phys. **105**, 1183 (1996)
10. U. Welp, J. Fendrich, W.K. Kwok, G.W. Crabtree and B.W. Veal: Phys. Rev. Lett. **76**, 4809 (1996)
11. H. Pastoriza, M.F. Goffmann, A. Arribere and F. de la Cruz: Phys. Rev. Lett. **72**, 2951 (1994)
12. E. Zeldov, D. Majer, M. Konczykowski, V.M. Geshkenbein, V.M. Vinokur and H. Shtrikman: Nature (London) **375**, 373 (1995)
13. T. Hanaguri, T. Tsuboi, A. Maeda, T. Nishizaki, N. Kobayashi, Y. Kotaka, J. Shimoyama and K. Kishio: Physica C **256**, 111 (1996)
14. T. Naito, T. Nishizaki, F. Matsuoka, H. Iwasaki and N. Kobayashi: Czech. J. Phys. **46-S3** 1585 (1996)
15. T. Sasagawa, K. Kishio, Y. Togawa, J. Shimoyama and K. Kitazawa: Phys. Rev. Lett. **80**, 4297 (1998); T. Sasagawa, Y. Togawa, J. Shimoyama, K. Kitazawa and K. Kishio: *The 11th Int. Symp. on Supercond., Fukuoka, 16-19 Nov., 1998, Advances in Superconductivity XI* (Springer-Verlag, Tokyo, 1999) p. 617
16. A. Schilling, R.A. Fisher, N.E. Phillips, U. Welp, D. Dasgupta, W.K. Kwok and G.W. Crabtree: Nature (London) **382**, 791 (1996); A. Schilling, R.A. Fisher, N.E. Phillips, U. Welp, W.K. Kwok and G.W. Crabtree: Phys. Rev. Lett. **78**, 4833 (1997)
17. A. Junod, M. Roulin, J.Y. Genoud, B. Revaz, A. Erb and E. Walker: Physica C **275**, 245 (1997)

18. M. Roulin, A. Junod, A. Erb and E. Walker: Phys. Rev. Lett. **80**, 1722 (1998)
19. R.E. Hetzel, A. Sudbøand D.A. Huse: Phys. Rev. Lett. **69**, 518 (1992)
20. S. Ryu, S. Doniach, G. Deutscher and A. Kapitulnik: Phys. Rev. Lett. **68**, 710 (1992); S. Ryu and D. Stroud: Phys. Rev. B **54**, 1320 (1996); S. Ryu and D. Stroud: Phys. Rev. Lett. **78**, 4629 (1997)
21. R. Šášik and D. Stroud: Phys. Rev. Lett. **75**, 2582 (1995)
22. H. Nordborg and G. Blatter: Phys. Rev. Lett. **79**, 1925 (1997)
23. X. Hu, S. Miyashita and M. Tachiki: Phys. Rev. Lett. **79**, 3498 (1997); X. Hu, S. Miyashita and M. Tachiki Phys. Rev. B **58**, 3438 (1998)
24. M.P.A. Fisher: Phys. Rev. Lett. **62**, 1415 (1989)
25. R.H. Koch, V. Foglietti, W.J. Gallagher, G. Koren, A. Gupta and M.P.A. Fisher: Phys. Rev. Lett. **63** 1511 (1989)
26. H. Safar, P.L. Gammel, D.A. Huse, D.J. Bishop, W.C. Lee, J. Giapintzakis and D.M. Ginsberg: Phys. Rev. Lett. **70**, 3800 (1993)
27. M. Daeumling, J.M. Seuntjens and D.C. Larbalestier: Nature (London) **346**, 332 (1990)
28. N. Kobayashi, K. Hirano, T. Nishizaki, H. Iwasaki, T. Sasaki, S. Awaji, S. Watanabe, H. Asaoka and H. Takei: Physica C **251**, 255 (1995)
29. N. Chikumoto, M. Konczykowski, N. Motohira and A.P. Malozemoff: Phys. Rev. Lett. **69**, 1260 (1992)
30. K. Kadowaki and T. Mochiku: Physica C **195**, 127 (1992)
31. T. Kimura, K. Kishio, T. Kobayashi, Y. Nakayama, N. Motohira, K. Kitazawa and K. Yamafuji: Physica C **192**, 247 (1992)
32. L. Krusin-Elbaum, L. Civale, V.M. Vinokur and F. Holtzberg: Phys. Rev. Lett. **69**, 2280 (1992)
33. A.A. Zhukov, H. Küpfer, H. Claus, H. Wühl, Kläser and G. Müller-Vogt: Phys. Rev. B **52**, R9871 (1995)
34. T. Tamegai, Y. Iye, I. Oguro and K. Kishio: Physica C **213**, 33 (1993)
35. K. Kishio, J. Shimoyama, Y. Kotaka and K. Yamafuji: *Proc. 8th Int. Workshop on Critical Currents in Superconductors, Kitakyushu, 27-29 May, 1996* (World Scientific, Singapore,1994) p. 339
36. T. Nishizaki, T. Naito T and N. Kobayashi: Phys. Rev. B **58**, 11 169 (1998); T. Nishizaki, T. Naito and N. Kobayashi: Physica C **282–287**, 2117 (1997); T. Nishizaki and N. Kobayashi: Supercond. Sci. Technol. **13**, 1 (2000)
37. B. Khaykovich, E. Zeldov, D. Majer, T.W. Li, P.H. Kes and M. Konczykowski: Phys. Rev. Lett. **76**, 2555 (1996)
38. D. Giller D, A. Shaulov, R. Prozorov, Y. Abulafia, Y. Wolfus, L. Burlachkov, Y. Yeshurun, E. Zeldov, V.M. Vinokur, J.L. Peng and R.L. Greene: Phys. Rev. Lett. **79**, 2542 (1997)
39. T. Naito, T. Nishizaki, Y. Watanabe and N. Kobayashi: *Advances in Super-conductivity* IX (Springer-Verlag, Tokyo 1997) p. 601
40. T. Naito, T. Nishizaki and N. Kobayashi: Physica C **293**, 186 (1997)
41. G.W. Crabtree, W.K. Kwok, U. Welp, J.A. Fendrich and B.W. Veal: J. Low Temp. Phys. **105**, 1073 (1996)
42. J.A. Fendrich, U. Welp, W.K. Kwok, A.E. Koshelev, G.W. Crabtree and B.W. Veal: Phys. Rev. Lett. **77**, 2073 (1996)
43. D. Giller, A. Shaulov, Y. Yeshurun and J. Giapintzakis: Phys. Rev. B **60**, 106 (1999)
44. N. Yeshurun, N. Bontemps, L. Burlachkov and A. Kapitulnik: Phys. Rev. Lett. **68**, 2672 (1994)

45. G. Yang, P. Shang, S.D. Sutton, I.P. Jones, J.S. Abell and E.E. Gough: Phys. Rev. B **48**, 4054 (1993)
46. T. Giamarchi and P. Le Doussal: Phys. Rev. Lett. **72**, 1530 (1994); T. Giamarchi and P. Le Doussal: Phys. Rev. B **55**, 6577 (1997)
47. D. Ertas and D.R. Nelson: Physica C **272**, 79 (1996)
48. R. Ikeda: J. Phys. Soc. Jpn. **65**, 3998 (1996)
49. G. Blatter, M.V. Feigelman, V.B. Geshkenbein, A.I. Larkin, and V.M. Vinokur: Rev. Mod. Phys. **66**, 1125 (1994)
50. T. Nishizaki, K. Shibata, T. Naito, M. Maki and N. Kobayashi: J. Low Temp. Phys. **117**, 1375 (1999); T. Nishizaki T. Naito and N. Kobayashi: Physica C **317-318**, 645 (1999)
51. K. Shibata, T. Nishizaki, T. Naitoand N. Kobayashi: Physica C **317-318**, 540 (1999)
52. T.K. Worthington, M.P.A. Fisher, D.A. Huse, J. Toner, A.D. Marwick, T. Zabel, C.A. Field and F. Holtzbelg: Phys. Rev. B **46**, 11854 (1992)

3 Magnetic Ordering and Superconductivity in La-based High-T_c Superconductors

T. Fukase and T. Goto

The La-based high-T_c cuprates $La_{2-x}Ba_xCuO_4$ (LBCO) and $La_{2-x}Sr_xCuO_4$ (LSCO) show a local minimum of T_c around the specific hole concentration of $x \sim 1/8$. In LBCO, the sharp dip appears in the T_c versus x curve around $x \sim 0.125$. With lowering temperature, the structural transformation from the orthorhombic phase (MTO, space group $Cmca$) to the tetragonal phase (LTT, space group $P4_2/ncm$) occurs at $T_{d2} \sim 70\,K$ [1,2] and the magnetic ordering in Cu-3d spins occurs at $T_N \sim 35\,K$ [3–6]. Since the region of the concentration where these two phase transformations exist is close to $x \sim 1/8$, it has been suggested that they are related in a way to the anomalous suppression in T_c. For the Nd-doped system $La_{2-x-y}Nd_ySr_xCuO_4$ ($x \sim 1/8$), which undergoes a structural phase transition to the LTT phase at $\sim 70\,K$ [7], neutron experiments report the existence of charge ordering below $\sim 65\,K$ and a modulated antiferromagnetic-like magnetic ordering below $\sim 50\,K$. The model is proposed that a strip-like charge ordering is pinned to the crystal structure of $P4_2/ncm$, and hence is stabilized [7].

In LSCO, a small dip appears in the T_c versus x curve around $x \sim 0.115$. While magnetic ordering occurs [9,8,10–12] similarly to that for LBCO, no evidence for the macroscopic transformation to the TLT phase has been reported so far at low temperatures down to $4.2\,K$ [13]. In elastic neutron-diffraction measurements on orthorhombic $La_{1.88}Sr_{0.12}CuO_4$, magnetic superlattice peaks were observed [14,15] suggesting the existence of a modulated magnetic order similar to that observed in $La_{1.66-x}Nd_{0.40}Sr_xCuO_4$ in the LTT phase. However, it is unknown whether the type of spin ordering of the materials, which have different crystal structures, is the same or not and whether the spin ordering really coexists with the superconductivity in both crystal structures. In the present paper we report the relation between the type of the spin ordering, the crystal structure and the superconducting state.

3.1 Sound Velocity and Effects of the Magnetic Field on the Crystal Lattice

Figure 3.1 shows the temperature dependence of the sound velocity, V_s, of $La_{2-x}Ba_xCuO_4$, $La_{2-x}Sr_xCuO_4$ and $La_{1.96-x}Y_{0.04}Sr_xCuO_4$ ($x \sim 1/8$). V_s rapidly decreases with decreasing temperature and shows a sharp bend at the temperature T_{d1} where the structural transformation from the high-temperature tetragonal phase (HTT: space group $I4/mmm$) to the MTO phase occurs. LBCO shows an upturn of V_s at the transformation temperature ($\sim 65\,K$) from the MTO phase to the LTT phase. In the case of $La_{1.84}Y_{0.04}Sr_{0.12}CuO_4$, another transition temperature, $T_{d\alpha}$, from the MTO phase to the LTO phase (space group $Pccn$) occurs, which is defined at the upturn of V_s. The transition temperature from the LTO phase to the LTT phase, $T_{d\beta}$, is defined to be $\sim 57\,K$ [16] as the mid-point of the s-shaped variation in the velocity-temperature curve. These two definitions have been confirmed to be consistent with the X-ray analysis [17]. In the temperature range between the local maximum and minimum of the s-shaped variation, the LTO and the LTT phases coexist. While in LSCO ($x \sim 1/8$) the transformation to the LTT phase is not observed down to 4.2 K, the upturn of the sound velocity is observed at low temperatures below $\sim 10\,K$, the phenomenon of which is similar to that observed at the structural transformation to the TLT phase in LBCO [18]. It is suggested that the fluctuation of the structural transformation is extremely large in LSCO around $x \sim 1/8$. This velocity increase at low temperatures in LSCO ($x \sim 1/8$) is enhanced with the application of a high magnetic field [19]. The amount of the increase under the field of 15 T is twice that with zero field at 4.2 K. The onset temperature of the rapid increase is also raised from $\sim 10\,K$ at zero field to $\sim 20\,K$ at 15 T. This phenomenon suggests that the structural phase transformation is enhanced by the strong magnetic field. This enhancement is not observed for LBCO and $La_{1.96-x}Y_{0.04}Sr_xCuO_4$ ($x \sim 1/8$), in which the TLT phase is inherently stable.

3.2   La-NMR and Antiferromagnetic Spin Ordering

The spin structure of La214 type cuprates couples with the crystal structure through the Dzyaloshinskii-Moriya interaction which comes from the small distortion of the CuO_2 plane in the OMT and TLT structures. The end member of the cuprate La_2CuO_4 is antiferromagnetic below $T_N \sim 310\,K$, where the spin direction is approximately along $[010]_o$ in the orthorhombic notation with a small canting up out of $(001)_o$ due to the Dzyaloshinskii-Moriya interaction. The ordered $3d$-spin brings the hyperfine field of $H_{La}^{\parallel} \sim 1\,kOe$ to the La-site [20] so that the resonance position of La-nuclei belonging to the different sublattice shifts in the opposite direction. When the magnetic field H is applied parallel to the spin-aligned axis $[010]_o$, the field-swept spectrum shows two peaks with a splitting width $\delta H_{La} = 2H_{La}^{\parallel}$, corresponding to spins parallel or anti-parallel to H. The splitting width between the two peaks suddenly decreases to nearly zero for magnetic fields above 10 T, as shown in Fig. 3.2. The disappearance of the split at high field is due to spin-flop in the canted-spin system with the Dzyaloshinskii-Moriya interaction, by which the direction of the ordered spins rotates from $[010]_o$ to $[100]_o$ [21].

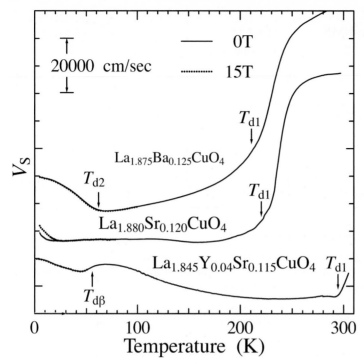

Fig. 3.1. Temperature dependence of the longitudinal sound velocity V_s at 0 T and 15 T in $La_{2-x}Sr_xCuO_4$, $La_{1.96-x}Y_ySr_xCuO_4$ and $La_{2-x}Ba_xCuO_4$ ($x \sim 1/8$). Solid lines and dotted lines represent V_s at 0 T and 15 T, respectively

In a sample of $La_{2-y}Y_yCuO_4$ ($y = 0.04$), the structural transformation to the LTT phase occurs at $T_{d2} \sim 37$ K. The resonance line width shows a significant increase below the antiferromagnetic transition temperature $T_N \sim 200$ K and the splitting of the resonance peak is observed at 4.2 K. The hyperfine field $\delta H_{La}^{\parallel}$ is evaluated from this splitting width of the resonance peaks to be approximately 600 Oe. In the MTO phase above $T_{d2} \sim 37$ K, the hyperfine field shows a step-like decrease above $H \sim 11$ T, as shown in Fig. 3.3, indicating a spin-flop similar to the one in La_2CuO_4. On the other hand, in the LTT phase below T_{d2}, the hyperfine field shows no field dependence, suggesting a spin structure without spin canting. According to Koshibae et al. [22], there are two possibilities of the spin direction for the LTT phase, depending on the competition between the pseudodipolar interaction and the Dzyaloshinskii-Moriya interaction. One possible spin direction is along $[1\bar{1}0]_o$, perpendicular to the tilt axis of CuO_6 octahedra with a small canting up out of $(001)_o$, and the other is along $[110]_o$, parallel to the tilt axis of CuO_6 octahedra without spin canting. Our results show that the spin direction in the LTT phase is along $[110]_o$ without canting similar to that of La_2CoO_4, while the spin direction of orthorhombic $La_{1.96}Y_{0.04}CuO_4$ is approximately along $[010]_o$ with

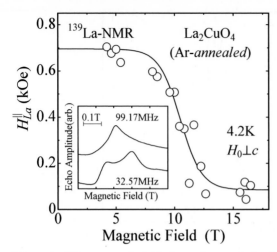

Fig. 3.2. The external field dependence of the hyperfine field at the La-site in La_2CuO_4. Inset: typical spectra of the c-axis-aligned powder sample in the magnetic field applied perpendicular to the aligned c-axis

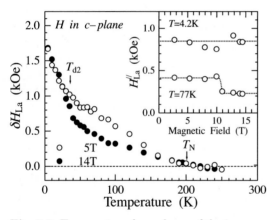

Fig. 3.3. Temperature dependence of the increase of the resonance line-width δH_{La} in the c-axis-aligned powder sample $La_{1.96}Y_{0.04}CuO_4$ in the magnetic field applied perpendicular to the aligned c-axis. Inset: the external field dependence of the hyperfine field at the La-site

a small canting up out of $(001)_o$ which is essential to the spin-flop, similar to the case of La_2CuO_4.

3.3   Cu/La-NMR in $La_{2-x}Ba_xCuO_4$ $(x = 0.125)$

In Ba-doped samples LBCO $(x \sim 1/8)$, the spectra are rather broad, because the random substitution of Ba^{2+} for La^{3+} causes a distribution in

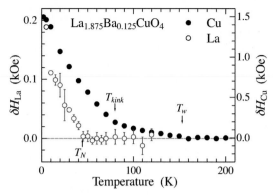

Fig. 3.4. Temperature-dependent increase of the resonance line width of Cu/La-NMR ($I_z = \pm 1/2$ transition) in the c-axis-aligned powder sample in the magnetic field applied perpendicular to the aligned c-axis, around 6 T for La and 15 T for Cu

the electric field gradient. Although no splitting of the resonance peak appears in the spectra, even in the magnetically ordered state, the resonance line width shows a significant increase below the magnetic transition temperature. In LBCO ($x = 0.125$), the transformation to the LTT phase and the antiferromagnetic-like order occurs at $T_{d2} \sim 70\,$K and $T_N \sim 35\,$K, respectively. The charge ordering may occur as in the case of $La_{1.66-x}Nd_{0.40}Sr_xCuO_4$ ($x \sim 1/8$) in the LTT phase. Since both the charge order and the magnetic order, including the long-period structure, contribute to the width of NMR resonance lines, one must detect these two contributions separately. The temperature dependence of the NMR line width at high fields for the La site and the Cu site allows the successful separation between the charge order and the magnetic order.

Figure 3.4 shows the temperature dependent increase of the resonance line width δH for Cu and La. The La-width stays constant above $T_N \sim 40\,$K and shows a significant increase below it [9,10]. This increase is confirmed [8] to be due to the hyperfine field at the La site. The increment of the width at 4.2 K is approximately 190 Oe [10]. No difference is observed between the temperature dependence under 5 and 14 T and no spin-flop is observed up to 20 T.

The temperature dependence of the Cu width is quite different from that of La. With decreasing temperature, it starts to increase gradually from $T_{onset} \sim 160\,$K, and shows a steep increase with a kink at around $T_{kink} \sim 75\,$K, which is very close to the structural phase transformation temperature at 70 K. The increment of the width at 4.2 K is 1500 Oe, about 8 times larger than that for the La width.

The difference between the temperature dependence of the La and Cu widths clearly demonstrates that the broadening in the Cu spectra in the higher temperature region does not originate from magnetic order, suggesting the existence of charge order. The gradual increase in the width from 160 K is considered to be from fluctuations of the charge order and the steep increase from 75 K is considered to be from its stabilization by being pinned to the $P4_2/ncm$ structure developing around this temperature. Insensitiveness of the La width to the charge order is explained by the confinement of the charge order within the CuO_2 plane.

The observed increment of the Cu width at 4.2 K is too small compared to that of the La width if the ratio of hyperfine coupling constants $A_{Cu}/A_{La} \simeq 80$ is assumed to be the same as that for La_2CuO_4 [10]. The ratio cannot be much smaller than 80 to explain the observed ordered moment $\mu_{Cu} \simeq 0.1 \sim 0.3 \mu_B$ [3,10]. The hyperfine field at the Cu site produced by the magnetic order at 40 K is considered to make little contribution to the Cu width. The large hyperfine field may shift the Cu spectra far from the zero-shift position.

3.4   La-NMR and Spin Ordering in $La_{2-x}Sr_xCuO_4$ and $La_{1.96-x}Y_{0.04}Sr_xCuO_4$ ($x \sim 1/8$)

The resonance line width for La at a low field shows a significant increase below the magnetic transition temperature T_N ($\sim$ 40 K), from which one can estimate the hyperfine field at the La site [10] to be approximately $H_{La}^{\parallel} \sim 120$ Oe at 4.2 K for $x = 0.12$. Line-shape analysis for the central transition peak ($I_z = \pm 1/2$) at 4.2 K indicates that the magnetic ordering occurs throughout most of the crystal. At a high field of 12.5 T, the increment of the resonance line width δH_{La} below T_N shows a complicated temperature dependence, as shown in Fig. 3.5. With decreasing temperature δH_{La} increases at first, and then instead decreases down to $\sim$ 0 Oe at $T \sim 15$ K, and below $\sim$ 15 K, rapidly increases again. At the lowest temperature of 4.2 K, δH_{La} at 14 T is smaller than that at 8.8 T. Below $\sim$ 20 K, the external field dependence of the hyperfine field of $H_{La}^{\parallel} = \delta H_{La}/2$ shows a step-like decrease above 11 T as shown in the inset of Fig. 3.5. For $T \sim 15$ K it decreases to almost zero, indicating a spin-flop similar to the one in La_2CuO_4. In the case of $T \sim 4.2$ K δH_{La} decreases to 50% of that at low fields. At the same field the sound velocity shows a steplike increase, indicating a change of the crystal lattice structure. There exists a fundamental difference between observed results for LSCO at 4.2 K and for La_2CuO_4. While the splitting in the resonance line in La_2CuO_4 and the broadening δH_{La} in LSCO at $\sim$ 15 K vanishes completely, the appreciable broadening still remains in LSCO at 4.2 K even at 14 T. Consistent with theoretical argument [22], we propose a model of a change of the spin and the lattice structure in LSCO [21,23]. In low field, the crystal structure is MTO and the spins are stable in the $[010]_o$ direction. With the application of a high field along the spin direction $[010]_o$, spin-flop occurs because of the existence of the spin canting in the MTO phase. When considering the spin energy alone, one expects that spins must rotate by 90°, being analogous to the case of La_2CuO_4. However, the instability of the LTT

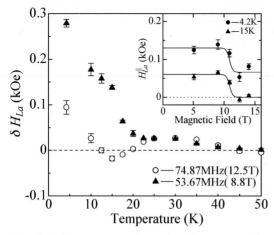

Fig. 3.5. Temperature-dependent increase of the resonance line width of the La-NMR $I_z = \pm 1/2$ transition in LSCO ($x = 0.125$) single crystal. Inset: the external field dependence of the hyperfine field at the La site

phase is large at low temperatures $< 15\,\mathrm{K}$ and the spin direction of $[110]_o$, which is $45°$ away from $[100]_o$, favors the crystal structure of the LTT phase, as mentioned for $La_{1.96}Y_{0.04}CuO_4$. Therefore, the sum of the free energy for the spin and the lattice system is expected to be lowest when the spins rotate to $[110]_o$, simultaneously provoking the structural phase transformation to the LTT phase. If the system falls into this ground state, the direction of the spins is now $45°$ away from that of the external field, where the contribution of the hyperfine field to the total field at the La site is reduced to $(1/2)^{1/2}$ rather than to zero.

In the case of $La_{1.96-x}Y_{0.04}Sr_xCuO_4$ (LYSCO) and LBCO in the LTT phase, no difference is observed between the temperature dependence under 5 and 14 T and no spin-flop is observed up to 20 T, suggesting that there is no spin canting in the ordered state. Therefore, out of the two possible spin structures proposed by Koshibae [22], the latter case with no canting is likely to be realized for the modulated spin ordering of $x \sim 1/8$ in the LTT phase. This means that the Dzyaloshinskii-Moriya interaction, which comes from the small distortion in the CuO_2 plane, is weaker than the pseudodipolar interaction, and so the spins are allowed to be parallel to the Dzyaloshinskii-Moriya interaction vector. We summarize our model in Fig. 3.6 on the change in the spin direction and the lattice with application of the magnetic field.

3.5   Elastic Properties of the Flux-Line Lattice and Superconductivity

Taking into account the layer structure of high-T_c superconductors, uniaxial anisotropy is assumed. There are three independent directions of flux motion.

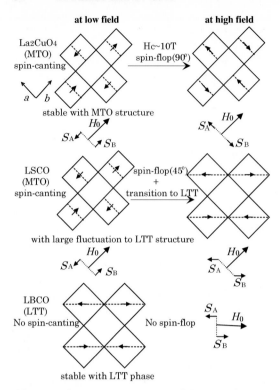

Fig. 3.6. A schematic drawing of the spin-flop and the structural transformation in La$_2$CuO$_4$, LSCO, LYSCO and LBCO. The configuration between the applied field H_0 and the spins belonging to each of the sublattices S_A and S_B is also shown

For magnetic fields applied along the c-axis, the direction of the flux motion is unique and is always perpendicular to the c-axis. For magnetic fields applied perpendicular to the c-axis, two independent directions of flux motions are possible. One of those is parallel to the c-axis and the other is perpendicular to both the applied field and the c-axis. Hereafter, patterns of motion are referred to as CA, AC and AB motion, respectively. For example, in the case of CA motion, the first letter C denotes the direction of the magnetic field and the second letter A denotes the direction of the flux motion. The AC motion is related to the intrinsic pinning mechanism [24] due to the layered crystal structure.

In a temperature region above the characteristic depinning temperature T_p, pinning of the flux-line lattice (FLL) effectively vanishes and one can regard the FLL as almost independent of the crystal lattice (CL). Therefore, in this temperature region, the elastic constant measured by ultrasonic measurements only reflects the elastic constant of the CL. At temperatures below T_p, since the FLL is pinned to the CL, the sum of the elastic constants of the FLL and CL should be observed. Consequently, a step and a peak at T_p

appear in the temperature dependence of the sound velocity and the sound attenuation coefficient, respectively [25]. The height of the step in the temperature dependence of the sound velocity reflects the elastic constant of the FLL, and the shapes of the step and the peak are governed by the activation energy U which is necessary to depin the FLL.

The sound velocity V_s and the ultrasound attenuation coefficient α in the mixed state can be written as

$$\rho V_\mathrm{s}(H,T)^2 = c_{ij}^c(H,T) + \Delta c^f(H,T) \tag{3.1}$$

$$\alpha(H,T) = \alpha^c(H,T) + \Delta \alpha^f(H,T), \tag{3.2}$$

where, ρ is the mass density. Here, c_{ij}^c and α^c are the elastic constant of the CL and the sound attenuation coefficient of the CL, respectively. In (3.1) and (3.2), Δc^f and $\Delta \alpha^f$ are the contributions from the FLL and include information about U. Within the framework of the thermally assisted flux-flow (TAFF) model, the motion of the FLL against the CL is described by a diffusion-type equation with a relaxation constant Γ. Solving the equation of motion, one gets the following formulas for Δc^f and $\Delta \alpha^f$ for $\mathbf{u} \perp \mathbf{H}$:

$$\Delta c^f = c_{ii}^f \frac{\omega^2}{\omega^2 + (c_{ii}^f \Gamma k^2)^2} \tag{3.3}$$

$$\Delta \alpha^f = \frac{\omega^2}{2\rho V_\mathrm{s}(H,T)^3} c_{ii}^f \frac{c_{ii}^f \Gamma k^2}{\omega^2 + (c_{ii}^f \Gamma k^2)^2}. \tag{3.4}$$

Here, ω is the angular frequency of the ultrasound. For $\mathbf{u} \| \mathbf{H}$, the FLL does not couple to sound waves: $\Delta c^f = 0$ and $\Delta \alpha^f = 0$. Γ is related to the Ohmic TAFF resistivity σ^{-1} caused by the flux motion with the relation

$$\sigma^{-1} = \Gamma \frac{c^2}{4\pi} \frac{H}{B} c_{44}^f, \tag{3.5}$$

where c is the velocity of light. The activation energy U is related to σ^{-1} by the Arrhenius-like formula

$$\sigma^{-1} = \rho_0 \exp\left(-\frac{U}{k_\mathrm{B}T}\right). \tag{3.6}$$

Here, ρ_0 is the proportionality coefficient and k_B is the Boltzmann constant. Equations (3.5) and (3.6) represent a relation between Γ and U.

There are three independent components in the elastic constant tensor of the FLL [26], namely, c_{11}^f as the compression modulus, c_{44}^f as the tilt modulus and c_{66}^f as the shear modulus. These moduli are expressed as [27,28]

$$c_{11}^f - c_{66}^f = \frac{B^2}{4\pi} \frac{\mathrm{d}H}{\mathrm{d}B} \tag{3.7}$$

$$c_{44}^{f} = \frac{BH}{4\pi} \tag{3.8}$$

$$c_{66}^{f} \simeq \frac{H_{c2}^{2}}{4\pi} \frac{(H/H_{c2})(1 - H/H_{c2})^{2}}{8\kappa^{2}}. \tag{3.9}$$

Here, B is the magnetic induction, H is the applied magnetic field, H_{c2} is the upper critical field and κ is the Ginzburg–Landau parameter. Elastic constants $c_{11}^{f} - c_{66}^{f}$ and c_{44}^{f} are determined from thermodynamic arguments. On the other hand, c_{66}^{f} is affected by the structure of the FLL and goes to zero if the FLL melts. When the applied field is much larger than the lower critical field and the magnetic penetration depth is long enough, one can assume the magnetic flux density in the superconductor is almost uniform and nearly equal to the applied field. In such a situation, c_{11}^{f} and c_{44}^{f} are at least 10^{3} times larger than c_{66}^{f}. Therefore, to a good approximation one can rewrite these elastic constants as

$$c_{66}^{f} \ll c_{11}^{f} \simeq c_{44}^{f} \simeq \frac{H^{2}}{4\pi}. \tag{3.10}$$

The temperature dependence of the attenuation coefficient α and the velocity V_{s} of $La_{1.88}Sr_{0.12}CuO_{4}$ ($T_{c} = 29.9\,\mathrm{K}$) in the c_{44} crystal mode were measured under the configuration of $\mathbf{k} \parallel [100]_{0}$, $\mathbf{u} \parallel [100]_{0}$ and $\mathbf{H} \parallel [100]_{0}$ in the orthorhombic notation. In this configuration, sound waves couple to the AC flux motion via an intrinsic pinning mechanism. The elastic constant of the FLL is the c_{44}^{f} tilt modulus. The FLL contribution of the sound attenuation $\Delta\alpha^{f}(H)$ can be extracted by subtracting $\alpha(0)$ from $\alpha(H)$. The temperature dependence of $\Delta\alpha^{f}$ exhibits a peak anomaly at a temperature below T_{c}, as is expected from (3.4), (3.5) and (3.6). The excess velocity $\Delta V_{s}(H)$ is extracted by subtracting $V_{s}(0)$ from $V_{s}(H)$ and converted to the excess elastic constant $\Delta c^{f}(H)$ using the relation of $\Delta c^{f}(H) = 2\rho V_{s}\Delta V_{s}(H)$. The temperature dependence of the excess elastic constant Δc^{f} induced by the FLL is shown in Fig. 3.7. Δc^{f} exhibits a steplike anomaly at a temperature below T_{c} and comes to saturation in the low-temperature region, as is expected from (3.3), (3.5) and (3.6). The agreement between experimental results and calculated curves using the same parameters obtained in the analysis of $\Delta\alpha^{f}$ is fairly good. Here the temperature dependence of U is assumed to be $U(T) = U(0)(1 - T/T_{C})^{n}$ with $n = 1.5$ as in the case of the optimally doped sample $La_{1.85}Sr_{0.15}CuO_{4}$ [29]. The activation energy $U_{AC}(0, H)$ deduced from the ultrasonic measurements is listed in Table 3.1. $U_{AC}(0, H)$ of $La_{1.88}Sr_{0.12}CuO_{4}$ for magnetic order decreases to about half the value in the optimally doped sample $La_{1.85}Sr_{0.15}CuO_{4}$. The saturated value of Δc^{f} at low temperatures is the elastic modulus of the FLL c^{f} itself because the FLL is nearly completely pinned to the crystal lattice. At a high magnetic field $H_{c1} \ll H < H_{c2}$, the tilt modulus is expressed as $c_{44}^{f} \sim H^{2}/4\pi$. The experimental value of c_{44}^{f} is proportional to H^{2} and the absolute value is about 91% of $H^{2}/4\pi$, as shown in Fig. 3.8. The reduction factor of 91% is not

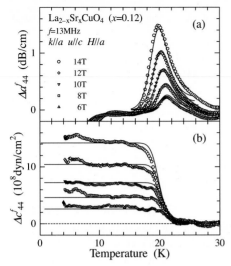

Fig. 3.7. The FLL-induced excess attenuation $\Delta\alpha^f$ (**a**) and excess elastic constant Δc^f (**b**) in $La_{1.88}Sr_{0.12}CuO_4$. Solid curves are fitted curves

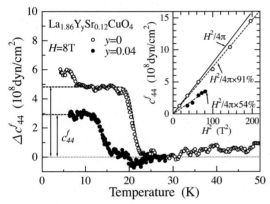

Fig. 3.8. The FLL-induced excess elastic constant Δc^f in $La_{1.88}Sr_{0.12}CuO_4$ (open circles) and $La_{1.84}Y_{0.04}Sr_{0.12}CuO_4$ (solid circles). Dotted curves are fitted curves. Inset: the plot of Δc^f against H^2. The solid line is the theoretical line, $c_{44}^f \sim H^2/4\pi$

smaller than the value of the optimally doped sample of $x = 0.15$ [29] and indicates that the volume fraction of the superconducting state is more than 91%.

The temperature dependence of the velocity V_s of $La_{1.84}Y_{0.04}Sr_{0.12}CuO_4$ ($T_c = 18\,K$) in the LTT phase was measured under the same configuration as $La_{1.88}Sr_{0.12}CuO_4$. The temperature dependence of the excess elastic constant Δc^f induced by the FLL is also shown in Fig. 3.8. For $La_{1.84}Y_{0.04}Sr_{0.12}CuO_4$ ($T_c = 18\,K$) in the LTT phase, the experimentally obtained value of the

Table 3.1. Activation energy $U_{AC}(0, H)$ in $La_{2-x-y}Y_ySr_xCuO_4$ deduced from ultrasonic measurements

| H | $x=0.15, y=1$ | $x=0.12, y=0$ | $x=0.12, y=0.04$ |
(T)	$U(0)/k_B$ (K)	$U(0)/k_B$ (K)	$U(0)/k_B$ (K)
6	1140	606	–
8	–	544	177
10	–	513	–
12	–	483	–
14	972	483	–

depinning activation energy $U_{AC}(0, H)/k_B = 177$ K is about 1/3 of that of $La_{1.88}Sr_{0.12}CuO_4$ and the FLL contribution of the elastic modulus c_{44}^f is 54% of $H^2/4\pi$, as shown in Fig. 3.8. This value is smaller than 91% of LSCO in the MTO phase but indicates that the superconducting state exists in the majority of the crystal. Notwithstanding that the degree of degradation of T_c and the degradation of the depinning activation energy are large for the LTT phase and small for the MTO phase, the superconducting state seems to coexist with antiferromagnetic-like ordering in the both phases and two kinds of carriers seem to exist: one contributes to the superconductivity and the other to the magnetic ordering.

3.6 Conclusion

The spin structure and the superconductivity in La214-type cuprates have been studied by La-NMR and ultrasonic measurements for the modulated magnetic ordering with a hole doping of $x \sim 1/8$. The spin direction has been revealed to depend on the crystal structure. That is, it is along the $[100]_o$ direction with small spin canting out of the CuO_2 plane in the orthorhombic phase and along the $[110]_o$ direction without spin canting in the low-temperature tetragonal phase. The volume fraction of the superconducting state of magnetically ordered samples with $x \sim 1/8$ has been evaluated by ultrasonic measurements and the superconductivity has been revealed to coexist with the magnetic ordered state both in orthorhombic $La_{2-x}Sr_xCuO_4$ and in tetragonal $La_{1.96-x}Y_{0.04}Sr_xCuO_4$ ($x \sim 1/8$), notwithstanding that the degree of the degradation of T_c is small for the orthorhombic phase and large for the tetragonal phase.

Acknowledgment

This work was mainly carried out at the High Field Laboratory for Superconducting Materials, IMR, Tohoku University and partially supported by the Grant-in-Aid for Scientific Research from the Ministry of Education, Science and Culture of Japan. The authors would like to thank to Prof. K. Yamada (Kyoto University) for valuable discussions.

References

1. J.D. Axe, A.H. Moudden, D. Hohlwein, D.E. Cox, K.M. Mohanty, A.R. Moodenbaugh and Y. Xu: Phys. Rev. Lett. **62**, 2751 (1989)
2. K. Kumagai, H. Matoba, N. Wada, M. Okaji and K. Nara: J. Phys. Soc. Jpn. **60**, 1448 (1991)
3. N. Wada, S. Ohsawa, Y. Nakamura, K. Kumagai: Physica, B **165-166**, 1345 (1990); N. Wada, Y. Nakamura and K. Kumagai: Physica C **185**, 1177 (1991)
4. G.M. Luke, L.P. Le, B.J. Sternlieb, W.D. Wu, Y.J. Uemura, J.H. Brewer, T.M. Riseman, S. Ishibashi and S. Uchida: Physica C **185**, 1175 (1991)
5. H. Tou, M. Matsumura and H. Yamagawa: J. Phys. Soc. Jpn. **61**, 1477 (1992); **62**, 1474 (1993)
6. K. Miyagawa, T. Goto and T. Fukase: Physica B **194-196**, 2175 (1994)
7. J.M. Tranquada, B.J. Sternlieb, J.D. Axe, Y. Nakamura and S. Uchida: Nature **375**, 561 (1995)
8. S. Ohsugi: J. Phys. Soc. Jpn. **64**, 3656 (1995)
9. T. Goto, S. Kazama and T. Fukase: Physica C **235–240**, 1661 (1994)
10. T. Goto, S. Kazama, K. Miyagawa and T. Fuakse: J. Phys. Soc. Jpn. **63**, 3494 (1994)
11. K. Kumagai, K. Kawano, H. Kagami, G. Suzuki, Y. Matsuda, I. Watanabe, K. Nishiyama and K. Nagamine: Physica C **235–240**, 1715 (1994)
12. I. Watanabe, K. Nishiyama, K. Nagamine, K. Kawano and K. Kumagai: Hyperfine Interact. **86**, 603 (1994)
13. Y. Maeno, A. Odagawa, N. Kakehi, T. Suzuki and T. Fujita: Physica C **173**, 322 (1991)
14. T. Suzuki, T. Goto, K. Chiba, T. Shinoda, T. Fukase, H. Kimura, K. Yamada, M. Ohashi and Y. Yamaguchi: Phys. Rev. B **57**, R3229 (1998)
15. H. Kimura, K. Hirota, H. Matsumura, K. Yamada, Y. Endoh, S.H. Lee, C.F. Majkrzak, R.W. Erwin, G. Shirane, M. Greven, Y.S. Lee, M.A. Kastner and R.J. Birgeneau: Phys. Rev. **59**, 6517 (1999)
16. T. Fukase, H. Geka, T. Goto, K. Chiba and T. Suzuki: J. Low Temp. Phys. **117**, 491 (1999)
17. T. Suzuki, M. Sera, T. Hanaguri and T. Fukase: Phys. Rev. B **49**, 12392 (1994)
18. T. Fukase, T. Nomoto, T. Hanaguri, T. Goto and Y. Koike: Physica B **165-166**, 1289 (1990), Jpn. J. Appl. Phys. Series 7, 213 (1992)
19. T. Suzuki, K. Chiba, T. Goto and T. Fukase: Czechoslovak J. Phys. **46**, S3-1237 (1996)
20. H. Nishihara, H. Yasuoka, T. Shimizu, T. Tsuda, T. Imai, S. Sasaki, S. Kanbe, K. Kishio, K. Kitazawa and K. Fueki: J. Phys. Soc. Jpn. **56**, 4559 (1987)
21. T. Goto, K. Chiba, M. Mori, T. Suzuki, K. Seki and T. Fukase: J. Phys. Soc. Jpn. **66** 2870 (1997)
22. W. Koshibae, Y. Ohta and S. Maekawa: Phys. Rev. **50** 3767 (1994)
23. K. Chiba, T. Goto, M. Mori, T. Suzuki, K. Seki and T. Fukase: J. Low Temp. Phys. **117**, 479 (1999)

24. M. Tachiki and S. Takahashi: Solid State Commun. **70**, 291 (1989), **72**, 1083 (1989)
25. J. Pankert; Physica C **168** 335 (1990), Physica B **165-166**, 1272 (1990)
26. A.M. Campbell and J.E. Evetts: Adv. Phys. **21**, 199 (1972)
27. R. Labusch: Phys. Status Solidi. **32** 439 (1969)
28. E.H. Brandt: J. Low Temp. Phys. **26** 735 (1977)
29. T. Hanaguri, T. Fukase, I. Tanaka and H. Kojima: Phys. Rev. B **48**, 9772 (1993)

4 Flux-Pinning Properties for CVD Processed YBa$_2$Cu$_3$O$_7$ Films

S. Awaji, K. Watanabe, N. Kobayashi and T. Hirai

Since the discovery of high-temperature superconductors (HTSC), their practical applications at high temperatures such as at liquid nitrogen temperature have been widely expected. Many groups have successfully developed Ag-sheathed Bi-Sr-Ca-Cu-O (BSCCO) and recently a magnetic field of 7 T was generated by a superconducting magnet using Bi$_2$Sr$_2$Ca$_2$Cu$_3$O$_{10}$ tapes under a cryocooling condition [1]. However, the BSCCO system shows a poor magnetic field dependence of the critical current density J_c in a high-temperature region above 30 K [2]. The YBa$_2$Cu$_3$O$_7$ (YBCO) system is, on the other hand, expected to see practical use as a superconducting wire of HTSC because of its high J_c properties even at high temperature and high magnetic field. This means that strong pinning centers exist in the YBCO system. Fine precipitates such as CuO [3] and Y$_2$O$_3$ [4], oxygen deficiency [5], Y211 large particles [6] and so on were pointed out as effective pinning centers in the YBCO system. Moreover, Tachiki and Takahasi have proposed a new concept for the flux pinning in HTSCs [7]. Superconducting layers can work as strong pinning centers intrinsically, due to a layer structure and a short coherence length along the c-axis of HTSCs. Therefore, this model is called an intrinsic pinning model and is characterized by a two-dimensional feature due to the large structure. Since the intrinsic pinning is closely related to the modulation of the superconducting energy along the c-axis, the relevant pinning strength depends on the anisotropy of the system. The conventional flux pinning by point defects, precipitates, grain boundaries and so on is called extrinsic pinning as opposed to intrinsic pinning. It is considered that both pinning mechanisms work competitively in a HTSC. Since the intrinsic pinning works mainly for the application of a magnetic field perpendicular to the crystal c-axis ($B \perp c$), high J_c characteristics are obtained in high magnetic fields up to 30 T even at 77.3 K for $B \perp c$. The J_c properties for $B \parallel c$ are more important for practical applications, because the J_c properties for $B \parallel c$ are much worse in comparison with those for $B \perp c$ at high temperature due to large anisotropy. The temperature-scaling law of the global pinning force is a good tool for characterizing the flux-pinning properties [8]. If the pinning force $F_p = J_c \times B$ and applied magnetic field are normalized by the maximum value F_{pmax} and the irreversibility field B_i, respectively, all the F_p versus B curves taken at various temperatures must be scaled and then the features of the scaled curve are determined. Such a temperature-scaling law was found not only for conventional superconductors [8] but also for HTSCs [9] and the features of the B_i properties for various kinds of HTSCs were reported by many groups [10]. Most HTSCs indicate concave curves for the temperature-dependent B_i values in the low-temperature region. Shimoyama et al. pointed out that the temperature dependence of the B_i values for various HTSCs are superimposed on a universal line, if $\gamma^2 B_i$ is plotted as a function of $T/T_c - 1$, where γ^2 is an anisotropic parameter [11]. In the YBCO case, however, the B_i properties are evaluated only in the temperature region near the critical temperature because of the large B_i values. It is pointed out that superconductivity of a HTSC has d-wave symmetry and the upper critical field B_{c2} shows the upward temperature dependence [12]. B_{c2} properties also affect the flux-pinning properties. In this review, the details of the flux-pinning properties for YBCO films prepared by chemical vapor deposition (CVD) are described and discussed in terms of the intrinsic and extrinsic pinning mechanisms. Moreover, we evaluate the B_i and J_c properties for $B \parallel c$ based on the temperature-scaling law. In the last part of this review, the capacities of YBCO for practical applications are discussed using the superconducting critical surface.

4.1   Critical Current Measurement

The bridges on the films were made using a Pt mask $20\,\mu m$ in thickness for critical current measurements. The typical size of the bridge was about $1\,mm$ in length and $500\,\mu m$ in width. Four lead wires for transport property measurements were attached by a mechanical pressure contact using indium tips. A heater and carbon glass and Pt thermometers were mounted inside a Cu block used as a sample holder. A capacitance thermometer and a sample were set on the same Cu block using GE7051 adhesive. This Cu block sample holder can be rotated with $0.5°$ angular resolution. The capacitance thermometer, which is hardly affected by the magnetic field, was used for controlling the sample temperature in a magnetic field. The sample holder was inserted into a cryostat in which the temperature could be controlled by flowing He gas. Critical currents were measured using a pulsed critical current measurement system [13] in order to avoid Joule heating at the terminals. The currents were monitored through a shunt resistance which was connected to the sample in series. The sample current and voltage were measured using a low-pass filter, an amplifier and a digital oscilloscope. The gain of the amplifier was $10\,000$ to get above $0.1\,\mu V$ sensitivity. Both current and voltage signals were measured as a function of time and these data were converted into current-voltage characteristics. The critical currents were obtained from the current-voltage relation with a voltage criterion of $2\,\mu V/cm$.

4.2   Characteristics of CVD-YBCO Films

YBCO films were deposited on SrTiO$_3$ (100) by a hot-wall type CVD process using β-diketonates of Y(thd)$_3$, Ba(thd)$_2$ and Cu(thd)$_2$, where thd is a 2, 2, 6, 6-tetramethyl-3, 5-heptanedionate [14]. The chemical compositions of films can be controlled by changing the vaporizer temperatures for the CVD process. The microstructures and superconducting properties strongly depend on the chemical composition [15]. The obtained films with Cu-rich composition show an excellent performance of the critical current densities in fields up to 30 T perpendicular to the c-axis, as shown in Fig. 4.1 [14]. The J_c values at 77.3 K are about 4.7×10^5 A/cm^2 at 0 T and 1.7×10^4 A/cm^2 at 23 T for $B \perp c$ in this case. However, J_c for $B \parallel c$ is degraded drastically by applying a few tesla. This is due to the large anisotropic properties for HTSCs. Figure 4.2 shows a typical surface morphology of a high J_c film measured by a scanning electron microscope (SEM). Many particles with a size of a few micrometers were observed in Fig. 4.2. An electron-probe-microanalysis (EPMA) indicates that these particles are CuO and are grown from substrates [15]. The concentration and size of the precipitates vary with the film composition. The number of particles increases with increasing excess Cu and the J_c values also tend to be improved with the number of CuO particles. Figure 4.3 indicates the plane view of the transmission electron microscopy

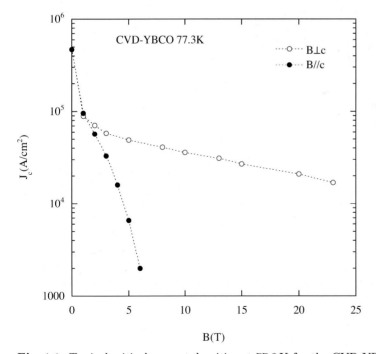

Fig. 4.1. Typical critical current densities at 77.3 K for the CVD-YBCO film

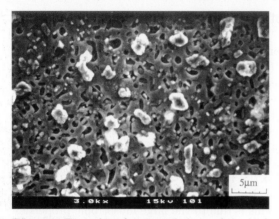

Fig. 4.2. Typical surface morphology of the high-J_c film measured by a scanning electron microscope (SEM)

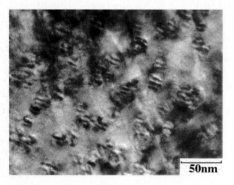

Fig. 4.3. Plane view of the transmission electron microscopy (TEM) image

(TEM) image. The numerous strain contrasts are observed as a moiré pattern and are considered to be fine precipitates [15]. The typical size of these fine precipitates is 20–30 nm in diameter and 1 nm in thickness. Since this sample was prepared in the Cu-rich compositional area, the fine precipitates are probably CuO. It was reported that fine Y_2O_3 precipitates of less than 5 nm in size were observed in sputtered YBCO films prepared in the Y-rich region [4]. It is considered that these fine precipitates are effective pinning centers for YBCO films with high J_c values.

4.3 Crossover from Extrinsic to Intrinsic Pinning

When a magnetic field is applied perpendicular to the c-axis, pinning of an intrinsic nature appears. In particular, most of the angular dependence of J_c for HTSCs was described by the intrinsic pinning model [16]. A few groups pointed out, however, that intrinsic pinning does not work at high tempera-

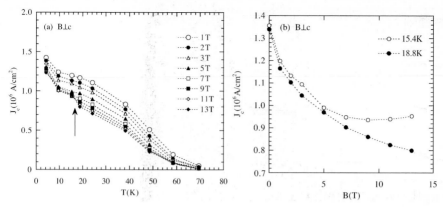

Fig. 4.4. Temperature dependence (**a**) and field dependence (**b**) of J_c for $B \parallel c$

tures around 77.3 K [17]. This contradiction can be understood as the competition between intrinsic and extrinsic pinning [18]. Figure 4.4a shows the temperature dependence of J_c. An anomalous kink in the temperature dependence of J_c is observed around 16 K at high magnetic fields above 7 T for $B \perp c$, as shown by the arrow, although J_c increases monotonically without any anomaly for $B \parallel c$ with decreasing temperature. The field dependence of J_c for $B \perp c$ is drastically changed around 16 K, as shown in Fig. 4.4b. The J_c values decrease monotonically with magnetic field at 18.8 K but become insensitive to magnetic fields above 7 T at 15.4 K. These anomalies are understood as a result of the crossover from extrinsic pinning at high temperatures to intrinsic pinning at low temperatures [19].

In order to confirm that intrinsic pinning is dominant at low temperature, the angular dependence of J_c is evaluated as shown in Fig. 4.5, where θ is the angle between the field direction and the c-axis. According to the intrinsic model, the angular dependence of J_c is determined by the c-axis component of the applied magnetic field. Therefore, the angular dependence of J_c is estimated from the field dependence of J_c for $B \parallel c$. The solid lines in Fig. 4.5 are adopted from the data of the field dependence of $J_c(B)$ for $B \parallel c$ and are compared with the measured angular dependence of J_c. The estimated values agree well with the measured ones below 10.5 K, but, above 18.8 K, the former deviate from the latter at high angles. These results indicate that intrinsic pinning is dominant below 10.5 K and does not work above 18.8 K. Hence, the effective pinning mechanisms are different in the high- and low-temperature regions.

The J_c values calculated by our competition model reproduce the directly observed results qualitatively and the anomalous kink appears to be due to the different temperature dependence of J_c. The crossover point is determined by the ratio between intrinsic and extrinsic pinning. In other words, the crossover point varies with the pinning strength. This prediction is confirmed

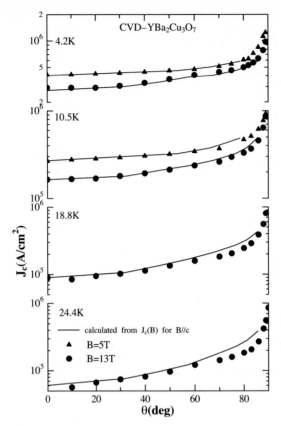

Fig. 4.5. Angular dependence of J_c at various temperatures. Observed data are compared with the calculated results from $J_c(B)$ for $B \perp c$

by experimental results that indicate that the crossover point is related to the magnitude of J_c [18]. For example, the pinning crossover does not appear for the high-J_c sample since the intrinsic pinning is masked.

The intrinsic pinning shows the problems associated with a large anisotropy of J_c. However, this model suggests that not only an improvement of the flux pinning but also a reduction of the anisotropy can be achieved by the introduction of strong extrinsic pinning centers.

4.4   Temperature-Scaling Law and Irreversibility Field

When the magnetic field is applied parallel to the c-axis, the J_c values are rapidly reduced by the magnetic field. For application of HTSCs, therefore, the characterization and improvement of J_c for $B \parallel c$ are most important. Since intrinsic pinning does not work for $B \parallel c$, we can consider only extrinsic pinning such as that from normal precipitates. In order to investigate the

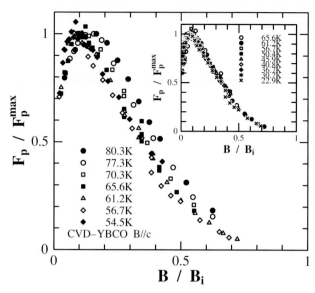

Fig. 4.6. Scaling plot of F_p at various temperatures. Inset shows the scaling plot focusing on low temperatures below 65 K

flux-pinning characteristics, we analyze the J_c properties in terms of the temperature-scaling law of the global pinning force F_p [8]. The temperature-scaling law is written as the following equation:

$$F_p = \left(\frac{B}{B_i}\right)^p\left(1 - \frac{B}{B_i}\right)^q,\tag{4.1}$$

where p and q are scaling parameters and B_i is the irreversibility field. This equation means that the scaling characteristics are obtainable if F_p and B are normalized by the maximum value of F_p and B_i, respectively. It is well known that the scaling parameters are affected by the flux pinning mechanism. Figure 4.6 shows the scaling plot of F_p calculated from the measured J_c values. The normalized F_p curves at various temperatures are not scaled in the high-temperature region above 65 K, but are superimposed on a universal curve in the low-temperature region below 65 K, as shown in the inset. In order to examine the temperature dependence of the scaling characteristics of F_p, the ratio of the magnetic field, B_{Fpmax}, at which the F_p value becomes a maximum, to the irreversibility field, B_{Fpmax}/B_i, is shown as a function of temperature in Fig. 4.7. If the global pinning force is scaled against the normalized irreversibility field, the B_{Fpmax}/B_i values become independent of temperature. However, in this case they decrease with temperature at high temperature and then saturate at about 0.1 at low temperatures below 65 K. As a result, the temperature scaling of F_p occurs only in the low-temperature region below 65 K [13]. Two possible reasons can be considered for the temperature dependence of B_{Fpmax}/B_i as shown in Fig. 4.7. One is a flux-creep

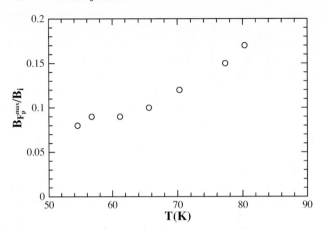

Fig. 4.7. Temperature dependence of B_{Fpmax}/B_i

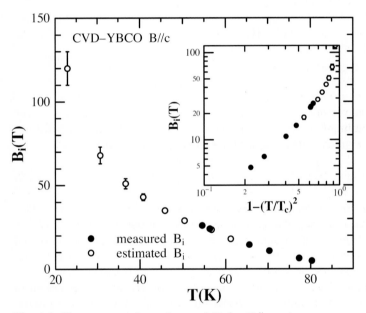

Fig. 4.8. Temperature dependence of B_i for $B \parallel c$-axis

effect and another is a change of the flux pinning. In the case of flux creep, the scaling behavior of F_p is strongly dependent on the flux-pinning strength because the competition between the thermal energy and pinning potential is important. However, the results as shown in Fig. 4.7 indicate that the temperature dependence of B_{Fpmax}/B_i for two different samples, whose J_c values are about one order of magnitude different from each other, trace the same line. Therefore, it is suggested that the flux-pinning characteristics change at high temperature around 65 K.

Although the B_i value at low temperatures is too large to be measured, the B_i value can be estimated from the F_p^{max} values, assuming the scaling relation between F_p/F_p^{max} and B/B_i. The scaling behavior of F_p/F_p^{max} at various temperatures in the low-temperature region below 65 K is shown in the inset of Fig. 4.6. The normalized curves are obtained using the estimated B_i at low temperature and the measured B_i at high temperature. All curves are well scaled, whether B_i is the estimated value or not. The peak position of the normalized pinning force appears at $B/B_i \simeq 0.1$ in the low-temperature scaling region. The scaling curve of F_p obtained in this study is different from previous studies by other groups [20,21]. This suggests that the CVD process with a Cu-rich chemical composition could introduce different pinning centers from those produced by other processes such as a physical vapor deposition [22]. However, the origin of the strong flux pinning still remain unsettled. The comparative study of microstructures and the J_c properties is necessary in order to solve this problem.

Figure 4.8 represents the temperature dependence of B_i for $B \parallel c$-axis. The closed and open symbols are the measured B_i values and the estimated ones, respectively. B_i follows the power-law dependence of $(1-(T/T_c)^2)^n$ with $n \simeq 1.5$ in the high-temperature region and then increases rapidly below 60 K with decreasing temperature, as shown in the inset of Fig. 4.8. This temperature is almost in agreement with the temperature at which the B_{Fpmax}/B_i saturates, as mentioned above. The B_i characteristics also indicate the change of the flux pinning mechanisms. A similar temperature dependence of B_i was observed for layered superconducting systems such as Bi-Sr-Ca-Cu-O [23] and κ-(BEDT-TTF)$_2$Cu(NCS)$_2$ [24]. These superconductors are characterized by the appearance of the 2nd peak in magnetization hysteresis at the so-called dimensional crossover field B_{cr}, which was proposed by Glazman and Koshelev [25]. In this case, the dimensionality change of the flux gives rise to an enhancement of the flux-pinning efficiency. As a result, the J_c and B_i values are enhanced. Recently, Koyama and Tachiki calculated B_{c2} in two-dimensional d-wave superconductors [12]. They found an enhancement of B_{c2} in the low-temperature and high-field region and it is similar to the temperature dependence of the irreversibility field obtained in this study. Therefore, the enhancement of B_i may be related to the temperature dependence of B_{c2} for two-dimensional d-wave superconductors.

4.5   Critical Surface of YBCO

Using our data, the typical critical surface of the CVD-YBCO film for $B \parallel c$ is presented in Fig. 4.9. In this figure, the obtained J_c and B_i data are shown as solid lines. J_c decreases drastically with magnetic field at high temperatures but depends weakly on magnetic field at lower temperatures. For high-temperature use at 77.3 K, immersed in liquid N$_2$, low-field applications up to about 5 T, such as for power cables, energy storage and small superconduct-

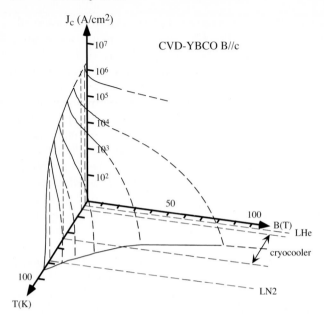

Fig. 4.9. Superconducting critical surface of CVD-YBCO for $B \parallel c$

ing magnets, can be developed. In the recent development of power cables, BSCCO wires are utilized in the self-field. However, YBCO is superior to BSCCO for applications in magnetic fields, because the high-field properties of YBCO are much better than those of BSCCO at high temperature [2]. An YBCO superconducting magnet operating at liquid nitrogen temperature has a capacity of generating fields up to about 5 T. The development of practical YBCO wires is now in progress by many groups. A convenient superconducting magnet cooled by liquid N_2 could be demonstrated by using YBCO wires in the future.

On the other hand, it is important that the superconducting region for practical use expands to very high magnetic fields, above 100 T, at low temperatures below 30 K, as seen in Fig. 4.8. Moreover, the degradation of J_c becomes small in magnetic fields up to at least 30 T below 50 K. Therefore, we can consider very high magnetic field applications of YBCO in cryocooling conditions and when immersed in liquid He, as shown in Fig. 4.9. The superconducting magnets employing conventional low-temperature superconductors are generally used immersed in liquid He. However, a new type of superconducting magnet, which is called a cryogen-free superconducting magnet, has been successfully developed [26]. A cryogen-free superconducting magnet has recorded the highest field, 15.1 T, so far [27]. One of advantages of the cryogen-free superconducting magnet is no use of liquid He. The available cooling power of the Gifford-McMahon (GM) cryocooler has become about

1 W at 4.2 K, recently, and the HTSC current leads play a role in the drastic reduction of heat flow to the superconducting coil through the current
leads. These two technologies are very important for the cryogen-free superconducting system. Since most of these magnets are made of the conventional
superconductors NbTi or Nb$_3$Sn, the operating temperature should be kept
to around 5 K. However, if we apply YBCO wire to the cryocooled superconducting magnet, the operating temperature can rise. In fact, Kato et al.
have developed a cryogen-free superconducting magnet wound with BSCCO
tapes which operates at about 25 K [1]. In this case, the technical merits are
a large heat capacity and a large cooling power due to the higher operating
temperature. Making good use of these technical advantages, they succeeded
in performing a repetitive operation with fast sweeping. Usually, the heating
resulting from ac losses increases upon increasing the energizing rate and this
causes serious problems for operation in the case of a conventional superconducting magnet. In HTSC magnets, however, since the temperature increase
due to ac losses is small enough in comparison with the critical temperature,
repetitive operation with fast sweeping can be carried out stably. The high
field properties of YBCO are still better than those of BSCCO in this temperature range. If we design a magnet for operation at 30 K, YBCO has a
capacity of generating fields up to 30 T, as shown in Fig. 4.9.

Furthermore, there is a promising application for HTSCs for use as insert
coils for very high field superconducting magnets because the J_c and B_i
values are extremely high at 4.2 K, as shown in Fig. 4.9. By immersion in
liquid He, conventional superconducting magnets can generate about 21.3 T
at 2.2 K [28]. HTSCs can be utilized as an innermost coil for a very high field
superconducting magnet generating over 30 T.

4.6 Conclusion

YBa$_2$Cu$_3$O$_7$ (YBCO) films have demonstrated excellent performance in achieving a high J_c value in high magnetic fields up to 30 T, even at a high temperature of 77.3 K. It is found that the crossover point of the pinning appears
as a kink anomaly in the temperature and field dependence of J_c. According to the competition model of two pinning mechanisms, the crossover from
extrinsic to intrinsic pinning is affected by the strength of the extrinsic pinning. For magnetic fields parallel to the c-axis, the temperature-scaling law
of the global pinning force was obtained at low temperatures but not at high
temperatures above 65 K. Although the irreversibility field B_i for $B \parallel c$ is
proportional to $(1 - T/T_c)^{3/2}$ at high temperatures, it is extremely enhanced
at low temperatures below 60 K. It is considered that this temperature dependence of B_i is related to the feature of the upper critical field of the
two-dimensional d-wave superconductor. Experimental results show that the
critical surface of superconductivity for YBCO expands to a very high field
region at low temperature.

Acknowledgements

We would like to thank Professor H. Yamane for a large contribution to this study. This work was performed at the High Field Laboratory for Superconducting Materials, the Institute for Materials Research, Tohoku University. We wish to express our gratitude to Messrs. M. Kudo, K. Sai, Y. Ishikawa and Y. Sasaki for operating the hybrid magnets and Messrs. S. Otomo, H. Miura, S. Tanno and K. Hosokura for supplying liquid helium.

References

1. K. Kato, K. Ohkura, M. Ueyama, K. Ohmatsu, K. Hayashi and K. Sato: *Int. Conf. Magnet Technology, Proc. 15th Magnet Technology Conf.* L. Lin, G. Shen and L. Yan eds., Beijin, 10-19 November, 1997 (Science Press, 1998) p. 793
2. K. Watanabe, S. Awaji: Sci. Rep. RITU **A42**, 371 (1996)
3. K. Watanabe, T. Matsushita, N. Kobayashi, N. Kawabe, H. Aoyagi, E. Hiraga, H. Yamane, T. Hirai and Y. Muto Appl. Phys. Lett. **56**, 1490 (1990)
4. P. Lu, Y. Q. Li, J. Zhao, S. Chern, B. Gallios, P. Norris, B. Kear and F. Cosandey: Appl. Phys. Lett. **60**, 1265 (1992)
5. M. Daeumiling, J.M. Seuntjents and D.C. Larbalestier: Nature **346**, 332 (1988)
6. M. Murakami, T. Oyama, H. Fujimoto, S. Gotoh, K. Yamaguchi, Y. Shinohara, N. Koshizuka and S. Tanaka: IEEE Trans. Magn. **27**, 1479 (1991)
7. M. Tachiki and S. Takahashi: Solid State Commun. **72**, 1083 (1989)
8. W.A. Fietz and W.W. Webb: Phys. Rev. **178**, 657 (1969)
9. H. Yamasaki, K. Endo, S. Kosaka, M. Umeda, S. Yoshida and K. Kajimura: Phys. Rev. Lett. **70**, 3331 (1993)
10. For example, K. Kishio, J. Shimoyama, T. Kimura, Y. Kotaka, K. Kitazawa, K. Yamafuji, Q. Li and M. Suenaga: Physica **C 235–240**, 2775 (1994)
11. J. Shimoyama, K. Kitazawa and K. Kishio: Physica **B 280**, 249 (2000)
12. T. Koyama and T. Tachiki: Physica C **203**, 25 (1996)
13. S. Awaji, K. Watanabe, N. Kobayashi, H. Yamane and T. Hirai: Jpn. J. Appl. Phys. **32**, L1532 (1992)
14. K. Watanabe, H. Yamane, H. Kurosawa, T. Hirai, N. Kobayashi, H. Iwasaki, K. Noto and Y. Muto: Appl. Phys. Lett. **54**, 575 (1989)
15. H. Yamane, T. Hirai, H. Kurosawa, A. Suhara, K. Watanabe, N. Kobayashi, H. Iwasaki, E. Aoyagi, K. Hiraga and Y. Muto: *Proc. 2nd Int. Symp. Superconductivity* T. Ishiguro and K. Kajimura eds., Tsukuba, 14-17 Nov, 1989 (Springer, Berlin, 1989) *Adv. Supercond. II* p. 767
16. M. Tachiki and S. Takahashi: Solid State Commun. **70**, 291 (1989)
17. B. Roas, L. Schultz and G. Saemann-Ischenko: Phys. Rev. Lett. **64**, 479 (1990)
18. S. Awaji, K. Watanabe, N. Kobayashi, H. Yamane and T. Hirai: Cryogenics **39**, 569 (1999)
19. S. Awaji, K. Watanabe, N. Kobayashi, H. Yamane and T. Hirai: Jpn. J. Appl. Phys. **32**, L1795 (1993)
20. T. Nishizaki, T. Aomine, I. Fujii, K. Yamamoto, S. Yoshii, T. Terashima and Y. Bando: Physica C **181**, 223 (1991)
21. J. N. Li, F.R. de Boer, L.W. Roeland, M.J. Menken, K. Kadowaki, A.A. Menovsky, J.J.M. Franse and P. Kes: Physica C **169**, 81 (1990)

22. K. Watanabe, S. Awaji and T. Fukase: Synthetic Metals **71**, 1885 (1995)
23. K. Kishio, J. Shimoyama, K. Kotaka and K. Yamafuji: *Proc. 7th Int. Workshop Critical Currents Supercond.*, Alpbach,1993, (World Sci. Publ. 1994) p. 339
24. T. Nishizaki, T. Sasaki, T. Fukase and N. Kobayashi: Synthetic Metals **85**, 1497 (1997)
25. L.I. Glazman and A.E. Koshelev: Phys. Rev. **B43**, 2835 (1991)
26. K. Watanabe, Y. Yamada, J. Sakuraba, F. Hata, C.K. Chong, T. Hasebe and M Ishihara: Jpn. J. Appl. Phys. **32**, L488 (1993)
27. K. Watanabe, S. Awaji, M. Motokawa, Y. Mikami, J. Sakuraba and K. Watazawa: Jpn. J. Appl. Phys. **37**, L1148 (1998)
28. R. Hirose, T. Kamikado, O. Ozaki, M. Yoshikawa, T. Hase, M. Shimada, K. Kawate, K. Takabatake, M. Kosuge, T. Kiyoshi, K. Inoue and H. Wada: *Int. Conf. Magnet Technology, Proc. 15th Magnet Technology Conf.* L. Lin, G. Shen and L. Yan eds., Beijin, 10-19 November, 1997 (Science Press, 1998) p. 874

5 Practical Application
of High Temperature Superconductors

K. Watanabe and M. Motokawa

NbTi alloy superconductors with a critical temperature of about 9 K, which are called the low temperature superconductors, have produced excellent superconducting magnets. For instance, a large helical device (LHD) was constructed for a nuclear fusion experimental reactor [1], and in basic research a large dipole superconducting magnet for a particle accelerator was very successful [2]. Further, a NbTi superconductor fundamentally contributed to an industrial application as a medical magnetic resonance imaging (MRI) superconducting magnet [3], and is now beginning to be industrialized for a magnetic levitation train [4]. Although Nb_3Sn compound superconductors with a critical temperature of 18 K are used for high magnetic field generation over 10 T, which NbTi superconductors cannot achieve, they are not finding wide use like NbTi, because of weak mechanical strength and large strain sensitivity. Recently, a 20 T superconducting magnet employing Nb_3Sn superconducting wires has been chosen for NMR experiments due to their usefulness as a powerful tool for structure analyses of protein [5]. On the other hand, high-temperature superconductors have achieved critical temperatures 130 K as high as fourteen years old after their discovery. It is important to reconsider what kinds of practical application for high-temperature superconductors have been demonstrated so far. The critical temperature of high-temperature superconductors is very high in comparison with low-temperature superconductors, and this results in very short coherence lengths for high-temperature superconductors. Since thermal fluctuations increase at high temperature, it is very difficult to observe a clear superconducting phase transition at liquid nitrogen temperature due to the thermal effect. High-temperature superconductors exhibit large anisotropy in superconducting properties, because the electrical conduction plane is two-dimensional due to the crystal structure. These features are disadvantageous from the viewpoint of superconducting applications, but do not become fatal. It is possible to use both high- and low-temperature superconductors in the desired temperature and field regions. In high-temperature superconductors, a manufacturing method called powder-in-tube, using a Ag sheath, has succeeded in wire-drawing of a long length tape for Bi-system high-temperature superconductors. It is very important to achieve the wire drawing, since a long length wire is indispensable for power applications. Until now, the following applications have been investigated by using Ag-sheathed Bi-system high-temperature superconducting tape: a superconducting magnet for high-field generation which exceeds 20 T at 4.2 K and a cryogen-free superconducting magnet operating at 20 K. Moreover, high-temperature superconducting bulk materials are already used for actual current leads, and are expected to be used for bulk magnets such as a magnetic levitation carrier device, a magnetic bearing, a flywheel, and a motor. In the High Field Laboratory for Super-conducting Materials, Tohoku University, we have investigated the critical current characteristics of high-temperature superconductors in fields up to 30 T. The history effect in the critical current, which is the characteristic feature of high-temperature superconducting bulk [6], and the high performance of CVD-processed $YBa_2Cu_3O_7$ (Y123) films at 77 K [7] have been reported. In addition, we have been the first to demonstrate a practical cryogen-free superconducting magnet using high-temperature superconducting current leads [8].

In this article, the research on practical applications of high-temperature superconductors carried out at the High Field Laboratory for Superconducting Materials HFLSM is described, and the future programme of this laboratory for high-temperature superconducting applications is presented.

5.1   Critical Surface and Critical Current Density Characteristics for High-Temperature Superconductors

The use of superconductors for power applications is carried out within the critical surface consisting of the critical temperature, upper critical field and critical current density. In high-temperature superconductors, the irreversibility field is utilized instead of the upper critical field for practical usefulness. Figure 5.1 shows the critical surfaces of NbTi, Nb_3Sn, Y123 and $Bi_2Sr_2Ca_2Cu_3O_{10}$ (Bi2223) for comparison. The value of the critical current density at 4.2 K in the absence of a magnetic field is approximately $J_c = 10^8 \, A/cm^2$ for low-temperature as well as high-temperature superconductors, whose value is almost close to the Cooper-pair-breaking current density. The characteristic point is that high-temperature superconductors have an extremely large irreversibility field. It is expected that the irreversibility field of Y123 high-temperature superconductors will reach several 100 T at 4.2 K [9].

Figure 5.2 shows the magnetic field dependence of the critical current density for NbTi, Nb_3Sn, and $Bi_2Sr_2CaCu_2O_8$ (Bi2212). Since high-temperature superconductors have large anisotropy in the critical current properties, the critical current densities for B/c are shown. The overall critical current density for a NbTi wire is decreased by about $1/3$ in comparison with the superconducting-core current density, even if the multifilamentary wire structure in which the superconducting cores are embedded in stabilized copper is adopted. On the other hand, the overall current density for practical Nb_3Sn compound superconductors is reduced due to bronze and diffusion barriers by more than one order of magnitude in comparison with the superconducting-

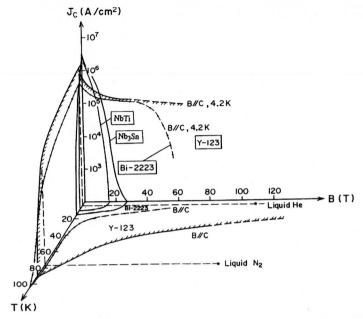

Fig. 5.1. Critical surfaces of NbTi, Nb₃Sn, Y123 and Bi2223 superconductors

core current density. When the practical overall current density of 10^4 A/cm^2 is selected for magnet design, one notes that the critical current density for high-temperature superconductors is very high at 4.2 K.

Although there is a problem whether or not high-temperature superconductors with mechanical reinforcement can achieve an overall current density over 10^4 A/cm^2, their potential for practical applications is considerably high.

5.2   High-Temperature Superconducting Applications to Current Leads

5.2.1   Y123 Current Leads and a Critical-Current Measurement Holder

In the application of high-temperature superconductors to current leads, the temperature-dependence of the thermal conductivity is to be given as a fundamental property. Thermal conductivity is largely different between c-axis and ab-plane directions, due to the very large anisotropy of high-temperature superconductors. Y123 bulk materials used for current leads are fabricated by extrusion or by the melt-growth process. The thermal conductivity in the ab-plane for melt-growth-processed Y123 bulk is presented in Fig. 5.3. Since the density of the electrons involved in the thermal conduction is one order of magnitude smaller in high-temperature superconductors than in a normal

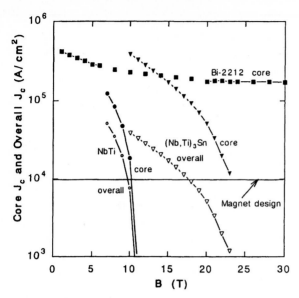

Fig. 5.2. Overall and core critical current densities in fields at 4.2 K for NbTi, Nb₃Sn, and Bi2223 superconductors

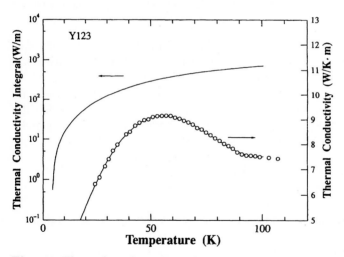

Fig. 5.3. Thermal conductivity and its integral value as a function of temperature for melt-growth-processed Y123 bulk

metal, the phonons are the main contributors to the thermal conduction. The thermal conductivities of copper, bronze and Y123 are 480, 65 and 20 W/m·K at 77 K, respectively. One can notice that high-temperature superconductors have very small thermal conductivity. This results in a large cross-section and a very short length for current leads to achieve a large transport current ca-

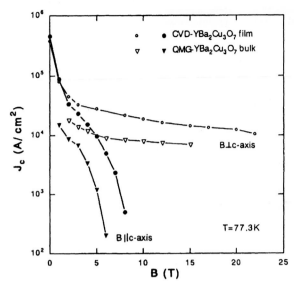

Fig. 5.4. Critical current densities versus magnetic field for melt-growth-processed Y123 bulk and CVD-processed Y123 thin film superconductors at 77.3 K

pability. These characteristics permit the utilization of new superconducting current leads where copper current leads cannot be used.

In critical current measurements, a sample holder with many copper leads causes a large consumption of liquid helium. Y123 current leads are available for the reduction of thermal inputs into liquid helium. In the case of Y123 current leads, bulk Y123 itself has to be covered with fiber reinforced polymer (FRP), whose thermal conduction is as low as that of high-temperature superconductors, because Y123 has an affinity for water and its superconducting properties are largely degraded by water. Bulk Y123 fabricated by the melt-growth process is cut into a block of dimensions $7 \times 7 \times 70 \,\mathrm{mm}^3$, and both ends are directly connected with copper electrodes. Although a compressive cooling stress is applied to the Y123 due to the slightly larger thermal contraction of the FRP cover, the critical-current characteristics of Y123 do not change in compressive stress.

Figure 5.4 shows the superconducting properties at 77.3 K for melt-growth-processed bulk Y123. The anisotropic behavior of the critical current density for a CVD-processed Y123 film is also shown for comparison. The critical current density for bulk Y123 is lower than for a Y123 film. However, it is possible to make Y123 current leads with a transport current capacity of 1500 A in fields up to 4 T for $B//c$-axis. This current lead has been used practically in the High Field Laboratory for Superconducting Materials (HFLSM) in Sendai as a critical-current measurement holder at 1.8 K, as shown in Fig. 5.5 [10].

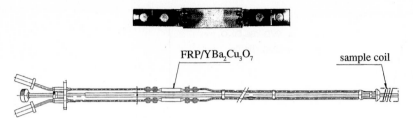

FRP/YBa$_2$Cu$_3$O$_7$ sample coil

Fig. 5.5. 1000 A–1.8 K sample holder using Y123 current leads

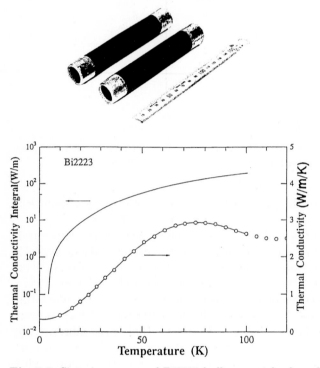

Fig. 5.6. Sintering-processed Bi2223 bulk current leads and their thermal conductivity characteristics

5.2.2   Bi2223 Current Leads and a Cryogenfree Superconducting Magnet

Figure 5.6 shows sintering-processed Bi2223 bulk current leads and the thermal conductivity integral. The integral value for Bi2223 sintered-bulk is 1 W/cm from 77 to 4 K, and is 1/500 and 1/3 in comparison with copper and stainless steel, respectively. Wesche and Fucks [11] compared bulk materials with Ag-sheathed tapes as high-temperature Bi-system superconducting current leads. When Bi-system superconducting current leads with 1000 A capacity are focused on, the characteristic point is that under the same heat-

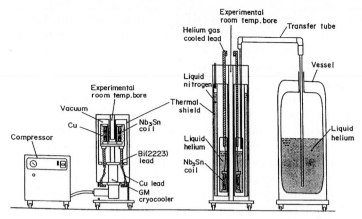

Fig. 5.7. Comparison between a cryogen-free superconducting magnet (*left*) and a pool-boiling superconducting magnet immersed in liquid helium (*right*)

load condition the bulk current lead with $1\,\mathrm{cm}^2$ cross-section and $10\,\mathrm{cm}$ length corresponds to the Ag-sheathed tape current lead with $0.1\,\mathrm{cm}^2$ cross-section and $100\,\mathrm{cm}$ length, where the critical current densities for the bulk and the tape current lead are 10^3 and $10^4\,\mathrm{A/cm^2}$ at $77\,\mathrm{K}$, respectively. This means that the cryostat can be made compactly using the Bi-system bulk current leads. Although a tiny cryocooler has been developed commercially, the refrigeration power is only 0.5–$1.0\,\mathrm{W}$ at most at low temperatures such as $4\,\mathrm{K}$. This small refrigeration power cannot be applied to the ordinary design using copper current leads for a cryocooled superconducting magnet. On the other hand, high-temperature superconducting current leads enable us to achieve a small heat injection of $0.1\,\mathrm{W}$, even if the very short length of $10\,\mathrm{cm}$ is used.

In 1992 we succeeded in realizing a practical cryogen-free superconducting magnet combined with a tiny Gifford-McMahon (GM)-cryocooler for the first time, which generated $4\,\mathrm{T}$ at $400\,\mathrm{A}$ in a $38\,\mathrm{mm}$ room-temperature bore. One notes how compact the developed cryogen-free superconducting magnet is in comparison with a traditional superconducting magnet immersed in liquid helium, as shown in Fig. 5.7. This is because a cryogen-free superconducting magnet using high-temperature superconducting current leads no longer need a long-length current lead. The high-temperature superconducting current leads that were used were sintering-processed bulk Bi2223, and they were hollow cylinders in form with $23\,\mathrm{mm}$ outer diameter, $20\,\mathrm{mm}$ inner diameter and $140\,\mathrm{mm}$ length. Although the critical current density for bulk Bi2223 is small and is only about $10^3\,\mathrm{A/cm^2}$ at $77\,\mathrm{K}$ in a zero magnetic field, a large transport current of $1000\,\mathrm{A}$ is obtainable by selecting a cross-section of $1\,\mathrm{cm}^2$, for instance. It is very difficult for copper current leads with good thermal conductivity to increase the cross-section. So far, we have successfully demonstrated a $15.1\,\mathrm{T}$ high-field cryogen-free superconducting magnet with a $52\,\mathrm{mm}$ room-temperature experimental bore [12] and a $7.4\,\mathrm{T}$ wide-bore cryogen-free superconducting magnet with a 220 mm room-temperature experimental bore [13]. Moreover, a 5.0 T split-pair cryogen-free superconducting magnet with a 50 mm room-temperature vertical bore and a 10 mm room-temperature horizontal gap was developed for X-ray diffraction [14]. Figure 5.8 shows new practical applications of developed cryogen-free superconducting magnets. A high-temperature heat-treatment equipment in magnetic fields realizes the long-term heat-treatment condition at 1200°C and at 15 T for 10 000 h. A chemical reaction process in fields up to 7 T is carried out at a controlled room temperature, using a wide-bore cryogen-free superconducting magnet. An X-ray diffraction investigation in magnetic fields enables the direct observation of lattice parameter changes for the field-induced phase transformation at temperatures ranging from 7.5 to 300 K in fields up to 5 T. Now, one notes that high-temperature superconducting current leads have become a key technology for producing a cryogen-free superconducting magnet combined with a tiny GM-cryocooler.

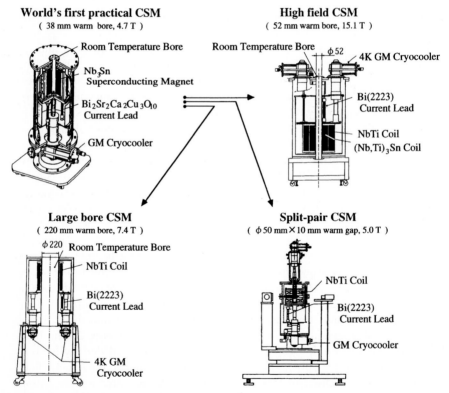

Fig. 5.8. Various cryogen-free superconducting magnets developed at the High Field Laboratory for Superconducting Materials, IMR, Tohoku University

5.3   Developmental Research
of a High-Temperature Superconducting Coil

The science using the high magnetic field is making drastic progress, and in particular a steady-state high magnetic field exceeding 23 T has been required for experiments like NMR. Therefore, it is very useful for a cryogen-free superconducting magnet to improve the high magnetic field generation. It is believed that the electromagnetic stress for a 20 T superconducting magnet is of the order of 150 MPa. For the development of a 20 T cryogen-free superconducting magnet employing high-temperature superconducting tapes, we have to know how the huge electromagnetic stress can be overcome and how the superconducting properties are affected by that stress.

As a reinforcement which withstands 150 MPa, the co-winding method using a Bi2212 tape made by the powder-in-tube method, a reinforcing tape and an insulating tape was attempted. The double-pancake test coil was co-wound with a multifilamentary Ag-sheathed Bi2212 tape ($5 \, \mathrm{mm}^w \times 0.3 \, \mathrm{mm}^t$), a reinforcing Ag-50 wt.% Cu tape ($5 \, \mathrm{mm}^w \times 0.1 \, \mathrm{mm}^t$), and a insulating Al_2O_3 paper tape ($5 \, \mathrm{mm}^w \times 0.1 \, \mathrm{mm}^t$), where w and t mean width and tickness, respectively. It was an 11-turn coil with a 280 mm outer diameter. In addition, the copper outermost layer ring was fitted onto the coil and was soldered to a current terminal. Finally, the test coil was impregnated with epoxy resin.

In order to evaluate the superconducting and mechanical properties in the large electromagnetic stress state, the double-pancake test coil for a high-temperature superconducting insert of a 20 T cryogen-free superconducting magnet was set into a large-current testing equipment. This equipment provided transport currents up to 1500 A in fields up to 12 T at 4.2 K in a large bore of 308 mm, as shown in Fig. 5.9. The performance test was carried out at 10 T and 4.2 K. The critical current of the test coil determined by a resistivity criterion of $10^{-11} \, \Omega \cdot \mathrm{cm}$ was 240 A. Transport currents up to 360 A above the critical current were applied to the test coil at 10 T. Hoop stress estimated by the relationship to B_{jr}, where B is the magnetic field, j is the critical current density, and r is the coil radius, was 200 MPa at 360 A and 133 MPa at 240 A. Stable operation for the large electromagnetic stress state of 200 MPa was confirmed for the AgCu-reinforced Ag/Bi2212 coil [15]. Since the AgCu reinforcing tape has good thermal conductivity as well as high strength, it is found that the co-winding method using a AgCu reinforcing tape is very useful for the insert coil of a 20 T cryogen-free superconducting magnet system. In the next stage, a high-temperature superconducting insert coil which will generate 7 T in a 52 mm room-temperature bore is planned and will be combined with a 13 T backup coil. A cryogen-free backup coil has been designed for an operating has been temperature of 5.5 K, and an insert coil will be operated at 8–10 K. In the near future, a cryogen-free superconducting magnet will achieve the outstandingly high field of 20 T.

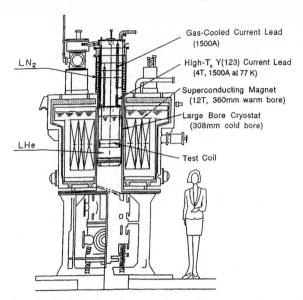

Fig. 5.9. 1500 A current testing equipment with a wide bore of 308 mm in liquid helium and in fields up to 12 T

5.4   Concluding Remarks

The High Field Laboratory for Superconducting Materials, (HFLSM), In-stitute for Materials Research, Tohoku University has evaluated the critical current densities in high fields for high-temperature superconductors, and carried out their practical applications. Melt-growth-processed $YBa_2Cu_3O_7$ bulk materials were utilized for high-temperature superconducting current leads in high fields at 77 K, where the Bi-system high-temperature superconductors are never applied. Sintering-processed $Bi_2Sr_2Ca_2Cu_3O_{10}$ bulk superconductors were adopted as the key component to realize a cryogen-free superconducting magnet (CSM). Since HFLSM succeeded in making the world's first practical cryogen-free superconducting magnet, various kinds of CSMs have been developed. In the next stage, a high-temperature superconducting insert coil employing $Bi_2Sr_2CaCu_2O_8$ tapes is being investigated for a 20 T cryogen-free superconducting magnet.

Acknowledgments

Cryogen-free superconducting magnets are being developed by collaborative research with Sumitomo Heavy Industries. The investigation for Ag-sheathed Bi-system superconductors is financially supported by Grant-in-Aid for scientific research from the Ministry of Education, Science and Culture, Japan.

References

1. T. Satow S. Imagawa, N. Yanagi, K. Takahata, T. Mito, S. Yamada, H. Chikaraishi, A. Nishimura, S. Satoh and O. Motojima: IEEE Trans. Appl. Supercond. **10**, 600 (2000)
2. L.R. Evans: IEEE Trans. Appl. Supercond. **10**, 44 (2000)
3. I. Nano: *Proc. 11th Int. Conf. on Magnet Technology*, ed. by T. Sekiguchi and S. Shimamoto, Tsukuba, 1989 (Elsevier Applied Science, New York, 1990) p. 1
4. J. Fujiie: *Proc. 11th Int. Conf. on Magnet Technology*, ed. by T. Sekiguchi and S. Shimamoto, Tsukuba, 1989 (Elsevier Applied Science, New York, 1990) p. 9
5. W.D. Markieuicz, L.A. Bonney, I.M. Eyssa, C.A. Swenson and H.J. Schneider-Muntau: Physica **B 216**, 200 (1996)
6. K. Watanabe, K. Noto, H. Morita, H. Fujimori, K. Mizuno, T. Aomine, B. Ni, T. Matsushita, K. Yamafuji and Y. Muto: Cryogenics **29**, 263 (1989)
7. K. Watanabe, H. Yamane, H. Kurosawa, T. Hirai, N. Kobayashi, H. Iwasaki, K. Noto and Y. Muto: Appl. Phys. Lett. **54**, 575 (1989)
8. K. Watanabe, Y. Yamada, J. Sakuraba, F. Hata, C. K. Chong, T. Hasebe and M. Ishihara: Jpn. J. Appl. Phys. **32**, L488 (1993)
9. S. Awaji, K. Watanabe, N. Kobayashi, H. Yamane and T. Hirai: IEEE Trans. Magn. **32**, 2776 (1996)
10. K. Watanabe, S. Awaji and K. Kimura: *Proc. 10th Int. Symp. on Supercond., Gifu, 27-30, Oct., 1997, Adv. Supercond. X*, ed. by K. Osamura, I. Hirabayashi (Springer, Hong Kong, 1998) p.1389.
11. R. Wesche and A.M. Fuchs: Cryogenics **34**, 145 (1994)
12. K. Watanabe, S. Awaji, M. Motokawa, Y. Mikami, J. Sakuraba and K. Watazawa: Jpn. J. Appl. Phys. **37**, L1148 (1998)
13. K. Goto, S. Iwasaki, N. Sadakata, T. Saitoh, S. Awaji, K. Watanabe, K. Jikihara, Y. Sugizaki and J. Sakuraba, to be presented at ASC 2000, Virginia-Beach
14. K. Watanabe, Y. Watanabe, S. Awaji, M. Fujiwara, N. Kobayashi and T. Hasebe: Adv. Cryog. Eng. **44**, 747 (1998)
15. S. Awaji, K. Watanabe, T. Wakuda and M. Okada: *Proc. 11th Inter. Symp. on Supercond., Fukuoka, 17-19 Nov., 1998, Adv. Supercond. XI*, ed. by N. Koshizuka and S. Tajima (Springer, Hong Kong, 1999) p. 923.

Part III

Conventional High-Field Superconductors

6 Highly Strengthened Nb₃Sn Superconducting Wires

6 Highly Strengthened Nb_3Sn Superconducting Wires

K. Katagiri, K. Noto and K. Watanabe

The superconducting wire in a large-scale high-field magnet experiences a large electromagnetic force. This force causes an axial tensile stress in the wire as well as a transverse compressive stress, if the radial displacement of the winding is limited by some constraints. There have been many attempts to reinforce superconducting wires in order to improve the mechanical properties and strain tolerance of I_c [1–4]. In wires for compact high-field magnets made by the wind-and-react technique, the choice of the reinforcing material is important because the reinforcing material and the superconducting composites are heat-treated collectively. In order to avoid a decrease in overall critical current density, simple addition of reinforcing material as an internal reinforcing method is not preferable. Replacement of copper in the superconducting composite wire by a material with high strength as well as high electric conductivity such as Ta [5], alumina-particle-dispersed Cu [6–9] and Cu-Nb microcomposite [10–18] is promising, although the stability of the wire is reduced to some extent [5,14,18].

The present authors have established that bronze-processed Nb_3Sn wires with Cu-Nb reinforcing stabilizer are applicable as high-strength and high-I_c superconductors [13–18]. The replacement by Cu-Nb reinforcing stabilizer changes not only mechanical properties such as the 0.2% proof stress but also the tensile-strain characteristics of the critical current I_c, especially ϵ_m, the strain corresponding to the I_c peak, I_{cm}, and the reversible strain limit, ϵ_{irr}, where permanent damage such as cracking starts to take place in the Nb_3Sn filaments [15–17]. Nb_3Sn wire fabricated by a tube process using alumina-dispersion-strengthened copper (alumina-Cu) has been proved to be another candidate for a compact superconducting magnet [8,9]. Changes in the strain characteristics of I_c similar to that in Cu-Nb reinforcing stabilized wires have been reported [9].

From the viewpoint of further improvement of the tensile strain/transverse compressive stress characteristics of I_c, results on some of the high-strength practical Nb_3Sn multifilamentary superconducting wires examined in magnetic fields up to 15 T at a temperature of 4.2 K are reviewed. Correlation between the tensile-strain dependence and the transverse compressive-strain dependence of I_c in the wires is briefly described.

6.1 Experimental

The multifilamentary Nb_3Sn superconducting wire specimens examined are various kinds of bronze-processed Cu-Nb reinforcing stabilized wires and tube-processed wires with alumina-dispersion-strengthened copper, the diameter ranging from 0.46 to 1.0 mm. For reference, conventional copper-stabilized wires with the same wire parameters, excluding the reinforcing material, were also examined. The specifications together with the heat-treatment conditions and the cross-sectional views of the wires are given below. If a small amount of Ti is added to Nb_3Sn, it is referred to as $(Nb,Ti)_3Sn$.

The axial-strain dependence of I_c as well as the transverse compressive stress dependence of I_c have been measured at 4.2 K, using two measuring probes inserted in the 15 T superconducting magnet at the Materials Research Institute, Tohoku University. The details of the testing apparatus have been reported elsewhere [19,20]. The accuracy of strain in the tensile testing apparatus using a high-compliance extensometer is 0.05% strain. It should be mentioned that the stress is underestimated by 10-20%, depending on the load level of the measurement at a strain of about 0.3%, where the error is largest. The I_c was determined by a four-probe method using a nominal electrical field criterion of 1 μV/cm based on the actual tap separation of 5 mm within the 17.5 mm gauge length. In the transverse compressive stress tests, the width of the pressure block was 1.5 or 3 mm. Thus, the voltage-sensing leads were connected to the test wire outside of the compressed region. The difference between I_c based on the actual tap separation and I_c based on the effective separation (block width) is less than 3%. Generally, more than two samples were used for each measurement.

Table 6.1. Specification of Cu-Nb reinforcing stabilized $(Nb,Ti)_3Sn$ wire and reference wire

Superconductor	$(Nb,Ti)_3Sn$	
Sheath	CuNb	Cu
Wire diameter (mm)	0.8	
Superconductor area		
Matrix	Cu-13 wt .% Sn	
Filament diameter (μm)	3.4 (Nb-1.2 wt.% Ti)	
Number of filaments	5587(151 x 37)	
Bronze ratio (bronze/Nb)	4.09	
Barrier	Ta	
Cu (+Cu-Nb) to superconductor ratio		
Pure copper	0.14	0.78
CuNb composite	0.63	—
	(20 wt.% Nb)	
Heat treatment	670 °C x 192 h	

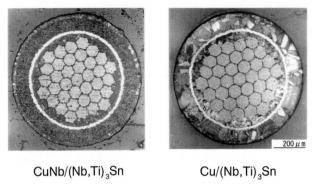

CuNb/(Nb,Ti)$_3$Sn   Cu/(Nb,Ti)$_3$Sn

Fig. 6.1. Cross-sectional view of Cu-Nb reinforcing stabilized wire and Cu stabilized wire

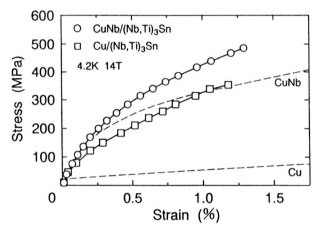

Fig. 6.2. Stress-strain curves of Cu-Nb/(Nb,Ti)₃Sn wire, reference wire and (reinforcing) stabilizer materials at 4.2 K and 14 T

6.2   Stress/Strain Characteristics of Wires

6.2.1   Bronze-Processed Nb₃Sn Wire Reinforcing-Stabilized with Cu-Nb Composite

Axial Tensile Strain Effect on Critical Current. The in-situ Cu-Nb composite rods adopted for the reinforcing stabilizer were fabricated from an ingot of Cu-Nb, which was made by induction-melting in a CaO crucible. Table 6.1 and Fig. 6.1 show the specification and the cross-sectional views of Cu-20 wt% Nb reinforcing-stabilized (Nb,Ti)₃Sn wire and Cu-stabilized (Nb,Ti)₃Sn wire [17].

The stress-strain curves of the superconducting wires at 14 T and 4.2 K are shown in Fig. 6.2.

Replacing the Cu stabilizer with the Cu-Nb reinforcing stabilizer increases the 0.2% proof stress from 170 MPa to 250 MPa. In Fig. 6.2, the stress-strain curves of constituting materials for fabrication of the superconducting wires, Cu-Nb (20 wt% Nb) and Cu wires, are also shown. Significant differences in the curves, resulting in the marked reinforcing effect of the composites, can be seen.

The strain dependence of I_c in the Cu-Nb/(Nb,Ti)$_3$Sn wire is compared with that in the Cu/(Nb,Ti)$_3$Sn (Fig. 6.3a). ϵ_m is shifted from 0.29% to 0.43%, when the Cu is replaced by the Cu-Nb. This is attributed to an increase in the proof stress of the stabilizer, which generates higher compressive thermal strain in the (Nb, Ti)$_3$Sn filaments on cooling from the temperature of the reaction heat treatment down to 4.2 K. Permanent damage in the (Nb,Ti)$_3$Sn filaments occurs when ϵ_{irr} is larger than about 0.95%; then I_c on unloading no longer falls on the loading curve. This is shifted by ϵ_m of the reinforcement. The strain sensitivity of I_c for both wires is identical, as shown in Fig. 6.3b, where the intrinsic strain obtained by subtracting the compressive thermal strain ϵ_m from the applied strain ϵ, $\epsilon_0 = \epsilon - \epsilon_m$, is used as the abscissa.

The strain dependence of the bulk average upper critical field B_{c2}^* in the two (Nb,Ti)$_3$Sn wires is shown in Fig. 6.4. B_{c2}^* is estimated by using the Kramer plotting technique based on I_c in the magnetic field range of 10 to 14.8 T. The estimate gave higher B_{c2}^* by about 2.5 T at $\epsilon=0$ as compared with that based on I_c in magnetic fields up to 23.1 T. There was a rather large scatter in the B_{c2}^* of about 1.5 T, depending on the sample. The broken line in the figure shows the analytically approximated empirical relation regarding the strain scaling law [21]:

$$B_{c2}^*(\epsilon)/B_{c2m}^* = 1 - a \mid \epsilon_0 \mid^u, \tag{6.1}$$

where $u = 1.7$ and $a = 900$ for compressive strain ($\epsilon_0 < 0$) and $a = 1250$ for tensile strain ($\epsilon_0 > 0$). No appreciable difference is seen between the data on the (Nb,Ti)$_3$Sn wires with Cu-Nb reinforcement and those without Cu-Nb. These data fit the empirical line fairly well.

Effect of Heat-Treatment Temperature. As the heat-treatment temperature increases from 973 K to 1023 K, the 0.2% proof stress decreases by 14% in the Cu-Nb reinforcing-stabilized Nb$_3$Sn wire (no Ti addition), as shown in Fig. 6.5, while no appreciable change in ϵ_m and the strain sensitivity of I_c are found [15]. Decrease in the proof stress is due to the decrease in the Sn concentration in the bronze with increase of volume fraction of Nb$_3$Sn. This coincides with an increase of I_c from 29 A to 34 A. Another possible reason is the decrease of the 0.2% proof stress of the Cu-Nb reinforcing stabilizer. This is likely to be due to recovery of the deformed microstructure in the Nb filaments in the Cu-Nb and the change of their morphology, i.e. spheroidization resulting in sausaging (diameter modulation along the filament) as evidenced in [15].

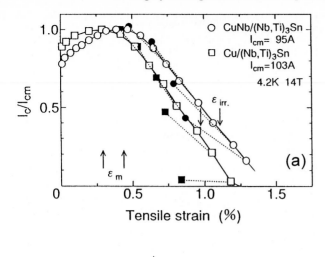

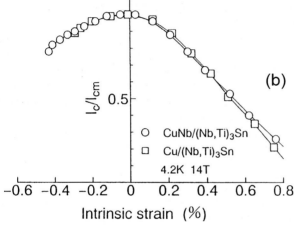

Fig. 6.3. Strain dependence (**a**) and intrinsic strain sensitivity (**b**) of I_c normalized to its peak, I_{cm}, in the Cu-Nb/(Nb,Ti)$_3$Sn wire and reference wire at 4.2 K and 14 T (solid symbols represent the data obtained on unloading)

Effect of Nb Content in Cu-Nb Composite. The effect of Nb content in the Cu-Nb composite in the superconducting wire has been examined using wires which contain Cu-20 wt% Nb, Cu-30 wt% Nb and Cu-40 wt% Nb as the reinforcing stabilizer. The increase of the Nb content leads to the improvement in the stress-strain relation. This is due to the increase in the volume fraction of Nb, the component with high strength, as well as the decrease in the spacing of the copper between the Nb filaments that serve as barriers to the dislocations in the Cu and increase the yield stress. Figure 6.6 shows the strain characteristics of I_c in the wires. By contrast to the replacement of Cu by Cu-Nb, in which the strength of the stabilizer is increased, ϵ_m decreases

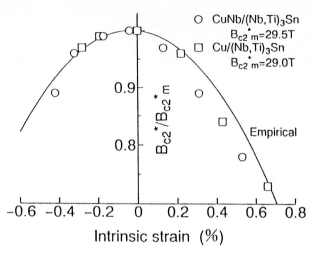

Fig. 6.4. Strain dependence of B_{c2}^* normalized to its peak, B_{c2m}^* in Cu-Nb/(Nb,Ti)$_3$Sn wire and reference wire

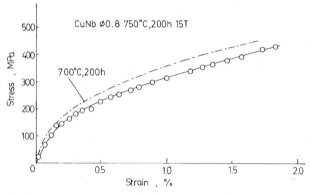

Fig. 6.5. Effect of heat-treatment temperature on the stress-strain curve of Cu-Nb/Nb$_3$Sn wire

with increasing strength as the Nb content is increased among the wires reinforced by Cu-Nb composite. This is due to the decrease in the coefficient of thermal expansion of Cu-Nb composite because that of Nb $(7.02 \times 10^{-6}\,\mathrm{K}^{-1})$ is smaller than that of Cu $(16.8 \times 10^{-6}\,\mathrm{K}^{-1})$. Figure 6.7 shows an example of the result of estimating the thermal residual strain induced in the constituents (ϵ_m in the case of (Nb,Ti)$_3$Sn) of the superconducting wire reinforcing stabilized by Cu-20 wt% Nb on cooling from reaction heat-treatment temperature down to 4.2 K. Here the law of mixture [16] and the thermal and mechanical parameters of the constituents [22] are used and yielding is taken into account by changing Young's modulus to the strain-hardening coefficient. It can be seen that Cu in the Cu/(Nb,Ti)$_3$Sn wire yields at higher tempera-

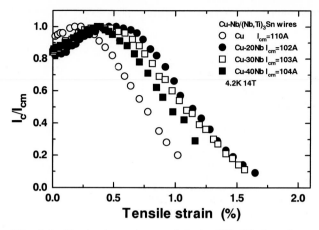

Fig. 6.6. Strain dependence of I_c in (Nb,Ti)$_3$Sn wires reinforcing stabilized by Cu-Nb with various Nb content

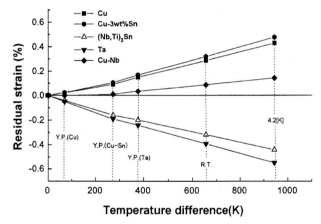

Fig. 6.7. Residual strains of constituents in Cu-20 wt% Nb/(Nb,Ti)$_3$Sn wire during cooling from 948 K via room temperature (R.T.) to 4.2 K (Y.P. is the yield point of the constituent)

tures while no yielding takes place in Cu-20 wt% Nb superconducting wire. The estimated residual strain in (Nb,Ti)$_3$Sn is 0.5%; this agrees fairly well with ϵ_m.

Because there is a limitation for increasing the Nb content in the Cu-Nb composite through the in-situ process, a jelly-roll method is adopted in order to increase the strength of the composite [23]. Increase in Nb content to 67% resulted in a 0.2% proof stress of 280 MPa at 4.2 K in the Cu-Nb/(Nb,Ti)$_3$Sn wire.

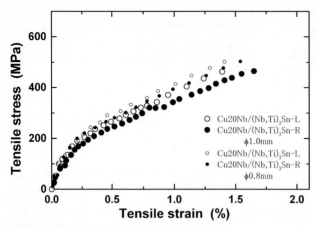

Fig. 6.8. Effect of Cu-Nb volume fraction and wire diameter on the stress-strain curve of Cu-20 wt% Nb/(Nb,Ti)₃Sn wire at 4.2 K and 14 T

Effect of Volume Fraction of Cu-Nb Composite and of the Wire Diameter. In order to increase the current-carrying capacity of the wire, the composition of a Cu-20 wt% Nb reinforcing-stabilized wire was modified. While the volume fraction of copper sheath and non-copper part was kept almost constant, 0.4 and 1 respectively, the volume fraction of Cu-Nb in the wire was decreased from 0.86 (Cu-20Nb/(Nb,Ti)₃Sn-R) to 0.63 (Cu-20Nb/(Nb,Ti)₃Sn-L), and thus the relative volume fraction of (Nb,Ti)₃Sn was increased. As a consequence, I_{cm} increased by 25%. Furthermore, it is seen that the 0.2% proof stress of Cu-Nb is increased from 210 MPa to 236 MPa, as shown in Fig. 6.8. This is because Nb₃Sn is a structural component with higher strength as compared to Cu-Nb composite; the Young's modulus is higher by a factor of 1.5 and the breaking strain is much higher than 0.2%. The strain dependence of the wire is shifted toward lower strain in the wires with lower Cu-Nb volume fraction due to the reduction of residual strain as discussed above.

In Fig. 6.8, data on wires with diameter of 1.0 mm and 0.8 mm are shown together. The flow stress is higher for the wires with small diameter. Since an increase in the area reduction ratio will decrease the Nb filament spacing, the flow stress becomes higher in thinner wires. The enhanced reaction of Nb into Nb₃Sn in thinner filaments will decrease through reduction of Sn content in the bronze matrix and increase through increase in volume fraction of (Nb,Ti)₃Sn with higher Young's modulus. These effects, however, were proved to be eventually cancelled because no size effect was found in the Cu stabilized wires. The strain dependence of I_c did not changed with the diameter of the wires that were examined.

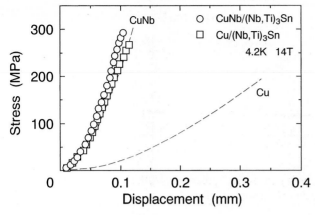

Fig. 6.9. Transverse stress versus displacement of Cu-Nb/(Nb,Ti)$_3$Sn wire, Cu/(Nb,Ti)$_3$Sn wire and (reinforcing) stabilizer materials

Cable Conductor. Cabling is a promising technique for obtaining higher current-carrying capacity and higher tolerance against bending deformation. Cables consisting of 6 strands of Cu-20 wt% Nb reinforcing-stabilized (Nb,Ti)$_3$Sn wires of 0.46 mm diameter were fabricated and tested [16]. It was established that the stress-strain characteristics and the I_c versus strain relation of the cable at 14 T and 4.2 K were almost the same as those in a single strand, ϵ_m being 0.36%. The I_{cm} value of the cable (131 A) was about 6 times as large as the I_{cm} value of the strand (23 A).

Transverse Compressive Stress Effect. Because of the difficulty in estimating the complicated strain state in the Nb$_3$Sn filaments in the superconducting wire with components in the elastic or plastic state, the effect of transverse loading is usually described by the compressive stress estimated by dividing the applied load by the projected compressive area (wire diameter × block width). Figure 6.9 shows the relations of transverse compressive stress versus displacement in the Cu-Nb/(Nb,Ti)$_3$Sn, Cu/(Nb,Ti)$_3$Sn, Cu-Nb and Cu wires [17]. In spite of the marked difference in the compressive behavior between Cu-Nb and Cu, as can be seen in Fig. 6.9, a pair of the superconducting composites stabilized with and without reinforcement shows rather small differences in the behavior resulting from both the deformations in the outer shell and those in the superconducting area. This is presumably due to the configuration of the reinforcing component Cu-Nb in the composite wire with regard to the loading direction and the anisotropy in the mechanical properties of the Cu-Nb composite.

The transverse compressive stress dependencies of I_c normalized by that at zero applied stress (I_{c0}) in the superconducting wires at a magnetic field of 14 T are shown in Fig. 6.10. I_c decreases monotonically with increasing stress. Degradation of I_c due to the transverse compressive stress occurs in

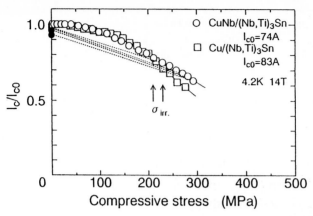

Fig. 6.10. Transverse stress dependence of I_c in Cu-Nb/(Nb,Ti)$_3$Sn wire at 4.2 K and 14 T

two stages [24], as in the case of the tensile-strain effect. All the constituents in the superconducting area behave almost elastically, and I_c recovers reversibly to I_{c0} on unloading in stage I and I_c does not recover in stage II. In Fig. 6.10, it can be seen that the stress sensitivity of I_c in stage I is almost the same for both wires with Cu-Nb and pure Cu. The solid circles in Fig. 6.10 show the data of I_c on unloading in Cu-Nb/(Nb,Ti)$_3$Sn wire. The boundary stress between stages I and II, σ_{irr}, is 230 MPa in the Cu-Nb/(Nb,Ti)$_3$Sn wire; this is higher by about 1.1 times as compared with 210 MPa in Cu/(Nb,Ti)$_3$Sn. Thus, we can conclude that the reinforcing stabilizer improves σ_{irr} and the stress sensitivity of I_c in the region of stage II. A more obvious effect has been reported for the wires with softer superconducting area (σ_{irr} being 190 MPa and 110 MPa for wires stabilized with and without reinforcement, respectively) [17]. An outer shell of Cu-Nb with high rigidity supports the transverse compressive load effectively and therefore the stress induced in the interior superconducting area is reduced. Although it is less evident compared to the tensile-strain effect as shown in Fig. 6.7, the effect of Nb content in Cu-Nb reinforcing stabilizer on the transverse compressive stress tolerance is clearly seen, in particular in stage II.

It has already been reported that the sensitivity of I_c in Nb$_3$Sn and Nb$_3$Al superconducting wire to the transverse compressive stress is higher than to that of axial tensile stress [25]. For wires that behave almost elastically, such as tube-processed Nb$_3$Al, however, it was suggested that there is intrinsically no large difference between the sensitivity of I_c to axial stress and to transverse stress, if the stress distribution in the cross-section of the wire is taken into account [26,27]. Using the geometric average strain, one representation of distortional strain, the present authors have shown that there is a large difference in the strain sensitivity for axial and transverse stresses, even in the highly strengthened bronze-processed wire [17]. Damages such as longitu-

Table 6.2. Specification of alumina-Cu reinforced (Nb,Ti)₃Sn wire

(Reinforcing) stabilizer	Cu-1.2 vol% Al₂O₃	Cu
Wire diameter (mm)	0.687	
Matrix	Cu-50 wt% Sn	
Fil. number	180	
Copper ratio	1.40	1.65
Cu/Al₂O₃-Cu	1/0.647	—
Heat treament	923 K x 50 h, 978 K x 50 h, 1023 K x 50 h	

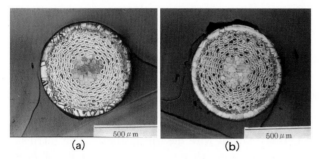

(a)   (b)

Fig. 6.11. Cross-sectional view of alumina-Cu reinforced (Nb,Ti)₃Sn wire (**a**) and reference wire (**b**)

dinal cracking in the filaments induced by the transverse compressive stress in this wire have already been described in detail [17,28]. Alumina-dispersion strengthened copper (alumina-Cu) reinforced Nb₃Sn wires fabricated using the tube process have been developed [7–9]. Table 6.2 and Fig. 6.11 show the specification and a cross-sectional view of the wire together with the reference wire with Cu tube instead of alumina-Cu.

6.2.2   Tube-Processed Nb₃Sn Wire Reinforcing Stabilized with Alumina-Dispersion-Strengthened Copper

Figure 6.12 shows stress-strain curves of the wire obtained at 4.2 K and 14 T. As compared to the reference wire, the flow stress at 0.3% strain increased to 205 MPa in the reinforced wire heat-treated at 923 K. The effect of reinforcement is less effective at a higher heat-treatment temperature of 1023 K, from 110 MPa to 153 MPa. Because of quench-like V–I behavior due to very high critical current density coupled with improperly high sweep rate, some of the I_c values of the wires heat-treated in the optimum condition are not accurate. I_{cm}, 122 A, in the reinforced wire heat-treated at 978 K is underestimated by about 10%, and, therefore, the strain dependence of I_c includes some uncertainty. Figure 6.13a shows the strain dependence of I_c. ϵ_m increased from 0.13–0.18% to 0.36% when Cu was replaced by alumina-Cu. The effect of heat-treatment temperature on the strength as well as on ϵ_m of the alumina-Cu reinforced (Nb,Ti)₃Sn superconducting wire is essentially

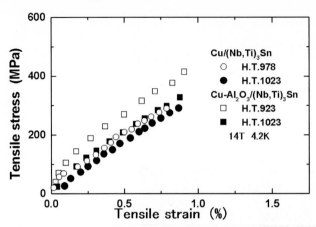

Fig. 6.12. Stress-strain curves of alumina-Cu/(Nb,Ti)₃Sn wires heat-treated at 978 K (H.T. 978) and 1023 K (H.T. 1023)

the same as that in the Cu-Nb/Nb₃Sn wire described above. The intrinsic strain behavior of I_c in Fig. 6.13b shows that the strain sensitivity of I_c in wire with/without alumina-Cu is essentially the same. Figure 6.14 shows the dependence of I_c on transverse compressive stress in the wires. The stress tolerance of the wires heat treated at both high and low temperatures is improved by the reinforcement, especially in the high stress region beyond 100 MPa.

The reinforcing-stabilized wires with high critical current density and strength are now applicable in practice and further development is in progress. A large-bore cryocooled magnet for a hybrid magnet system is under development using a Cu-Nb reinforcing-stabilized Nb₃Sn wire. The magnet can be fabricated by a react-and-wind and tension-winding method without need for a large-scale heat-treatment furnace and vacuum epoxy impregnation equipment [29]. Development of a 1 GHz superconducting NMR magnet is in progress where it is planned to use a Ta-reinforced wire [7–9,30].

6.3 Conclusion

In order to improve the mechanical properties as well as the stress/strain tolerance of the critical current I_c of Nb₃Sn multifilamentary superconducting wires, various attempts have been successfully developed. The stress/strain characteristics of some typical highly strengthened practical Nb₃Sn superconducting wires have been examined at the High Field Laboratory for Superconducting Materials, Institute for Materials Research, Tohoku University up to a magnetic field of 14 T and at a temperature of 4.2 K. Although the intrinsic axial tensile strain sensitivity of I_c does not basically change, the strain for peak I_c as well as the reversible strain limit change by thermal con-

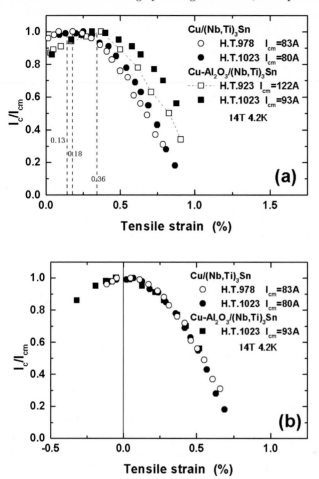

Fig. 6.13. Strain dependence of I_c (**a**) and intrinsic strain sensitivity of I_c (**b**) in alumina-Cu/(Nb,Ti)₃Sn wires at 4.2 K and 14 T

traction of the reinforcing material. Furthermore, the transverse compressive stress sensitivity of I_c decreases and the reversible-stress limit increases.

Acknowledgment

The authors wish to express their thanks to Dr. Kasaba, Mrs. Y. Shoji, technical official, and Y. Morii, H. Seto, N. Ebisawa, T. Takahashi and M. Ishizaki, graduate students, of Iwate University for their helpful assistance in the experiments. They appreciate helpful advice and assistance by Prof. S. Awaji and the staff in the High Field Laboratory for Superconducting Materials, Institute for Material Research, Tohoku University. Supply of valuable

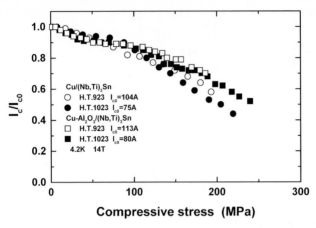

Fig. 6.14. Transverse compressive stress dependence of I_c in alumina-Cu/(Nb,Ti)$_3$Sn wires at 4.2 K and 14 T

superconducting wire specimens from Fujikura Ltd. and Toshiba Co. are acknowledged.

References

1. J.W. Ekin, R. Flükiger and W. Specking: J. Appl. Phys. **54**, 2869 (1983)
2. R. Flükiger, E. Drost and W. Specking: Adv. Cryog. Eng. **30**, 875 (1984)
3. K. Noto, N. Konishi, K. Watanabe, A. Nagata and T. Anayama: *Proc. Int. Sympl. Flux Pin. Electromagn. Prop. Superconds.*, eds. T. Matsushita, K. Yamafuji and F. Irie, Fukuoka, 11- 14 Nov. 1985 (Matsukura Press, Fukuoka 1985) p.272.
4. R. Flükiger and A. Nyilas: IEEE Trans. Magn. MAG-**21**, 285 (1985)
5. M. Matsukawa, K. Noto, C. Takahashi, Y. Saito, N. Matsuura, K. Katagiri, M. Ikebe, T. Fukutsuka and K. Watanabe: IEEE Trans. Magn. **28**, 880 (1992)
6. E. Gregory, L.R. Motowidlo, G.M. Ozeryansky and L.T. Summers: IEEE Trans. Magn. **MAG-27**, 2033 (1991)
7. S. Nakayama, S. Murase, K. Shimamura, N. Aoki and N. Shiga: Adv. Cryog. Eng. **38**, 279 (1992)
8. K. Koyanagi, S. Nakayama, S. Murase, S. Nomura, K. Shimamura and M. Urata: IEEE Trans. Appl. Supercond. **7**-2, 427 (1997)
9. S. Murase, S. Nakayama, T. Masegi, K. Koyanagi, T. Nomura, N. Shiga, N. Kobayashi and K. Watanabe: J. Japan Inst. Metals, **61**-9, 801 (1997)
10. C.C. Tsuei: J. Appl. Phys. **45**, 1385 (1974)
11. J. Bevk, J.P. Harbison and J.L. Bell: J. Appl. Phys. **49** , 6031 (1978)
12. C.V. Renaud, E. Gregory and J. Wong: Adv. Cryog. Eng. **32**, 443 (1986)
13. K. Watanabe, A. Hoshi, S. Awaji, K. Katagiri, K. Noto, K. Goto, M. Sugimoto, T. Saito and O. Kohno: IEEE Trans. Appl. Supercond. **3**, 1006 (1993)
14. K. Watanabe, S. Awaji, K. Katagiri, K. Noto, K. Goto, M. Sugimoto, T. Saito and O. Kono: IEEE Trans. Magn. **30**, 1871 (1993)

15. K. Katagiri, K. Watanabe, K. Noto, K. Goto, T. Saito, O. Kohno, A. Iwamoto, M. Nunogaki and T. Okada: Cryogenics **34**, 1039 (1994)

16. K. Watanabe, K. Katagiri, K. Noto, S. Awaji, K. Goto, N. Sadakata, T. Saito and O. Kohno: IEEE Trans. Appl. Supercond. **5**-2, 1905 (1995)

17. K. Katagiri, K. Watanabe, H.S.Shin, Y. Shoji, N. Ebisawa, K. Noto, T. Okada, K. Goto, T. Saito and O. Kohno: Adv. Cryog. Eng. **42**, 1423 (1996)

18. K. Watanabe, M. Motokawa, T.Onodera, K. Noto, K. Katagiri and T. Saito: Material Science Forum **308-311**, 561 (1999)

19. K. Katagiri, M. Fukumoto, K. Saito, M. Ohgami, T. Okada, A. Nagata, K. Noto and K. Watanabe: Adv. Cryog. Eng. **36**, 69 (1990)

20. K. Kamata, K. Katagiri, T. Okada, T. Takeuchi, K. Inoue, K. Watanabe, Y. Muto, T. Ogata and T. Tsuji: *Proc. 11th Int. Conf. Magnet Tech.*, ed by T. Sekiguchi and S. Shimamoto, Tsukuba, 28 Aug-1 Sept., 1989 (Elsevier Appl. Sci., London, 1989) p. 1231.

21. J.W. Ekin: Cryogenics **20**, 661 (1980)

22. D.S. Easton, D.M. Kroeger, W. Specking and C.C. Koch: J. Appl. Phys. **51**-5, 2748 (1980)

23. G.Iwaki, M. Kimura, H. Moriai, K. Asano, K. Watanabe and M. Motokawa: Adv. Cryog. Eng. **46**, 981 (2000)

24. T. Kuroda and H. Wada: Cryog. Eng. **28**, 439 (1993) (in Japanese)

25. J.W. Ekin: J. Appl. Phys. **62**, 4829 (1987)

26. T. Kuroda, H. Wada, S.L. Bray and J.W. Ekin: Fusion Eng. Des. **20**, 271 (1993)

27. K. Katagiri, T. Kuroda, H. Wada, H.S. Shin, K. Watanabe, K. Noto, Y. Shoji and H. Seto: IEEE Trans. Appl. Supercond. **5**, 1900 (1995)

28. K. Katagiri, K. Watanabe, T. Kuroda, K. Kasaba, K. Noto, T. Okada and O. Kohno: Sci. Rept. RITU **A42**, 381 (1996)

29. K. Watanabe, S. Awaji, M. Motokawa, S. Iwasaki, K. Goto, N. Sadakata, T. Saito, K. Watazawa, K. Jikihara and J. Sakuraba: IEEE Trans. Appl. Supercond. **9**-2, 440 (1999)

30. T. Kiyoshi, A. Sato, H. Wada, S. Hayashi, M. Shimada and Y. Kawate: IEEE Trans. Appl. Supercond. **9**-2, 559 (1999)

7   Development of Nb_3Al Superconductors

K. Inoue, T. Takeuchi, Y. Iijima, A. Kikuchi, N. Nakagawa, G. Iwaki,
H. Moriai and K. Watanabe

7.1   Variation of Fabrication Process
for Nb_3Al Superconductors

Stoichiometric Nb_3Al has higher values of T_c and H_{c2} than those of the commercial Nb_3Sn superconducting wire, but stoichiometric Nb_3Al is unstable except at high temperatures near 2000 °C, while non-stoichiometric Nb-rich Nb_3Al that has relatively low values of T_c and H_{c2} is stable at low temperatures [1]. In addition, the Nb_3Al formation rate through the solid-state diffusion reaction between Nb and Al is very slow.

In order to surmount the slow diffusion reaction by increasing the density of Nb/Al diffusion-couples in the microcomposite wire, various processes were proposed, such as a jelly-roll process [2], a powder metallurgy process [3], a rod-in-tube process [4] and a clad-chip extrusion process [5]. The combination of a low-temperature diffusion reaction and Nb/Al microcomposite wire can be used to produce Nb_3Al multifilamentary wires with large J_c values at low fields. However, the resulting wire has off-stoichiometric Nb_3Al filaments with lower T_c and H_{c2} (4.2 K) than those of commercial Nb_3Sn wire [6].

Six years ago we proposed the RHQT (Rapid Heating, Quenching and Transforming) process to fabricate Nb_3Al multifilamentary wire with near-stoichiometric composition [7]. In this process a Nb/Al multifilamentary microcomposite wire is rapidly heated up to about 2000 °C by resistive heating for 0.1-0.4 s, and then quenched into a molten-Ga bath at room temperature to form Nb-Al supersaturated bcc filaments in the Nb matrix, as shown in Fig. 7.1. The supersaturated bcc phase is quasi-stable and ductile at room temperature, and transforms into the Nb_3Al A15 phase of near-stoichiometric composition when annealed at 700–900 °C. The resulting Nb_3Al multifilamentary wire shows not only an excellent strain tolerance but also a critical current density J_c that is 2–5 times larger than that of the bronze-processed $(Nb,Ti)_3Sn$ multifilamentary wire [8]. Moreover, the RHQT-processed Nb_3Al wire shows maximum values of $T_c = 17.9$ K and $H_{c2}(4.2 K) = 25.8$ T; these values are nearly identical with those of commercial $(Nb, Ti)_3Sn$ multifilamentary wires.

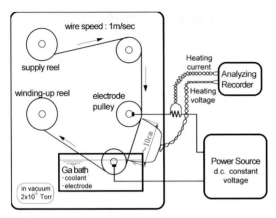

Fig. 7.1. Schematic diagram of the RHQT (Rapid Heating, Quenching and Transforming) treatment for the Nb/Al precursor wire

7.2   Nb/Al Microcomposite Precursor Wires for the RHQT Process

At first we studied the combination of the rod-in-tube processed precursor wire and the RHQT process, and succeeded in the fabrication of an Nb₃Al multifilamentary wire with very high-J_c [7] and excellent strain tolerance [10]. In order to industrialize the RHQT processed Nb₃Al wire, we studied the combination of the jelly-roll processed precursor wire and the RHQT process [9] as the next step, because the industrial fabrication technique of jelly-rolled Nb/Al precursor wire is well established [6]. Nearly the same results have been obtained for the jelly-roll processed precursor wire and for the rod-in-tube processed wire. On the other hand, the effects of adding a third element to the RHQT process have been studied by using the rod-in-tube processed precursor wire because of the easy fabrication of the initial Al-X alloy [11].

In the jelly-roll process, a monofilament billet, jelly-rolled with a thin laminated foil of Nb/Al on a Nb core, is extruded using a hydrostatic extrusion press and cold-drawn into a monofilament wire. The monofilament wire is cut into short pieces that are assembled into a Nb tube to fabricate a multifilamentary billet. The billet is extruded and then cold-drawn into multifilamentary precursor wires with diameters of 0.5–1.25 mm [12]. The volume ratio of Nb/Al in the filament was about 3.

In the rod-in-tube process, an Al-X (X = Ge or Cu) alloy rod is inserted into a Nb tube. The composite is cold-drawn into a monocore wire with a Nb/Al-X volume ratio of about 3. We assembled the monocore wires into a Nb tube. This was cold-drawn and the resulting composite wires were assembled again into a Nb tube in order fabricate a multifilamentary billet; this was cold-drawn into multifilamentary precursor wires with diameters of 0.5–0.9 mm.

Ordinarily the precursor wires were taken through the RHQ treatment to form the supersaturated bcc phase filaments, and then annealed at about 800 °C to transform these into A15 phase filaments. The superconducting properties of the wire were measured by a 4-probe resistance measurement. In this paper I_c is defined as the current at which the sample shows a potential drop of 100 μV/m in transverse magnetic fields. J_c is defined as I_c/S, where S is the total cross-sectional area of Nb/Al or Nb/Al-X microcomposite filaments, which is expected to be the maximum total cross-sectional A15-phase area formed in the sample.

7.3   Enhancements in Current Capacities

We have investigated the the enhancement of the critical current density, J_c, of RHQT-processed Nb₃Al wire by optimizing the fabrication conditions. The critical current density J_c of the Nb₃Al filaments increased monotonically with decreasing thickness of the Al layers in the jelly-rolled precursor wire, but the dependence became uncertain when the Al thickness was less than 300 nm.

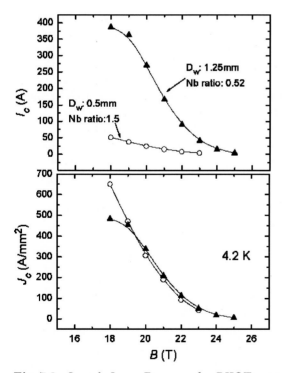

Fig. 7.2. I_c and J_c vs. B curves for RHQT-processed Nb₃Al wire with 0.5 mm diameter and an Nb ratio of 1.5, and for a wire with 1.25 mm diameter and an Nb ratio of 0.52 [12]

Fig. 7.3. Cross-section of a RHQT-processed Nb_3Al strand cable composed of 9 strands [12]

An excess of Nb in the precursor wire decreases the overall J_c, but is necessary to obtain enough mechanical strength during the RHQT-treatment. We could perform the RHQT-treatment for the precursor wire with a Nb/Nb_3Al ratio of 0.5–2.0 without encountering severe difficulty. Therefore the non-Cu overall J_c in a cross-sectional area excluding the Cu stabilizer can be increased up to $J_c/1.5$ with the reduction of the Nb matrix ratio [12].

With increasing precursor wire diameter, the cooling rate of the innermost region becomes slow and not sufficient due to the increase of both thermal resistance and heat capacity. Experimentally we could increase the precursor wire diameter up to 1.26 mm without causing any severe problems for the formation of the supersaturated bcc phase. However, when we used the precursor wire with 1.5 mm diameter, the as-quenched wire became brittle due to formation of the A15 phase. Experimentally, I_c (4.2 K, 21 T) for a monolithic Nb_3Al conductor could be enhanced up to 166 A from 15 A through the combination of increasing the wire diameter from 0.5 mm to 1.25 mm and decreasing the Nb/Nb_3Al ratio from 2.0 to 0.52, as shown in Fig. 7.2.

The as-quenched wire is composed of the Nb matrix and the Nb-Al supersaturated bcc filaments; both of these are ductile and deformable at room temperature. By making use of the ductility, we fabricated a compacted strand cable with 9 strands (Fig. 7.3), and then annealed the cable to transform the Nb-Al bcc phase to the A15 phase. I_c for the Nb_3Al strand cable is nearly 9 times the value of I_c for the strand.

7.4   Stabilization

The RHQT process includes a short heat treatment at about 2000 °C; this is much higher than the melting points of excellent electrical conductors such as Cu and Ag. For practical applications, the Nb_3Al conductors must be stabilized by incorporating an excellent electrical conductor. Simple electroplating with Cu after the RHQT-treatment could not stabilize the Nb_3Al wire sufficiently, because the Nb-oxide layer on the surface of the Nb matrix prevents thermal and electrical contact between the Cu and Nb. In order to break the Nb-oxide layer for sufficient stabilization, we have investigated several methods for incorporating the stabilizer into the RHQT-processed Nb$_3$Al wire, as follows.

7.4.1   Internal Ag Stabilization

To break the Nb-oxide layer through heavy cold-drawing of the Nb/Cu interface, it was attempted to incorporate the stabilizer into the precursor. In order to avoid melting down of the stabilizer during the RHQT-treatment, the stabilizer must be embedded into the precursor wire when the multifilamentary billet is assembled. In addition, a Nb film must be inserted between the stabilizer (Cu or Ag) and the Nb/Al microcomposite as a diffusion barrier to prevent the diffusion reaction between the Nb/Al microcomposite and the stabilizer. The Nb-oxide layer between the stabilizer and the diffusion barrier was broken by the heavy cold-drawing of the precursor wire. When the Cu stabilizer was embedded experimentally with the Nb barrier, a thick diffusion layer was formed between the Nb barrier and the Cu stabilizer during the RHQT-treatment, and the conductivity of the Cu stabilizer was reduced by the formation of Nb dendrites. On the other hand, when using Ag as the stabilizer, the thick diffusion layer is not formed between the Ag and the Nb, maintaining the good conductivity of the Ag without Nb dendrites. Therefore, internal Ag stabilization is one of the promising solutions for stabilizing the RHQT-processed Nb$_3$Al wire [13].

However, for the internally-stabilized wire, the Ag/non-Ag ratio should be lower than 0.3 [14], because the precursor wire must have sufficient yield strength in spite of melting of the Ag during the RHQT-treatment. We are planning to test a small coil wound with the Nb$_3$Al wire stabilized with internal Ag.

7.4.2   Mechanical Cladding of Cu

Mechanical cladding of the RHQT-processed wire with Cu has been investigated [15]. In this method the as-quenched wire is wrapped longitudinally with a Cu sheet and groove-rolled to ensure the mechanical bond between the Nb matrix and the Cu stabilizer. We have been successful in the fabrication of a Nb$_3$Al wire clad with a Cu layer with a ratio of 0.45, as shown in Fig. 7.4. The initial as-quenched wire of 1.26 mm diameter was plastically deformed to a nearly rectangular shape (1.2×0.62 mm) with a total area reduction of 42%.

As described before, J_c (21 T, 4.2 K) was increased from 150 A/mm^2 to 335 A/mm^2 by the cold drawing (Fig. 7.4). The shear deformation between Cu and Nb, caused by the round-to-rectangular deformation, should play an important role in breaking up the Nb-oxide layer. In the round-to-round deformation, we could obtain good bonding between Cu and Nb when the total reduction ratio is more than 90%. Heavy deformation with total area reduction ratio over 50% causes degradation of J_c. Therefore, this method is only useful for making a rectangular stabilized Nb$_3$Al wire. We are planning

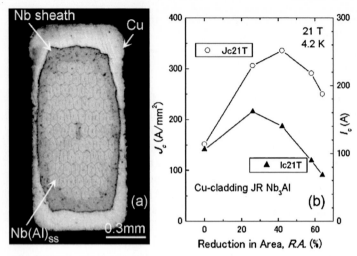

Fig. 7.4. Cross-section of the RHQT-processed wire clad mechanically with a Cu sheath; J_c and I_c vs. reduction ratio curves during the mechanical Cu cladding [15]

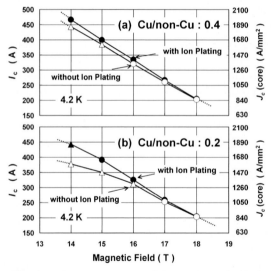

Fig. 7.5. I_c vs. B curves of Cu-stabilized Nb$_3$Al wires with and without Cu-ion plating, (**a**) Cu/non-Cu ratio: 0.4, (**b**) Cu/non-Cu ratio: 0.2 [16]

to test a small coil wound with the rectangular Nb$_3$Al wire stabilized with mechanical Cu-cladding [14].

7.4.3   Combination of Cu-Ion Plating and Cu Electroplating

Ion plating is one of the physical vapor deposition processes, and can be used to deposit a layer with a rapid deposition rate and firm bonding between the substrate and the layer. Therefore, we studied Cu-ion plating on the RHQT-processed Nb₃Al wire for stabilization. The combination of Cu-ion plating and Cu electroplating was found to be very effective for stabilizing the Nb₃Al wire, as shown in Fig. 7.5.

Cu-ion plating of 1 μm thickness on the RHQT-treated wire breaks up the Nb-oxide layer effectively and prevents re-oxidization in the air. The Nb₃Al wire, Cu-plated electrically with several tens of micrometers thickness after being Cu-ion plated, showed improved J_c at lower fields and an increase in the recovery current, indicating improved stability of the Nb₃Al wire. The duration required for the 1 μm thick Cu-ion plating is less than 20 s for the wire with a length of 150 mm and a diameter of 0.88 mm. Therefore, the cost of the stabilizing method is moderate. Moreover, this method is useful for stabilizing not only a rectangular wire but also a round wire, without any limitations on the Cu ratio.

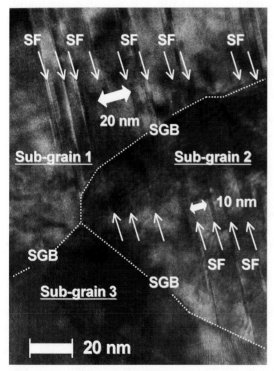

Fig. 7.6. Microstructure of an A15 subgrain in the RHQT-processed Nb₃Al wire. Many stacking faults are formed in the A15 subgrains in parallel with spaces of 10–20 nm [17]

7.5   Microstructure of RHQT-Processed Nb_3Al Wire

By using a focused-ion beam apparatus, RHQT-processed Nb_3Al wires were sliced into thin films of 100 nm thickness, to be used as the samples for transmission electron microscopy (TEM) [17]. The TEM analysis showed that the bcc supersaturated solid-solution phases consist of many crystal grains with a diameter of 2–4 μm that are surrounded by large-angle grain boundaries. All of the grain boundaries are simple flat planes. Some spherical voids, with about 0.1 μm diameter, are observed at the intra-grains and the inter-grains. By additional annealing at 800 °C, the Nb-Al bcc phases were transformed into A15 phases with 0.5-2 μm grain size. Any secondary phases were not observed in the A15 phase filaments. The grain boundaries of A15 phases show zigzag shapes unlike those of the Nb-Al bcc phases. Each grain of the A15 phase is an aggregation of subgrains with 80–150 nm diameter. The subgrain boundaries are small-angle boundaries. Moreover, we found that many parallel stacking faults were formed in the A15 subgrains with a spacing of 10–20 nm, as shown in Fig. 7.6. These numerous plane defects seem to be the main pinning centers in the RHQT-processed Nb_3Al wire.

7.6   Additional Effects of Ge and Cu on the Precursor Wire

In investigating the effects of adding a third element to the RHQT-processed Nb_3Al multifilamentary wire, we found that the addition of Ge or Cu is very effective for increasing both T_c and H_{c2}, with relatively high J_c at high fields [18,19]. Multifilamentary microcomposite precursor wires of Nb/Al-20 at%Ge and Nb/Al-2 at%Cu were fabricated through the rod-in-tube process.

We performed the normal RHQT-treatment using the precursor wires. According to the X-ray diffraction patterns of the as-quenched wires, disordered A15-phase filaments are formed directly in the Nb matrix of the Nb/Al-X composite wire during the RHQT treatment. The A15 phases in the as-quenched Nb/Al-Ge and Nb/Al-Cu wires show a T_c of about 15 K and about 11 K, respectively. With additional annealing at 750–800 °C, T_c of the Nb/Al-Ge and the Nb/Al-Cu wires were increased up to 19.4 K and 18.2 K, respectively. The additional annealing seems to increase T_c through the improvement of long-range ordering in the A15 crystal structure. It is well known that the T_c values of arc-melted Nb_3Al and $Nb_3(Al, Ge)$ are increased with the improvement of the long-range ordering in the A15 crystal structure by heat treatments at 750–800 °C. The maximum H_{c2} (4.2 K) values for the $Nb_3(Al, Ge)$ and $Nb_3(Al, Cu)$ wires were 39.5 T and 28.7 T, respectively; these values are much higher than those of the RHQT-processed Nb_3Al wire. We call the new process RHQO-(Rapid Heating, Quenching and Ordering).

J_c (4.2 K) versus B curves for the RHQO-processed Nb/Al-Ge wires and the RHQT-processed Nb/Al wire are shown in Fig. 7.7 as a function of the Al-alloy core diameter of the precursor wires. For the RHQT-processed Nb/Al

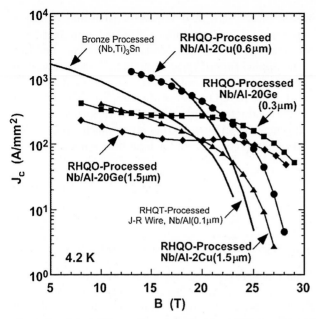

Fig. 7.7. J_c(4.2 K) vs. B curves for the RHQO-processed Nb/Al-X (X = Ge [18] and Cu [19]) wires and the RHQT-processed Nb/Al wire, as a function of the Al-X alloy core diameter in the precursor wires. For the RHQT-processed Nb/Al wire, the thickness of Al films is shown in parenthesis. The non-Cu overall J_c (4.2 K) vs. B curve for the bronze-processed (Nb,Ti)₃Sn wire [18] is shown for comparison

wire, the thickness of Al films is shown in parenthesis. The non-Cu overall J_c (4.2 K) versus B curve for commercial bronze-processed (Nb, Ti)₃Sn wire is also shown for comparison. With the reduction of the Al-alloy core diameter, J_c for the RHQO-processed Nb₃Al wires increased drastically. The J_c (4.2 K) values above 20 T for the Nb/Al-Ge wire with core size of 0.3 μm are very high and therefore of great practical interest.

If the Nb matrix ratio of RHQO-processed wire can be reduced down to 0.5 in such a way as to maintain the mechanical strength of the wire, $J_c/1.5$ becomes the optimized non-Cu overall J_c, which is very high in high fields. For example, J_c(4.2 K)/1.5 values above 150 A/mm² are obtained in fields up to 23 T for the RHQO-processed multifilamentary wires. If the RHQO-processed Nb₃Al-based wire can be commercialized, it will be possible to make a superconducting magnet generating fields up to 23 T at 4.2 K, and up to 26 T at 1.8 K.

7.7   Remark

High-T_c superconducting wires, such as Bi-2223 and Bi-2212 wires, are presently known to have comparably high J_c values in high fields at cryogenic temperatures below 20 K. However, these wires are very expensive and deficient in reliability, because large-scale fabrication under optimized manufacturing conditions is very difficult. The undesirable combination of high cost and low reliability has prevented the application of these high-T_c superconductors in large-scale magnets for the last decade. By contrast, the wire fabrication costs for the RHQT-process and for the RHQO-process are comparable to those of the commercial metallic superconductors that are used for various large-scale applications. Therefore, the new Nb_3Al conductors show much promise for becoming the next-generation superconductors for high-field applications.

References

1. J. L. Jorda, R. Flukiger and J. Muller: J. Less-Common Met. **75**, 227-239 (1980)
2. S. Ceresara, M.V. Ricci, N. Sacchetti and G. Sacerdoti: IEEE Trans. Magn. **MAG-11**, 263-265 (1975)
3. R. Akihama, R.J. Murphy and S. Foner: Appl. Phys. Lett. **37**, 1107-1109 (1980)
4. K. Inoue, Y. Iijima and T. Takeuchi: Appl. Phys. Lett. **52**, 1724-1726 (1988)
5. S. Saito, S. Ikeda, K. Ikeda and S. Hanada: J. Jpn. Inst. Met. **54**, 737-740 (1990)
6. N. Ayai, Y. Yamada, A. Mikumo, K. Takahashi, K. Sato, N. Koizumi, M. Sugimoto, T. Ando, Y. Takahashi, N. Nishi and H. Tsuji: IEEE Trans. Appl. Supercond. **5**, 893-896 (1995)
7. Y. Iijima, M. Kosuge, T. Takeuchi and K. Inoue: Adv. Cryog. Eng. Mater. **40**, 899-915 (1994)
8. K. Kamata, H. Moriai, N. Tada, T. Fujinaga, K. Itoh and K. Tachikawa: IEEE Trans. Magn. **MAG-21**, 277-280 (1985)
9. K. Fukuda, G. Iwaki, M. Kimura, S. Sakai, Y. Iijima, T. Takeuchi, K. Inoue, K. Kobayashi, K. Watanabe and S. Awaji: *Proc. ICEC/ICMC*, Kitakyushu, 20-24 May 1996 (Elsevier Science, Oxford, 1997) pp. 1669-1671.
10. T. Takeuchi, Y. Iijima, K. Inoue, H. Wada, B. ten Haken, H.H. ten Kate, K. Fukuda, G. Iwaki S. Saki and H. Moriai: Appl. Phys. Lett. **71**, 122-124 (1997)
11. Y. Iijima. A. Kikuchi, K. Inoue and T. Takeuchi: IEEE Trans. Appl. Supercond. **9**, 2696-270 (1999)
12. T. Takeuchi, K. Tagawa, T. Kiyoshi, K. Itoh, M. Kosuge, M. Yuyama, H. Wada, Y. Iijima, K. Inoue, K. Nakagawa, G. Iwaki and H. Moriai: IEEE Trans. Appl. Supercond. **9**, 2682-2687 (1999)
13. K. Tagawa, T. Takeuchi, T. Kiyoshi, K. Itoh, H. Wada and K. Nakagawa: to be published in Adv. Cryog. Eng. 46 (2000)
14. T. Takeuchi, K. Tagawa, N. Banno, T. Kiyoshi, S. Matsumoto, H. Wada, K. Aihara, Y. Wadayama, M. Okada and K. Nakagawa: IEEE Trans. on Supercond. **11**, 3972-3975 (2001)Å@
15. T. Takeuchi: to be published in IEEE Trans. Magn. **36** (2000)
16. A. Kikuchi, Y. Iijima, M. Fukutomi and K. Inoue: Cryog. Eng. **34**, 515-523 (1999)
17. A. Kikuchi, Y. Iijima and K. Inoue: IEEE Trans. on Supercond. **11**, 3938-3971 (2001)

18. A. Kikuchi, Y. Iijima, K. Inoue, T. Asano and M. Yuyama: *Proc. EUCAS'99*, Sitges, 14-17 September, 1999 (Institute of Physics Publishing, London, 2000)
19. Y. Iijima, A. Kikuchi and K. Inoue: Cryogenics, **40**, 345-348 (2000)

8 High-Field A15 Superconductors Prepared Via New Routes

K. Tachikawa, T. Kato, H. Matsumoto and K. Watanabe

A magnetic field of 18.8 T has been generated so far at 4.2 K using the bronze-processed $(Nb,Ti)_3Sn$ conductor [1]. The generation of fields over 20 T at 4.2 K may be realized by the development of Nb_3Al-based conductors with high upper critical field B_{c2} or by the improvement of the high-field performance of Nb_3Sn. In this study, $Nb_3(Al.Ge)$ conductors were fabricated by means of the reaction between intermediate σ-phase $Nb_2(Al,Ge)$ and Nb powders. Then, $(Nb,Ta)_3Sn$ conductors were prepared by a similar process using a $(Nb,Ta)_6Sn_5/Nb$ mixed powder core and a Ta sheath. Subsequently, $(Nb,Ta)_3Sn$ conductors were prepared by fabricating a new composite of Ta-Sn core and Nb (Nb-Ta) sheath. These conductors prepared through new routes gave sufficient performance to generate over 20 T at 4.2 K.

8.1 $Nb_3(Al,Ge)$ Superconductors Prepared from σ-Phase/Nb Mixed Powder Core

Since the bronze process is not applicable for the fabrication of Nb_3Al-based A15 compounds, direct diffusion between Nb and Al has been studied. However, in this route intermediate compounds richer in Al, e.g. $NbAl_3$ and Nb_2Al (σ-phase), are predominantly formed. To minimize the formation of intermediate compounds richer in Al, the Al should be fabricated into a very thin filament [2] or foil [3]. In the present study, the $Nb_3(Al,Ge)$ phase was synthesized through the reaction between the σ-phase and Nb since the stable phase between σ and Nb is only A15 $Nb_3(Al,Ge)$ [4]. $Nb_3(Al,Ge)$ is well known to have appreciably better high-field performance than Nb_3Al.

8.1.1 Experimental Procedure

The σ-phase $Nb_2(Al_{0.8},Ge_{0.2})$ starting material was prepared by conventional plasma-arc melting in an argon atmosphere. The obtained σ-phase buttons (usually 20 g in weight) were crushed into powders, and passed through a -330 mesh sieve. Subsequently, σ-powder was mixed with Nb powder of -325 mesh, the final Nb/Al-Ge composition ratio being 2.8–3.2. Furthermore, 1–10 vol.% of MgO was added to the σ/Nb mixed power. Some powders were crushed into finer size using a zirconia ball mill in open air, with a crushing period of 20 min. This powder is designated as "fine powder" in this paper.

The σ/Nb mixed powders were packed into a Nb tube with 8 mm outer diameter and 5 mm inner diameter. The composite tubes were groove-rolled into 2.5 mm square rods, and then flat-rolled into tapes 0.5 mm thick and 5 mm wide; the width and thickness of the σ/Nb core was 2.5 mm and 0.3 mm, respectively. The tape specimens were heat-treated at a temperature between 1200 °C and 1450 °C in a vacuum of 1×10^{-5} torr. Some specimens were subsequently annealed in vacuum at a temperature between 650 °C and 750 °C for 100 h.

Microstructures of the cross-section of specimens were observed using an optical microscope after anodic oxidization. X-ray diffraction (XRD) analysis was carried out in order to identify phases formed by the heat treatment. The transition temperature T_c was measured by a conventional four-probe resistive method using a calibrated Ge thermometer. The critical current I_c was measured at 4.2 K by a four-probe resistive method; this was defined as the current at which a voltage across the 10 mm length of the specimen reached 1 μV/cm. The magnetic field was applied parallel to the tape surface and perpendicular to the specimen current. The I_c measurement was limited up to a few hundreds of amperes due to the current capacity of the specimen holder. The critical current density J_c was obtained by dividing I_c by the cross-sectional area of the core.

8.1.2   Experimental Results

Figure 8.1 illustrates the XRD pattern of the σ/Nb (Nb ratio 2.9) tape heat-treated at 1300 °C for 1 h. In the XRD pattern taken before the heat treatment, the peaks from the σ-phase, Nb and a small amount of A15 phase are identified. The XRD pattern of the tape heat-treated at 1200 °C for 1 h consists of peaks from the major A15 phase and a minor residual σ phase. After the heat treatment at 1300 °C for 1 h, the XRD pattern is almost changed to that of a single A15 phase. These results imply that the reaction temperature of 1300 °C is required for the synthesis of the A15 $Nb_3(Al,Ge)$ phase from the σ/Nb composite.

The transition width at T_c of tape specimens is fairly sharp; it is less than 0.5 K. The T_c (mid-point) of specimens exceeds 17.0 K after the reaction at 1400 °C and subsequent annealing at 700 °C. The post-annealing at 700 °C enhances T_c by 0.5–1.0 K. An onset T_c of 18.0 K has been obtained in the fine powder tape that shows a slightly higher T_c than that of coarse powder tapes.

Figure 8.2 illustrates I_c versus magnetic field curves of different tape specimens reacted at 1400 °C. The specimens show a pronounced peak effect in the I_c-B performance. However, in the fine powder specimen the peak effect becomes less pronounced. The fine powder specimen also shows better high-field performance than that of the coarse powder specimen. The annealing at 700 °C after the reaction shifts the I_c-B curve of specimens to higher field by ~1.5 T corresponding to the increase in T_c. Regarding the effect of the Nb composition ratio, the $Nb_{2.9}(Al_{0.8},Ge_{0.2})$ tape seems to show the best

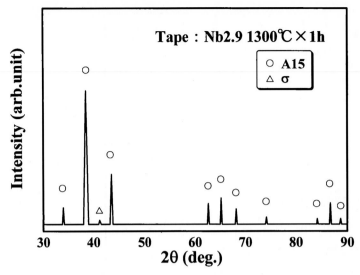

Fig. 8.1. XRD pattern of the σ/Nb core-Nb$_{2.9}$(Al,Ge) tape reacted at 1300 °C for 1 h

high-field performance. The J_c (core) of the fine powder tape reacted at 1400 °C for 1 h and then annealed at 700 °C is about 3×10^4 A/cm^2 at 21.5 T and 4.2 K.

The peak effect in the I_c-B curve is considered to originate from the degradation of J_c at low fields. The addition of MgO powder was then attemped in order to introduce artificial pinning centers effective at low fields, and to suppress the peak effect. The electron probe microanalysis (EPMA) pattern taken on the MgO-added tape revealed the fine dispersion of Mg throughout the cross-section of the tape. The MgO addition is effective for enhancing I_c at low fields, e.g. the I_c at 12 T is increased by a factor of 2–3 by the MgO addition. Thus the MgO powder addition of 5–10 vol% is found to be effective for eliminating the peak effect in the Nb$_3$(Al,Ge) specimens with large grain size reacted at high temperatures [4].

8.1.3   Features of this Conductor

A15 Nb$_3$(Al,Ge) has been synthesized through the reaction between σ-phase Nb$_2$(Al,Ge) and Nb. An onset T_c of 18.0 K and a J_c (core) of about 3×10^4 A/cm^2 at 4.2 K and 21.5 T have been obtained in the Nb$_3$(Al,Ge) tape with σ/Nb mixed-powder core. However, a reaction temperature of 1300 °C–1400 °C is required to obtain a single A15 Nb$_3$(Al,Ge) phase. The addition of MgO powder which may act as artificial pinning centers is effective for eliminating the peak effect in the I_c-B curve of Nb$_3$(Al,Ge) tapes reacted at high temperatures.

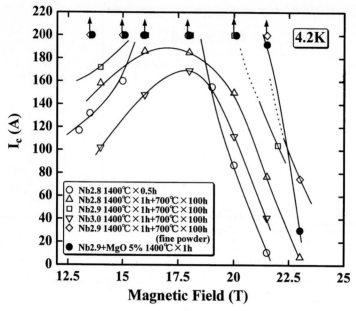

Fig. 8.2. I_c-B curves of different σ/Nb core-Nb$_3$(Al,Ge) tapes reacted at 1400 °C

8.2   Nb$_3$Sn and (Nb,Ta)$_3$Sn Superconductors Prepared from Intermediate Compound Powder

A Nb$_3$Sn superconductor with B_{c2} of about 24.7 T at 4.2 K has been prepared through a new process starting from an intermediate Nb$_6$Sn$_5$ compound. The Nb$_6$Sn$_5$ powder has been easily synthesized by the melt-diffusion process of Sn using Nb and Sn powders [5]. A small amount of Ti or Ta substitution for Nb is performed, which yields a further increase in B_{c2}. Substituting Ta for Nb makes the curvature of the I_c-B curve convex to upward, which results in a J_c (core) of 4×10^4 A/cm^2 at 23 T and 4.2 K after the reaction at 925 °C [6]. When Nb-Ta alloy powder is used in the specimen preparation, a more homogeneous structure and better high-field performance have been achieved [7].

8.2.1   Experimental Procedure

A mixed powder of Nb and Sn was heated in vacuum at 900 °C for 10 h using an alumina pot, where molten Sn reacted with Nb (melt diffusion, MD). The purity and the size of the Nb powder were 99.8% and under 325 mesh, respectively, while those of the Sn powder were 99.9% and under 350 mesh, respectively. The synthesized Nb$_6$Sn$_5$ powder was ground for 0.5 h by a planetary-type ball mill under Ar atmosphere. The Nb$_6$Sn$_5$ powder was mixed with Nb powder in the composition of Nb$_3$Sn using the ball mill for 3 h under Ar atmosphere. A small amount of Ti or Ta was substituted for Nb in the Nb_6Sn_5 powder using Ti or Ta powder at the time of the melt diffusion. The specimen, for example, where 5 at% Ta is substituted for Nb in the Nb_6Sn composition, i.e. $(Nb_{0.95}Ta_{0.05})_3Sn$, is denoted as 5Ta in this paper. Another way to substitute Ta for Nb has been performed in this study. Nb-4 at.% Ta alloy powder was prepared through the hydride-dehydride process, and was used for the preparation of $(Nb,Ta)_6Sn_5$ powder by the melt-diffusion reaction with Sn powder. The particle size of Nb-4 at.% Ta powder was under 325 mesh. Then the Nb-4 at.% Ta powder was mixed with $(Nb,Ta)_6Sn_5$ powder in the composition of $(Nb_{0.96},Ta_{0.04})_3Sn$. The resulting powder is designated as NT(4Ta). The resulting mixed powder of Nb_6Sn_5/Nb with or without Ti or Ta substitution was encased in a Ta tube, 8 mm in outer diameter and 5 mm in inner diameter, and then fabricated into a tape, with width and thickness of 5 mm and 0.5 mm, respectively. The resulting tape specimens were heat-treated in vacuum to form a Nb_3Sn core. The superconducting properties and structures of the specimens were characterized by the procedure described in 8.1.1.

8.2.2   Experimental Results

Nb_6Sn_5 powder prepared by the MD reaction exhibits a nearly round shape of 40–50 μm in diameter. After grinding for 0.5 h in the ball mill, the diameter of most of the Nb_6Sn_5 powders was distributed in the range of 1–2 μm. The XRD pattern revealed that the heat treatment at 900 °C for 20 h after the fabrication transformed the Nb_6Sn_5/Nb core almost completely into A15 Nb_3Sn. No diffusion of Sn from the core to Ta sheath was observed according to the EPMA analysis; this implies the preservation of stoichiometry of the core during the heat treatment. This is a difference between the present process and the ECN process [8], where the intermediate compound core reacts with the Nb sheath.

The T_c of pure Nb_3Sn specimens (NS) reaches its maximum value after the reaction at 900–925 °C. This is about 0.5 K higher than that of bronze-processed Nb_3Sn [9]. In bronze-processed Nb_3Sn, the amount of Sn supply is limited by the solubility of Sn in Cu, which may cause a slight degradation in T_c. More stoichiometric A15 Nb_3Sn may be formed in the present specimens. The T_c of these specimens is not sensitive to the amount of Ti substitution for Nb up to 3 at.%. The small amount of Ta substitution for Nb yields a further increase in T_c, and a T_c (mid) of 18.3 K has been obtained in the 5Ta specimen.

Figure 8.3 shows curves of I_c and J_c (core) versus field for different specimens in fields over 20 T. The Kramer plot indicates that the B_{c2} of NS tape is 24.7 T at 4.2 K, which is by $\sim$ 4 T higher than that of bronze-processed Nb_3Sn. The enhanced B_{c2} obtained in the NS specimen may be due to its large ρ_n value [6] and high T_c described above. The 2 at.% Ti and 5 at.% Ta substitution yields further increase in ρ_n and B_{c2} values [6]. The substitution of Ta for Nb changes the curvature of the log I_c-B curve from convex

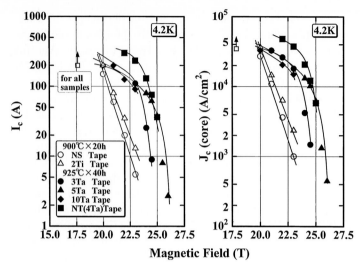

Magnetic Field (T)

Fig. 8.3. I_c-B and J_c-B curves of NS, 2Ti, 5Ta, and NT(4Ta) tapes prepared from intermediate compounds and reacted at quoted conditions

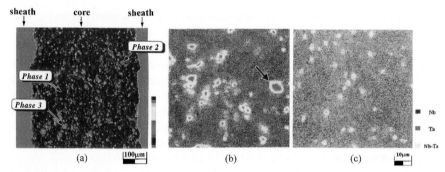

Fig. 8.4. (a) EPMA mapping of Ta on the cross-section of 5Ta tape. **(b)** Enlarged EPMA mapping of Ta on the core of 5Ta tape. **(c)** Enlarged EPMA mapping of Ta on the core of NT(4Ta) tape. All specimens were reacted at 925 °C for 40 h

to upward, which results in significant improvement of the high-field performance. Substitution of 5–7 at.% Ta for Nb yields the best result. 3 at.% Ta substitution is insufficient for the improvement of the high-field performance, while 10 at.% Ta substitution slightly degrades I_c as compared to 5 at.% Ta substitution. A J_c (core) of $2.4 \times 10^4 \text{A/cm}^2$ has been obtained at 23 T in the 5Ta tape reacted at 925 °C for 40 h. The NT(4 Ta) tape exhibits further improvement of I_c in high fields by comparison to 5Ta tape, as shown in Fig. 8.3. A J_c (core) of $4 \times 10^4 \text{A/cm}^2$ has been obtained at 23 T and 4.2 K, corresponding to a non-Cu J_c of nearly $1 \times 10^4 \text{A/cm}^2$.

Figure 8.4a is the EPMA mapping of Ta taken on the cross-section of a 5Ta specimen reacted at 925 °C for 40 h. The structure of the core with Ta

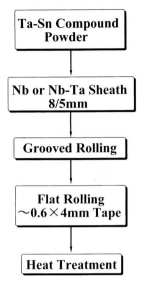

Fig. 8.5. Fabrication procedure of $(Nb,Ta)_3Sn$ tape using Ta-Sn core

substitution is composed of three different phases as indicated in the figure. Table 8.1 shows the results of EDX analysis for phases 1, 2 and 3 in Fig. 8.4a. The Sn concentration in phases 1 and 2 corresponds to $(Nb,Ta)_3Sn$, where the Ta concentration in phase 2 is about twice as large as that in phase 1. Phase 3 is enriched with Ta. The compositions of phase 1, 2 and 3 are not much changed by the amount of Ta substitution in the specimens. The increase in the amount of Ta substitution results in the increase in the areal fraction of phase 2 and phase 3. Thus, two $(Nb,Ta)_3Sn$ phases with different Ta concentrations are found in the Ta-substituted specimens. Phase 1 may be formed by the diffusion of Sn into Nb from $(Nb,Ta)_6Sn_5$, while phase 2 may be formed by the release of Sn from $(Nb,Ta)_6Sn_5$. Figure 8.4b, c is enlarged EPMA composition mappings of Ta on the cross-section of 5Ta and NT (4Ta) tapes reacted at 925 °C for 40 h respectively. In the 5Ta tape, aggregation of Ta is indicated by an arrow; phase 3 in Table 8.1 is observed. In the NT (4Ta) tape, the aggregation of Ta becomes less prominent. Ta is mixed as a powder in the 5Ta specimen at the time of the melt diffusion C, while in the NT (4Ta) specimen Ta is mixed in an atomic level using Nb-4 at.% Ta alloy. This may cause a finer dispersion of Ta in the $(Nb,Ta)_3Sn$ phase. The EDX analysis indicates that the average Ta concentration in the $(Nb,Ta)_3Sn$ phase of the NT (4Ta) specimen is 2.3 at.%; this is close to that of phase 2 in Table 8.1. Even the nominal concentration of Ta in the NT (4 Ta) specimen is lower than that in the 5Ta specimen. The finer dispersion of Ta in the NT (4Ta) core may result in an effective Ta incorporation into the $(Nb,Ta)_3Sn$ phase which attributes to the enhancement of I_c in high fields shown in Fig. 8.3.

Table 8.1. Result of EDX composition analysis for different phases in the 5Ta specimen shown in Fig. 8.4a

	Phase 1			Phase 2			Phase 3		
Element	Nb	Sn	Ta	Nb	Sn	Ta	Nb	Sn	Ta
at. %	77.3	25.0	1.3	72.4	24.9	2.7	16.8	5.2	78.0

Table 8.2. EDX analysis on the $(Nb,Ta)_3Sn$ layers formed in the Ta-Sn/Nb 3.3 at.% Ta tape specimens reacted at 900 °C for 80 h

at.%	Ta-Sn/Nb tape			Ta-Sn/Nb-3.3 at.% Ta tape		
	core side	middle	sheath side	core side	middle	sheath side
Nb	71.14	74.00	75.04	67.54	72.35	72.61
Ta	3.82	0.97	0.14	7.26	2.35	2.35
Sn	25.04	25.03	24.82	25.20	25.30	25.30

8.2.3   Features of this Conductor

Nb_3Sn superconductors have been prepared starting from Nb_6Sn_5 intermetallic compound powder that is easily synthesized by the melt-diffusion reaction of Nb and Sn powders at 900 °C. The mixed powder of Nb_6Sn_5 with or without Ti or Ta substitution and Nb was encased in a Ta tube, and fabricated into tapes without intermediate annealing. The Nb_3Sn tape prepared by the present process shows higher T_c than that of bronze-processed Nb_3Sn and a B_{c2} of 24.7 T at 4.2 K. No diffusion reaction takes place between the core and the sheath. Substituting a small amount of Ti or Ta for Nb still increases B_{c2}. Substituting Ta for Nb changes the curvature of the I_c-B curve from convex to upward, resulting in significant improvement of the high-field performance. Furthermore, Nb-4 at.% Ta alloy powder has been used for the preparation of Ta-substituted tape. The Nb-Ta alloy powder facilitates a much finer dispersion of Ta in the core compared to that prepared from conventional Ta powder. A J_c (core) of 4×10^4 A/cm^2, and a non-Cu J_c of nearly 1×10^4 A/cm^2 have been obtained at 23 T and 4.2 K after the reaction at 925 °C for 80 h.

8.3 $(Nb,Ta)_3Sn$ Superconductors Prepared from Ta-Sn Core

The $(Nb,Ta)_3Sn$ superconducting tape is fabricated using Ta-Sn core as shown in Fig. 8.5. Figure 8.6 is the EPMA composition mapping of Nb, Sn and Ta taken on the cross-section of the tape with 6/5 (Ta/Sn ratio) powder core and Nb sheath reacted at 900 °C for 80 h. One of the features of the present process is that a quite thick $(Nb,Ta)_3Sn$ layer is synthesized after the diffusion reaction. About 80 μm-thick and 100 μm-thick $(Nb,Ta)_3Sn$ layers are obtained in the 6/5 powder core and 2/3 powder core tapes, respectively. Even the 3/1(Ta_3Sn) core produces a nearly 30 μm-thick $(Nb,Ta)_3Sn$ layer.

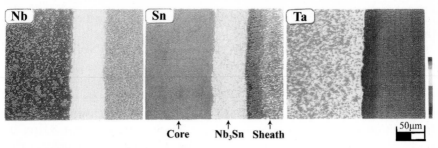

Fig. 8.6. EPMA composition mapping of Nb, Sn and Ta on the cross-section of Ta-Sn/Nb composite tape reacted at 900 °C for 80 h

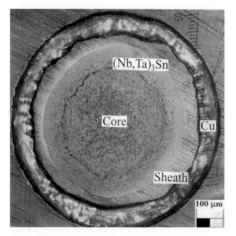

Fig. 8.7. Optical microstructure on the cross-section of the $(Nb,Ta)_3Sn$ wire reacted at 800 °C for 80 h with 2 wt.% Cu addition to the Ta-Sn core

Sn seems to have a much stronger affinity for Nb than for Ta; this results in a significant enhancement of the diffusion of Sn from the Ta-Sn core to the Nb sheath. The thickness of the $(Nb,Ta)_3Sn$ layer does not depend on the Ta concentration in the Nb-Ta sheath. It is worth noting that Nb is incorporated into the core from the sheath passing through the $(Nb,Ta)_3Sn$ layer. No void is formed in the core after the reaction, since Sn in the core and Nb in the sheath inter-diffuse with each other. In the Nb_3Sn wire fabricated by the Nb-tube process using a Cu-Sn core and a Nb sheath, Nb does not penetrate into the Cu-Sn core after the reaction [10]. The core after the reaction is composed of two different phases: these are Ta-Sn binary and Nb-Ta-Sn ternary phases.

An appreciable amount of Ta is incorporated into the $(Nb,Ta)_3Sn$ layer from the Ta-Sn core and the Nb-Ta sheath. Table 8.2 is the result of the EDX analysis on the $(Nb,Ta)_3Sn$ layer formed in the tape specimen with Nb and Nb-3.3Ta sheath, both reacted at 900 °C for 80 h. A gradient in Ta concentration is observed in the $(\text{Nb,Ta})_3\text{Sn}$ layer that is richer in Ta near the core side. The Ta concentration in the $(\text{Nb,Ta})_3\text{Sn}$ layer is increased by using a Nb-Ta alloy sheath. The Sn concentration in the $(\text{Nb,Ta})_3\text{Sn}$ layer is very close to that of the stoichiometric A15 composition. Ta seems to substitute for the Nb site in the A15 $(\text{Nb,Ta})_3\text{Sn}$ phase. A non-uniform and thin $(\text{Nb,Ta})_3\text{Sn}$ layer is formed between the core and the sheath after the reaction at 800 °C for 40 h. On the contrary, in the specimen with 2–5 wt.% Cu addition to the Ta-Sn core, a uniform Nb_3Sn layer of about 80 μm in thickness is obtained by the reaction at 800 °C. Figure 8.7 illustrates the cross-section of the wire specimen reacted at 800 °C with 2 wt.% Cu addition to the core.

The T_c versus reaction temperature curve indicates that T_c reaches the maximum value after the reaction at 900 °C. Specimens with Nb-3.3 at.% Ta and 4.2 at.% Ta sheaths show a T_c (mid-point) of 18.3 K, while those with a Nb sheath show a T_c of 18.0 K. The addition of 2–5 wt.% Cu to the core decreases the optimum reaction temperature to 750–800 °C with respect to T_c. A T_c (mid) of 18.3 K is also obtained in the specimen with 2 wt.% Cu addition after the reaction at 800 °C.

The Ta addition to the sheath yields an enhancement of I_c in high fields. This may be attributed to the increase of Ta concentration in the $(\text{Nb,Ta})_3\text{Sn}$ layer indicated in Table 8.2. The tapes with Nb-3.3 at.% Ta sheath and Nb-4.2 at.% Ta sheath exhibit almost the same high-field performance. The optimum Ta concentration in the sheath seems to be 3.3 at.%, since the Nb-3.3 at.% Ta sheath shows the most excellent workability. The increase of Sn concentration in the Ta-Sn core results in the increase of the $(\text{Nb,Ta})_3\text{Sn}$ layer thickness. The tape with the core composition of 6/5 shows nearly the same I_c as that with the core composition of 2/3. The optimum Ta-Sn core composition seems to be 6/5, since excess Sn exudes out from the open ends of the tape with the core composition of 2/3.

Figure 8.8 shows the effect of reaction temperature and Cu addition on the non-Cu J_c versus B curves of the specimen with 6/5 Ta-Sn core and Nb-3.3 at.% Ta sheath. The non-Cu J_c versus B performance at high fields is fairly sensitive to the reaction temperature. The I_c of the tape reacted at 900 °C for 80 h is about 280 A; the corresponding non-Cu J_c of the tape reaches 1.2×10^4 A/cm^2 at 23 T and 4.2 K. The B_{c2} of this specimen seems to be about 26.5 T and 4.2 K. The I_c of the specimen is depressed by the reaction at higher temperatures. The specimen reacted at 950 °C shows a peak effect in the non-Cu J_c versus B performance, the peak field being about 21 T. This may be caused by the grain growth of $(\text{Nb,Ta})_3\text{Sn}$ formed at temperatures above 900 °C. The specimen with 2 wt.% Cu addition and reacted at 800 °C shows a rapid increase in J_c with decreasing applied magnetic field, although the B_{c2} of this specimen is a little lower than that of the specimen without Cu addition. A small amount of Cu addition enhances the J_c value below 22 T as a result of the finer $(\text{Nb,Ta})_3\text{Sn}$ grain size synthesized at lower temperatures.

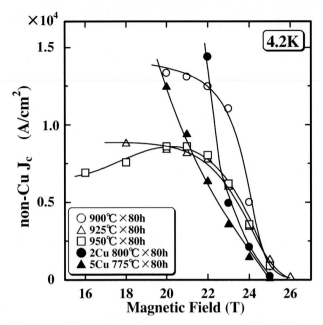

Fig. 8.8. Effects of reaction temperature and Cu addition on the non-Cu J_c versus B curve of $(Nb,Ta)_3Sn$ tapes with 6/5 Ta-Sn core and Nb-3.3 at%Ta sheath

The tape with 2 wt.% Cu added to the core and reacted at 800 °C shows a non-Cu J_c of about 1.4×10^4 A/cm^2 at 22 T and 4.2 K. Recently, the high-field performance of bronze-processed $(Nb,Ti)_3Sn$ conductors has been improved by increasing the Sn concentration in the bronze. Present $(Nb,Ta)_3Sn$ conductors show a non-Cu J_c of over 1×10^4 A/cm^2 at a field about 4 T higher than that reported for the improved bronze-processed $(Nb,Ti)_3Sn$ [11].

8.4   Conclusion

A J_c (core) of about 3×10^4 A/cm^2 at 4.2 K and 21.5 T has been obtained in the $Nb_3(Al,Ge)$ tape prepared from $Nb_2(Al,Ge)/Nb$ mixed powder core. However, a reaction temperature of 1300 °C–1400 °C is required to synthesize a single A15 phase. The $(Nb,Ta)_3Sn$ tape prepared from $(Nb,Ta)_6Sn_5/Nb$ mixed powder core shows a J_c (core) of nearly 4×10^4 A/cm^2 at 4.2 K and 23 T after the reaction at 925 °C, when Nb-4 at.% Ta alloy powder is used as the Ta source. A new $(Nb,Ta)_3Sn$ conductor with a Ta-Sn core exhibits a non-Cu J_c of 1.2×10^4 A/cm^2 at 4.2 K and 23 T after the reaction at 900 °C. Adding 2 wt.% Cu to the Ta-Sn core reduces the optimum reaction temperature to 800 °C; this is most promising for practical applications. Three processes require no intermediate annealing during the fabrication; this is a substantial advantage over the conventional bronze process for fabricating A15 conductors.

Acknowledgments

The authors deeply appreciate the contributions of Professor S. Awaji and other staff members of IMR of Tohoku University at the time of using the high-field magnet facilities for the I_c measurement. They are also very grateful to Mr. Miyamoto of Tokai University for his EPMA and EDX analyses of specimens.

References

1. M. Yosikawa, T. Kamikado, R. Hirose, S. Hayashi, Y. Kawate, M. Hamada and K. Takabatake: *Abstract 59th Meeting on Cryogenics and Supercond.*, Yamaguchi, 13-15 October 1998 (Cryogenic Association of Japan, Tokyo, 1998) p.56 (in Japanese)
2. T. Takeuchi, M. Kosuge, Y. Iijima, A. Hasegawa, T. Kiyoshi and K. Inoue: IEEE Trans. Magn. **MAD-27**, 2045 (1991)
3. T. Ando, Y. Takahashi, M. Nishi, Y. Yamada, K. Ohmatsu and M. Nagata: IEEE Trans. Magn. **MAD-27**, 1775 (1991)
4. K. Sakinada, M. Kobayashi, M. Natsuume and K. Tachikawa: Adv. Cryog. Eng. **40**, 923 (1994)
5. K. Tachikawa, M. Natsuume, Y. Kuroda and H. Tomori: Cryogenics **36**,113 (1996)
6. K. Tachikawa, Y. Kuroda, H. Tomori and M. Ueda: Adv. Cryog. Eng. **44**, 895 (1998)
7. K. Tachikawa, T. Yokoyama, T. Kato and H. Matsumoto: to be published in Adv. Cryog. Eng. **46** (2000)
8. W.L. Nijmeijer and B.H. Kolster: J. Less-Common Metals **160**, 161 (1990)
9. H. Sekine, K. Itoh and K. Tachikawa: J. Appl. Phys. **63**, 2167 (1988)
10. S. Murase, H. Shiraki, M. Tanaka, M. Koizumi, H. Maeda, I. Takano, N. Aoki, M. Ichihara and E. Suzuki: IEEE Trans. Magn. **MAG-21**, 316 (1985)
11. T. Miyazaki, Y. Murakami, T. Hase, M. Shimada, K. Itoh, T. Kiyoshi, T. Takeuchi, K. Inoue and H. Wada: IEEE Trans. Appl. Supercond. **9**, 2505 (1999)

Part IV

Magnetic and Optical Properties
in High Fields

9 Magnetic Properties of Rare-Earth Monopnictides in High Magnetic Fields

T. Sakon, Y. Nakanishi, T. Komatsubara, T. Suzuki and M. Motokawa

9.1 Introduction

9.1.1 Properties of Rare-Earth Compounds: Valence Fluctuations and Heavy Fermions

Since more than two decades, the physics of rare-earth compounds has been of particular interest because of the variety of their magnetic properties, classified as valence fluctuations and heavy fermions. In Ce compounds, the valence changes between $+3$ ($4f^1$) and $+2$ ($4f^0$) in some materials called valence fluctuation systems, although the valence is normally supposed to be $+3$ ($4f^1$). The very small valence excitation energy for Ce and Yb leads to an intermediate valence state by hybridization between the 4f states and neighboring atoms, and creates a resonance scattering band near the Fermi level E_F. The energy width of the band is about $\Delta = \pi N_F V^2$, where V is the hybridization amplitude and N_F is the density of states (DOS) of the conduction band near E_F. Therefore, the DOS at the Fermi surface is enhanced on the order of $1/\Delta$. Some of the Ce and Yb compounds show heavy-fermion behavior. A typical material is CeAl$_3$. As a result of the enhanced DOS at E_F, the effective mass is a hundred or a thousand times larger than that of the normal metals. At low temperatures, the mixing between the 4f orbital and the itinerant bands results in an excessively narrow quasi-particle band at the Fermi level. The Sommerfeld coefficient g is proportional to the DOS at E_F and $\gamma = 1.6\,\mathrm{JK^{-2}mol^{-1}}$ in CeAl$_3$ [1]. While in the typical normal metal, copper, γ is about $0.7\,\mathrm{mJK^{-2}mol^{-1}}$. Therefore, very strong mass enhancement occurs in CeAl$_3$. We can know only the amplitude of the total interactions between electrons from the specific heat measurements. The detail of the Fermi surface should be studied to understand these interactions more precisely and more correctly. The de Haas-van Alphen (dHvA) effect is the most effective experimental method to investigate the Fermi surface parameters such as the cyclotron mass m_C^* and the external cross-sectional area of the Fermi surface S. In this manuscript, after some introductory materials, some experimental results obtained at the IMR are summarized for rare-earth monopnictides using magnetization and dHvA techniques.

9.1.2   Competition Between Kondo Effect and Magnetic Ordering

Among the 4f electron systems, rare-earth monopnictides with simple cubic crystal structure have been attracting much attention for both experimentalists and theorists for their interesting physical phenomena such as the Kondo effect, metal-insulator transitions, valence fluctuation behavior and complicated magnetic phases. At temperatures higher than the Kondo temperature T_K, the scattering between 4f electrons and carriers occurs individually and locally at each atomic site, and the electric resistivity ρ shows a logarithmic temperature dependence on the temperature as [2]

$$\rho = -\ln T. \tag{9.1}$$

Below T_K, the scattering changes coherently and ρ is decreased by lowering the temperature. At low temperatures the 4f magnetic moment disappears due to the cancellation with the moment of the carriers, changing the quasiparticles that are formed by the enhanced extremely narrow band. The value of $k_B T_K$ is identical to the bandwidth Δ. For dense Kondo materials such as CeAl$_3$ and CeCu$_6$, T_K is only a few K and, therefore, g becomes very large. On the other hand, T_K of the intermediate valence materials is around 10 or 100 K and γ is not to be too large, on the order of 10–100 mJK^{-2}mol^{-1}.

The Kondo effect competes with the Ruderman-Kittel-Kasuya-Yoshida (RKKY) interaction in some materials. If the RKKY energy is larger than the Kondo energy, long-range magnetic ordering is stabilized. The hamiltonian of the RKKY interaction [3–5] is expressed as

$$H = \sum_{k\delta} \epsilon C_{k\delta}^* C_{k\delta} + J_{cf} \sum_i S_i s_i, \tag{9.2}$$

where S_i is a local spin at the ith site and s_i is the spin density of the carrier at the ith site. As indicated in (9.2), the local spin S_i and the carrier spin s_i interact through the cf-exchange interaction J_{cf}. The negative J_{cf} induces an antiferromagnetic correlation. The magnetic susceptibility $\chi(r)$ is written as a function of the distance r for temperatures above T_N,

$$\chi(r) = \frac{1}{N_0} \chi_p F(2k_F r), \tag{9.3}$$

where

$$F(2k_F r) = \frac{-2k_F r \cos(2k_F r) + \sin(2k_F r)}{(2k_F r)^4}, \tag{9.4}$$

and k_F is the Fermi wavelength. The susceptibility oscillates and decreases in magnitude with r. The negative cf-exchange interaction J_{cf} in (9.2) produces an antiferromagnetic long-range order at low temperatures below T_N

.

9.1.3   Magnetic and Electrical Properties of Low-Carrier Systems

In the rare-earth monopnictides the physical phenomena vary by depending both on the rare-earth and pnictogen elements. The rare-earth pnictides are semimetallic compounds. The hole band and the electron band overlap each other. Therefore, the carriers can move easily in electric fields. The carrier density of the semimetal is very low, 1/100 or 1/1000 of that in the metal. Within the framework of the free-electron model, the decrease of the electron density means the decrease of the Fermi energy, or in other words, the decrease of the kinetic energy of the carriers. In the rare-earth materials, the Coulomb interaction, magnetic exchange interaction, crystal field effect and quadrupolar interactions play an important role and consequently affect sensitively the magnetic properties of rare-earth monopnictides.

CeSb is the most typical compound and shows 14 magnetic phases in the $B-T$ diagram, as shown in Fig. 9.1 [6]. This phase diagram is confirmed from specific heat experiments. The magnetic structures have been determined by neutron diffraction experiments.

All the observed structures are formed by the combination of the (100) layers, with their magnetic moments orientated ferromagnetically in the [001] direction (denoted by up or down arrows) and completely disordered (100) layers (indicated by open circles). This is understood by the Axial Next Nearest Neighbor Ising (ANNNI) model [7]. In the rare-earth compounds, interactions between mostly localized f-electrons and carrier electrons at the Fermi level are very strong. The dHvA effect is one of the most important methods of investigation to understand about the conduction electrons near the Fermi surface. The reference material of CeSb is non-magnetic LaSb, which is a compensated semimetal and has no f-electrons. The Fermi surface of LaSb consists of three sheets, as shown in Fig. 9.2, and the angular dependence of the cross-sectional area is shown in Fig. 9.3. These sheets consist of electron pockets a, with the Lanthanum d-orbit character (t_{2g}) at the X points, and β and γ hole pockets with the Sb-p character at the Γ point [8]. The anisotropic effective mass of LaSb varies from 0.14 to 0.49 m_0 [9]. The shape of the Fermi surface of CeSb (Ce^{3+}, $4f^1$, $J = 5/2$) is basically similar to that of LaSb.

However, it has an additional characteristic β_4 hole pocket, as shown in Fig. 9.4. The external cross-section of the β_4 branch corresponds to 4.2 m_0 along the [100] direction [9,10]. This is almost 10 times larger than the maximum effective mass of LaSb. Sakai et al. performed the band calculation and predicted the large hole sheet with a heavy effective mass proposed by the p-f mixing model [11]. The cross-sectional area and effective mass of β_4 are in good agreement with values predicted by the theoretical calculation. The p-f mixing interaction is considered to be the dominant role for the magnetic properties of CeSb.

For other rare-earth elements from Pr ($4f^2$) to Tm ($4f^{12}$), the 4f state is more localized compared to the case for Ce or Yb. In these compounds, the quadrupolar interaction due to the anisotropic charge distribution plays an

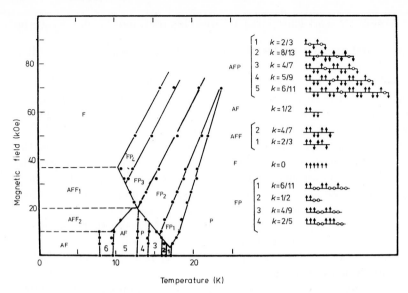

Fig. 9.1. The magnetic phase diagram of CeSb [6]. F, AF, P, AFP, AFF and FP indicate ferromagnetic phase, antiferromagnetic phase, paramagnetic phase, antiferro-paramagnetic phase, antiferro-ferromagnetic phase and ferro-paramagnetic phase, respectively

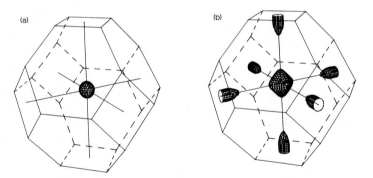

Fig. 9.2. The Fermi surface of LaSb [8]

important role. The typical example is HoP. It shows a ferro-quadrupolar ordering. The magnetic moment is aligned on the (100) plane and parallel and perpendicular to [100] by turns under magnetic field parallel to [100] axis, due to the strong quadrupolar interaction. Some compounds show antiferromagnetic order and quadrupolar order. It is important to investigate the magnetic and electric properties systematically.

In this paper, the dHvA effect and magnetization of four interesting rare-earth monopnictides are mainly described. A list of the dHvA effect measurements and magnetization measurements on these materials is shown in

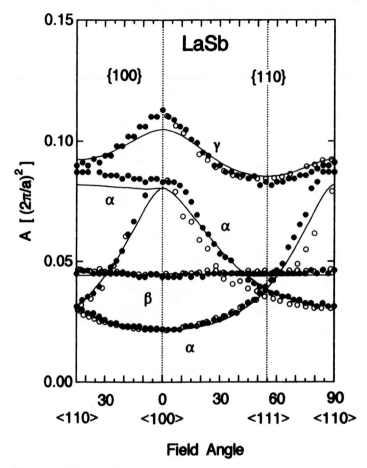

Fig. 9.3. The angular dependence of the cross sectional area of the Fermi surface of LaSb [9]

Table 9.1. The four materials are $Ce_{0.9}La_{0.1}Sb$ (Ce^{3+}, $4f^1$, $J = 5/2$), DySb (Dy^{3+}, $4f^9$, $J = 15/2$), GdAs (Gd^{3+}, $4f^7$, $L = 0$, $S = J = 7/2$) and TbSb (Tb^{3+}, $4f^8$, $J = 6$). DySb shows a large lattice distortion at T_N. Gd^{3+} has no orbital angular momentum ($L = 0$) and the magnetic moment comes only from the spin ($S = 7/2$). Gd monopnictides are expected to show typical properties and are considered to be good reference compounds in rare-earth monopnictides. In TbSb an antiferromagnetic transition and trigonal distortion appear simultaneously at T_N=15.5 K. We shall discuss the effective masses and an exchange interaction between conduction electrons and 4f-electrons in rare-earth monopnictides.

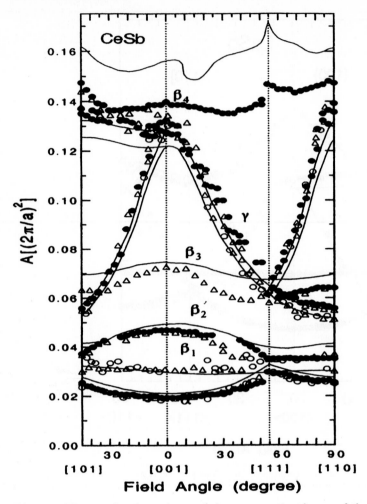

Fig. 9.4. The angular dependence of the cross sectional area of the Fermi surface of CeSb [9]

9.2   Investigation of the Fermi Surface

9.2.1   Theoretical Background of the dHvA Effect

The conduction electrons are quantized in a magnetic field and Landau levels are formed. The magnetic susceptibility changes drastically when each Landau level crosses the Fermi energy, and oscillatory behavior is observed as a function of the applied field according to the Landau level crossing to the Fermi surface. The oscillation is periodic with respect to the inverse of the external magnetic field, $1/B$. High magnetic fields are favorable to the dHvA experiment. The magnetization M due to the conduction electrons is given by

Table 9.1. List of the dHvA effect and magnetization measurements of the rare earth-Sb and rare earth-As compounds. Filled circles denote the compounds in this work. Open circles denote other compounds that have been investigated. Crosses denote compounds which have not been investigated. LaSb and LaAs are nonmagnetic compounds. dHvA: LaSb [9], LaAs [12], CeSb [9,10], CeAs [13]. Magnetization: CeAs [14], PrSb [15], PrAs [16], GdSb and GdAs [17], DyAs [18]

dHvA	La	Ce	Pr	Gd	Tb	Dy
Sb	○	○	●	✕	●	●
As	○	○	✕	●	✕	✕

Magnetization	La	Ce	Pr	Gd	Tb	Dy
Sb	nonmag.	●	○	○	●	●
As	nonmag	○	○	○	✕	○

Lifshitz and Kosevich [19]. The oscillation part of the magnetization, M_{OSC}, is written as

$$M_{OSC} = \sum_r A_r \sin\left(\frac{2\pi F}{B}\right) + \phi_r \,, \tag{9.5}$$

where A_r is

$$A_r = J_2(x) T B^{-1/2} \left|\frac{\partial^2 S}{\partial k_B^2}\right| \frac{\exp[-\alpha_c^*(T + T_D)/B]}{1 - \exp(-2\alpha_c^* T/H)} \cos(\pi g r m_c^*/2m_0) \,. \tag{9.6}$$

Here

$$\alpha = \frac{\pi c k_B}{e h} \,, \qquad x = \frac{2\pi F h_{OSC}}{B^2} \,, \tag{9.7}$$

where h_{OSC} is the modulation field strength applied to the sample, $J_2(x)$ is a Bessel function and r is a positive integer. M_{OSC} shows periodic behavior with respect to $1/B$. F is the dHvA frequency, written as

$$F = \frac{hS}{e} \,. \tag{9.8}$$

The dHvA frequency is proportional to the external cross-section S .

The cyclotron mass m_C^* can be estimated from the temperature dependence of the dHvA amplitude given in (9.6), namely, by the slope of $\ln(A(1 - \exp(-2\alpha m_C^* T/B))/T)$ versus T plot at constant external magnetic field and constant modulation field for $r = 1$.

9.2.2 Experimental Details

The single crystals were grown by the M. Kubota and D.X. Li in a high-frequency induction furnace by the Bridgman method or mineralization

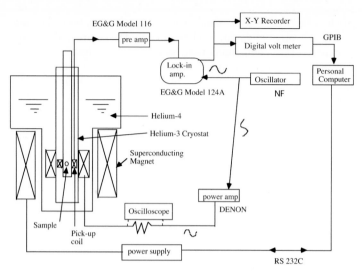

Fig. 9.5. The dHvA measurement system

method at Tohoku University. In order to investigate the "real" physical properties, high-quality samples are needed. The dHvA effects were observed for all samples, which were very high-quality single crystals. The experimental system was installed in the High Field Laboratory, Institute for Material Research, Tohoku University as shown in Fig. 9.5. This system was composed of a home-made ^{3}He cryostat, a ^{3}He handling system and measuring instruments. In order to observe the dHvA effect in detail, an AC modulation field method was used. It is better to apply a modulation field with a small amplitude and low frequency for avoiding eddy-current heating of the metallic sample. The amplitude of the modulation field is usually about 50 Oe and the frequency is 30–300 Hz. The sample was rotated to observe the angle dependence of the dHvA cyclotron mass m_C^* and the external cross-sectional area of the Fermi surface S.

A modulated magnetic field was generated in the sample space and the induction current from the sample was detected by the pickup coil (induction coil). The signal generated by the pickup coil was proportional to the oscillation amplitude of M_{OSC} shown in (9.1) and was transferred to a computer via an analog lock-in amplifier (PARC 124A). As an example, the measured dHvA oscillation of PrSb is shown in Fig. 9.6. dHvA oscillations were clearly observed. As shown by (9.1), M_{OSC} is a function of $1/B$ and the unit of the oscillation frequency is the tesla [T]. The Fast Fourier Transform (FFT) spectrum of the dHvA oscillation is shown in Fig. 9.6b. The horizontal axis indicates the dHvA frequency F, which is proportional to the external cross-section S. The detail of the Fermi surface was estimated from the dHvA oscillation measurements.

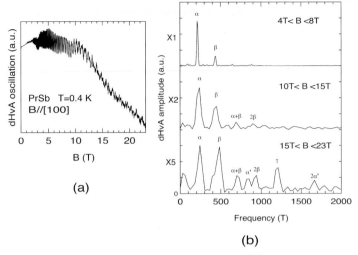

Fig. 9.6. (a) The dHvA effect of PrSb and (b) the FFT spectrum of the dHvA oscillation of PrSb

9.3   Negative Pressure Effect on CeSb

We present first the results for $Ce_{0.9}La_{0.1}Sb$ and compare them to those for CeSb. The magnetic structure of CeSb changes drastically under pressure [20]. The lattice constant increases with decreasing Ce concentration in $Ce_XLa_{1-X}Sb$, where the negative pressure effect can be realized by substitution of Ce ions by nonmagnetic La ions. Figure 9.7 shows the magnetization process for $Ce_{0.9}La_{0.1}Sb$ in static external magnetic fields. At 4.2 K, there are two sharp jumps at $B_{C1} = 2\,T$ and $B_{C2} = 4.8\,T$. The magnetization is saturated above 10 T. At 23 T, the magnetic moment is slightly larger than $2\mu_B/Ce$, but smaller than the moment expected from a Ce_{3+} ($J = 5/2$) ion. B_{C1} increases and B_{C2} decreases with decreasing temperature below 10 K.

The magnetic phase diagram of $Ce_{0.9}La_{0.1}Sb$ is presented in Fig. 9.8. The triangles are the transition points determined by the magnetization measurements and the squares are the points at which the specific heat shows a sharp peak. The closed circles are the transition temperatures obtained from AC susceptibility measurements. There are four magnetic phases: paramagnetic (P), ferro-paramagnetic (FP), antiferromagnetic (AF) and ferromagnetic (F) phases. In the FP phase, the magnetic moment is slightly larger than half the value of the moment in the ferromagnetic phase (F-phase). The FP-phase is considered to be a ferromagnetic and paramagnetic mixed phase and can be compared to the phase diagram of CeSb. The FP-phase of $Ce_{0.9}La_{0.1}Sb$ appears at lower temperatures than in CeSb and the area of the FP-phase is wider than that of CeSb. It is interesting to consider the reason why the area of the FP-phase of $Ce_{0.9}La_{0.1}Sb$ is larger than that of CeSb. From a neutron-scattering study of CeSb under pressure, Chadpadhayay et al. found that

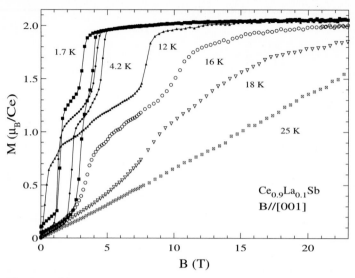

Fig. 9.7. The magnetization process of $Ce_{0.9}La_{0.1}Sb$

the FP-phase appears at higher magnetic fields and its area becomes narrow with pressure [20]. Therefore, it is concluded that the area of the FP-phase decreases by decreasing the lattice constant by pressure. As a result of the mixing of La, the lattice constant is expanded and the area of the FP-phase becomes large. We measured the dHvA effect in $Ce_{0.9}La_{0.1}Sb$ and concluded that the Fermi surface is almost the same as that of CeSb [21]. The area of the g surface at the X_Z point is about 10% smaller than that of CeSb, because of the lattice decrease due to the Ce ion density. This indicates that the p-f mixing interaction becomes weaker with decreasing Ce ion density. In the case of $Ce_{0.5}La_{0.5}Sb$, the Fermi surface was shown to be the same as in LaSb by the dHvA effect experiments.

9.4   Antiferromagnetic Order and Quadrupole Order in DySb

We next turn to a type-II antiferromagnet, DySb (Dy^{3+}, $4f^9$, $J = 15/2$, $T_N = 9.5\,K$) [22]. Below T_N, lattice distortion occurs. Magnetization and dHvA effect measurements were performed. Figure 9.9 shows the magnetization $M(B)$ along the [001] and [110] directions of DySb using a SQUID magnetometer with fields up to $5\,T$ and a pulsed field magnetometer from $5\,T$ to $26\,T$. For $B\,//[001]$, $M(B)$ shows two steps. One shows a step at $2.5\,T$ and increases to $5\mu_B/Dy$, and shows another step at $5\,T$ and saturates at $10.5\mu_B/Dy$ above $20\,T$, which is the same value as the saturation moment of the Dy^{3+} ion. With the magnetic field along the [001] axis, neutron scattering experiments below T_N indicate a transition from the type-II antiferromagnetic

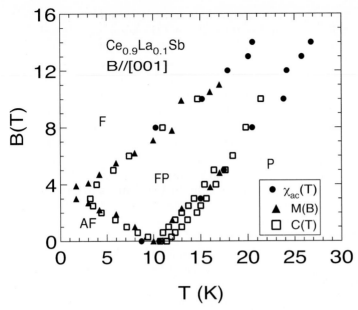

Fig. 9.8. Magnetic phase diagram of $Ce_{0.9}La_{0.1}Sb$. Filled circles, filled triangles and open squares denote the transition point determined by the AC susceptibility χ_{AC} (T), the magnetization $M(B)$ and the specific heat $C(T)$, respectively. P, AF, F and FP denote the paramagnetic phase, the antiferromagnetic phase, the ferromagnetic phase and ferromagnetic-paramagnetic phase, respectively

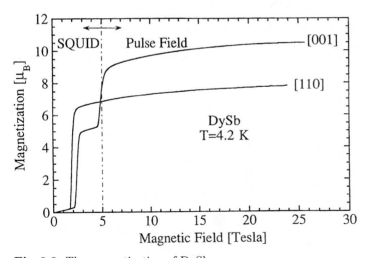

Fig. 9.9. The magnetization of DySb

state (A type) to the HoP-type intermediate ferrimagnetic phase (Q phase) at about 2.5 T, as illustrated in Fig. 9.10 [23], and the ferromagnetic ordering (F phase) occurs above 5 T. The magnetization in the Q phase is just half the value of the F phase, suggesting that the ratio of the magnetic moment parallel and perpendicular to the [001] axis is 1:1. Neutron scattering experiments suggest that the ratio of the moments parallel and perpendicular to the [001] axis is 1:1 and consistent with that of the magnetization. $M(B)$ for B //[110] shows a step around 2 T and reaches the value $7.7\mu_B$/Dy above 20 T. This is equal to $1/(\cos(45°))$ of the saturated moment along the [001] axis. The HoP-type intermediate ferrimagnetic phase (Q phase) is realized and the magnetic moment is strongly fixed along the [001] axis even in high magnetic fields along the [110] axis. Below T_N the cubic lattice distorts to a tetragonal one; this phenomenon has been understood to be driven by the quadrupolar interaction. The magnetic transition occurs at the same temperature. This quadrupolar interaction is so strong that the angular momentum generated by the cloud of 4f electrons of Dy^{3+} ion is fixed along the [001] axis. Furthermore, we have investigated the Fermi surface of DySb. The dHvA signal is shown in Fig. 9.11. The F-phase is stabilized within 25° of the [001] axis, and the Q-phase is stabilized beyond 25°. The Fermi surface of DySb is basically similar to that of LaSb. However a large splitting was observed for the α and

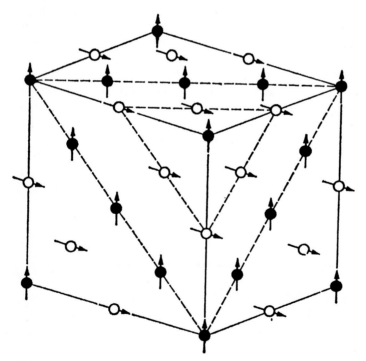

Fig. 9.10. The magnetic structure of HoP [23]

β branches. The α electron branch at the X_Z point splits into two respective magnetic phases, (α_1, α_2) and (α'_1, α'_2), as shown in Fig. 9.11. The Fermi surface shows nearly the same split and, in the Q-phase, shifts upward parallel to the F-phase. On the other hand, the large β hole splits into (β_1, β_2) in the F-phase and is compared to (β'_1, β'_2) in the Q-phase. As for the γ hole branch, the cross-sectional area A is nearly the same and constant (spherical surface). In other rare earth-Sb compounds [24–29] except CeSb and TbSb (Tb), Sb is mentioned in Sect. 9.6, there are no splittings in the α and β branches. The β hole branch of CeSb shows splitting, as shown in Fig. 9.4. This splitting can be explained by the p-f mixing model. For DySb, the exchange splittings are the same for both the electrons and holes. It is assumed that these exchange splittings originate from the intra-atomic d-f exchange interactions. The splitting of the a branch is larger than that of CeSb and is considered to be due to the difference in the localized spin value in these two compounds. The other problem in DySb comes from the strong decrement of the electron Fermi surface located at the X_Z point. The interaction between the quadrupole moment and the conduction electrons strongly affects the Fermi surfaces for DySb compared to other rare-earth antimonides.

9.5   Fermi Surface of GdAs

Next we discuss the Fermi surface of Gd monopnictides. For these compounds, the electric conductivity varies from semiconductive to metallic. GdN is ferromagnetic, while the other pnictides are antiferromagnetic. The antiferromagnetic ordering in the semimetallic materials is destroyed and changed to ferromagnetic ordering under external magnetic fields. It has been suggested that the Fermi surface might be affected under external fields due to the strong exchange interaction between the conduction electrons and the localized 4f electron spins [30]. GdAs with the NaCl crystal structure is antiferromagnetic below 18.7 K [17]. However, the spin structure has not been determined so far. Among the rare-earth monopnictide systems, Gd monopnictides are the simplest ones and the crystal electric field effects in GdX are considered to be fairly weak due to the S-state character of the Gd^{3+} ion. Actually, the magnetization of GdAs increases linearly and saturates at 17 T, while the magnetic structure does not change. There is little anisotropy, due to the weakness of the crystalline electric fields (CEF) effects [30].

In order to understand the electronic structure of GdAs, dHvA effect measurements were performed. Recently a high-quality single crystal of GdAs was successfully grown and dHvA signals were observed clearly. For the strongly correlated electron systems called heavy-electron compounds, it is interesting to investigate experimentally the mass-enhancement effect of the quasiparticles.

Single crystals of GdAs were grown by the mineralization method in tungsten crucibles [17]. The value of the residual resistivity r was 5.2 $\mu\Omega$ cm and

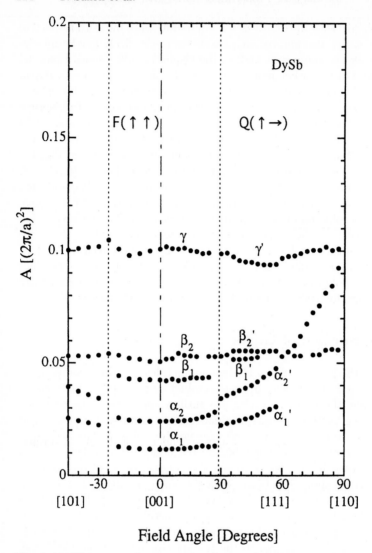

Fig. 9.11. The external cross sectional area of the Fermi surface of DySb

the residual resistivity ratio was 16.7. The dHvA effects were measured using the field modulation method. A ^{3}He cryostat and a superconducting magnet with fields up to 15 T were used. The temperature was changed from 0.5 K to 4.2 K by controlling the vapor pressure of helium gas. Data were taken from the (100) and (110) planes. A correction to the internal magnetic fields is made in the induced ferromagnetic ordered phase, because the 6% difference with respect to the external fields is not negligible.

Figure 9.12 shows a typical dHvA oscillation signal of GdAs at 0.5 K along the [100] axis and the corresponding FFT spectrum, respectively. The FFT analysis reveals three main peaks corresponding to the α-, β- and 2α-branches.

Figure 9.13 shows the angular dependence of the external cross-sectional area of the Fermi surface for GdAs. The shape of the Fermi surface is very similar to that of LaSb [8,9]. Hence, the oscillations originate from an electron pocket at the X-point and two hole pockets at the Γ-point. We analyzed the temperature dependence of the FFT spectrum and obtained the effective cyclotron mass m_C^*. For the α branch, m_C^* is 0.20 (0.14) m_0 for B$//$ [100] and 0.24 m_0 for [110]. For the β-branch, m_C^* is 0.26 (0.15) m_0 for [100] and 0.22 (0.23) m_0 for [111]. Parentheses denote the m_C^* value of LaSb measured by Settai et al. [9]. The m_C^* of GdAs for B$//$ [100] is larger than that of LaSb, but for B$//$[111] the m_C^* of GdAs has the same value as that of LaSb. This means that the interactions between quasi-particles are not so strong as those in CeSb. It is assumed that the orbital moment of the Gd ion is zero ($L = 0$) and the orbital of the carriers of the Gd ion is spherical. Therefore, the p-f mixing is very small and the effective mass is not so different from that of LaSb.

The field dependence of the external cross-section is directly shown by the number of filled levels n_f versus $1/B$, since n_f varies linearly with respect to $1/B$. From precise FFT analysis in several field regions, a very slight field dependence was detected in the α-, β- and 2α-branches up to 15 T. Below 15 T, GdAs shows a type-II antiferromagnetic state and its Fermi surface does not change with increasing magnetic field.

The shapes of the Fermi surfaces determined by the dHvA study are very similar to those of LaSb. From the angular dependence of the Fermi surface shown in Fig. 9.3, the carrier density of the α and β-branches are estimated to be 2.27×10^{20} cm^{-3} (0.0115/Gd) and 0.516×10^{20} cm^{-3} (0.0026/Gd), respectively. These values are consistent with those estimated from the Hall effect [17]. Since GdAs is a semimetallic compound with an equal number of electrons and holes, there should be another hole surface which corresponds to the γ- branch of LaSb, and its carrier density should be 1.76×10^{20} cm^{-3} (0.0089/Gd). However, we have not observed this branch in GdAs. It is known that the lattice constant depends on the carrier density. The shapes of Fermi surfaces and carrier density for LaAs, CeAs and YbAs were successfully determined using the dHvA effect. The carrier density increases linearly with decreasing lattice constant. This is different from rare-earth Sb compounds. The Fermi surfaces of the rare-earth As compounds are not considered to be influenced in the same way as those of rare-earth Sb compounds by interactions between the carriers and the localized 4f electrons spins in the rare-earth atom.

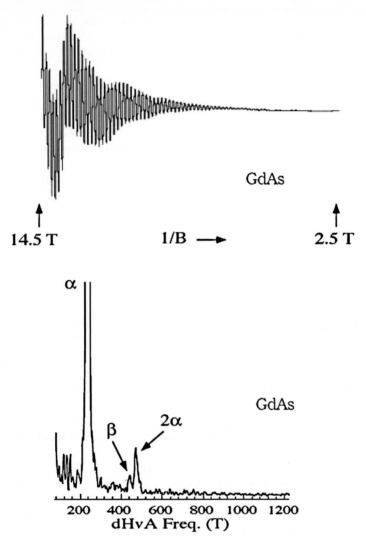

Fig. 9.12. Typical dHvA oscillations of GdAs in external magnetic fields for [100] at $T = 0.5$ K and the corresponding FFT spectrum

9.6   Magnetic Interaction and Fermi Surface of TbSb

Finally, we present and discuss results for TbSb, which shows an antiferromagnetic transition and trigonal distortion simultaneously. In the AF state, the magnetic moments are aligned ferromagnetically in the (111) plane and antiferromagnetically in the adjacent layers pointing along the [111] direction. The complicated magnetic structures are expected to be realized in a magnetic field due to the quadrupole interaction. In order to investigate the

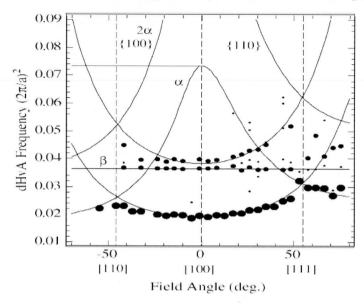

Fig. 9.13. The external cross-sectional areas of the Fermi surface of GdAs as a function of the direction of the magnetic fields. The α-branch is assumed to be three ellipsoids centered at the X-points. The β-branch is assumed to be a sphere centered at the Γ-points

magnetic properties and interactions between carriers and 4f electrons, magnetization and dHvA effect measurements were performed up to 30 T using pulsed magnetic fields and up to 23 T using the hybrid magnet, respectively. The results of the magnetization experiments on a single crystal TbSb at $T = 4.2$ K are shown in Fig. 9.14. Three sharp steps are clearly observed under an external field along the [111] axis. To make these steps clear, the differential magnetization dM/dB for $B//$[111] was measured and the results are shown in Fig. 9.15. The induced magnetization increases up to $9\mu_B$/Tb, which is supposed to be the full moment of Tb^{3+} ion, and saturates at 16 T. For the other two directions, only one step is observed. For $B//$[111], the highest critical field B_{C3} is constant up to 10 K, while B_{C1} and B_{C2} decrease slightly above 6 K, as shown in Fig. 9.16. Cooper and Buschbeck et al. studied the magnetization process of single crystal [31,32]. They observed an S-shape like anomaly at 7 T and 8.5 T for $B//$[100] and [111], respectively. These transitions were very broad and only one transition was observed as for $B//$[111]. They suggested that this is a second-order transition from type II-antiferromagnetism to paramagnetism. They described the magnetization process well by taking account of a mean field approximation model (MFA) in addition to the crystal field and Heisenberg exchange interactions, and also a quadrupolar coupling. On the other hand, the present result showed three sharp steps for $B//$[111] with clear hysteresis, as shown in Fig. 9.15.

Therefore, these transitions are first order. The high-quality sample, which was used for the dHvA effect in this study, clearly showed first-order transitions for each directions. These properties could not be described by the MFA model proposed by Buschbeck et al. Now we consider the magnetic structure in magnetic fields. When the field is applied along an easy axis, [111], the magnetic moments are aligned within the planes perpendicular to the field and antiferromagnetically in the adjacent layers. In this phase, the magnetization process is well described by the MFA model. On the other hand, in the III-phase between $B_{C1} = 7.8$ T and $B_{C2} = 9$ T at 4.2 K, the magnetization is about $4.5\mu_B$/Tb, as shown in Fig. 9.16. This value is just half of the magnetic moment of Tb^{3+} ion. The MFA model took the quadrupolar coupling into account, and DySb shows HoP-type quadrupolar order, as shown in Fig. 9.10.

Therefore, the magnetic structure in the III-phase is assumed to be the quadrupolar structure, where the magnetic moment is aligned parallel and perpendicular to the [111] axis, one by one. In the III-phase, the field region is only about 1 T and the magnetization was gradually increased. This is because the quadrupolar interaction, in the MFA model, is 0.018 K and quite small compared to the antiferromagnetic interaction (1.25 K). Consequently, the magnetic moments are assumed to move with increasing fields. In the II-phase ($B_{C2} < B < B_{C3} = 10$ T), the magnetization is about $6\mu_B$/Tb, which is 2/3 of the magnetic moment of Tb^{3+} ion. In the I-phase (paramagnetic or field-induced ferromagnetic phase), the magnetization increases linearly with increasing fields and saturates at 16 T. The magnetic structure is as-

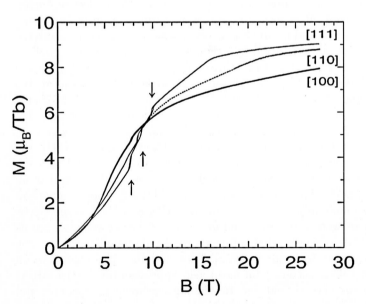

Fig. 9.14. The magnetization process of TbSb sample #1 at $T = 4.2$ K in a pulsed magnetic field. Arrows indicate the steps for $B//[111]$

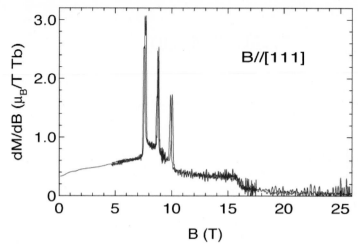

Fig. 9.15. The differential magnetization of TbSb sample #1 at $T = 4.2$ K

sumed to be a canted antiferromagnetic order. If the magnetic structure is the canted antiferromagnetism in the II-phase, the transition at B_{C3} cannot be explained. From the value of the magnetization, it is assumed that 2/3 of the total magnetic moments are aligned parallel to the [111] axis and the others are aligned perpendicular to the [111] axis. Neutron scattering studies in magnetic fields are needed to clarify this problem.

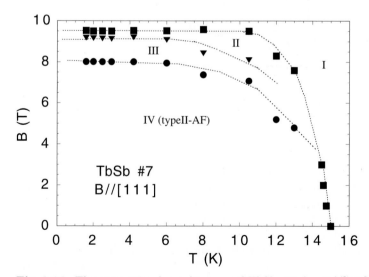

Fig. 9.16. The magnetic phase diagram of TbSb #7 for $B//[111]$. Filled circles, filled triangles, and filled squares indicate the transition fields, B_{C1}, B_{C2}, and B_{C3}, respectively

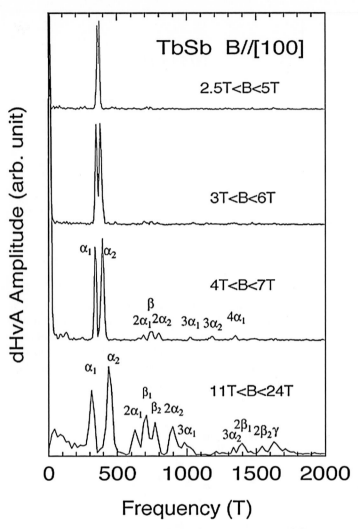

Fig. 9.17. The FFT spectra of dHvA oscillations in TbSb #1 at $T = 1.5\,\mathrm{K}$

The dHvA effect in TbSb was studied up to 23 T and the oscillations were observed above 1.5 T even at temperatures as high as 6 K, which means the sample quality is high. We clearly observed beats, which come from spin splitting. The FFT spectra of the dHvA oscillation are illustrated in Fig. 9.17. The width of the splitting of the α- and β-branches increases with increasing field. From the results of the dHvA effect and magnetization measurements, we can estimate the strength of the c-f exchange interaction as about 37 meV for $B//[100]$. The details of the calculation are described [33,34]. It reveals that the exchange interaction between carriers and 4f electrons is very strong and is comparable to that of TmSb [29]. On the other hand, the shape of the

Fermi surface of TbSb is different from that of DySb, which shows a large anisotropy due to a strong quadrupole interaction, and the Fermi surface of TbSb is rather similar to that of LaSb.

9.7 Fermi Surface of Rare-Earth Antimonides

We now compare the Fermi surfaces of rare-earth antimonides. The external cross-sectional area A of the α-electron Fermi surface at the X_Z point versus the lattice constant is shown in Fig. 9.18. As shown in Fig. 9.17, the $\alpha(X_Z)$ branch of TbSb shows a spin splitting and the splitting increases with increasing field, so the cross-sectional area A of $\alpha(X_Z)$ of TbSb is in the middle of the two branches. The most characteristic feature of the Fermi surface is that the α-electrons of DySb and CeSb are much smaller than that of the other rare-earth Sb compounds. These compounds show tetragonal distortion and quadrupole ordering at low temperatures. Figure 9.18 suggests the interaction between the conduction electrons and the 4f quadrupole moment strongly affects the Fermi surface of DySb and CeSb. As for TbSb, the cross-sectional area A of $\alpha(X_Z)$ is roughly the same as that of GdSb, which indicates that the quadrupole interaction is much smaller than that of DySb. This fact suggests that only the c-f exchange interaction affects the Fermi surface of TbSb, and consequently spin splitting occurs.

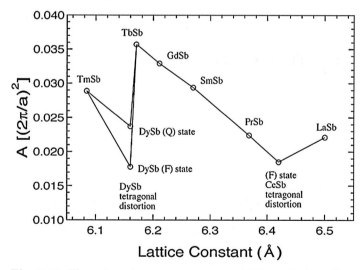

Fig. 9.18. The external-cross sectional area of the α-electron Fermi surface vs. the lattice constant for rare-earth Sb (rare-earth = La, Ce, Pr, Sm, Gd, Tb, Dy, Tm)

9.8   Conclusion

The Authors would like to thank Mr. A. Uesawa, Dr. Kubota and Mr. M. Ozawa for collaboration. The magnetic and electric properties of the rare-earth monopnictides of $Ce_{0.9} La_{0.1}Sb$, DySb, GdAs and TbSb have been investigated. It is clear that the magnetization and dHvA effects in high magnetic fields showed much valuable properties in characterizing these quite different and interesting behaviors.

Acknowledgment

This work was carried out at the High Field Laboratory for Superconducting Materials, IMR, Tohoku University and supported by a Grant-in-Aid for Scientific Research from the Ministry of Education, Science, and Culture of Japan.

References

1. F. Steglich: Physica B&C **130**, 145 (1985)
2. J. Kondo: Prog. Theor. Phys. **32**, 37 (1964)
3. M.A. Ruderman and C. Kittel: Phys. Rev. **96**, 99 (1954)
4. T. Kasuya: Prog. Theor. Phys. **16**, 45 58 (1956)
5. K. Yoshida: Phys. Rev. **106**, 893 (1957)
6. J. Rossat-Mignod, P. Burlet, H. Bartholin, O. Vogt and R. Lagnier: J. Phys. C **13**, 6381 (1980)
7. T. Kasuya, Y.S. Kwon, T. Suzuki, K. Nakanishi, F. Ishikawa and K. Takegahara: J. Magn. Magn. Mater. **90–91**, 389 (1990)
8. A. Hasegawa: J. Phys. Soc. Jpn. **54**, 677 (1985)
9. R. Settai, T. Goto, S, Sakatsume, Y.S. Kwon, T. Suzuki and T. Kasuya: Physica B **186–188**, 176 (1993)
10. H. Kitazawa, Y.S. Kwon, A. Oyamada, N. Takeda, H. Suzuki, S. Sakatsume, T. Satoh, T. Suzuki and T. Kasuya: J. Magn. Magn. Mater. **76–77**, 40 (1988)
11. O. Sakai, Y. Kaneta and T. Kasuya: Jpn. J. Appl. Phys. **26**, Suppl. **26–3**, 477 (1987)
12. K. Morita, T. Goto, H. Matsui, S. Nakamura, Y. Haga, T. Suzuki and M. Kataoka: Physica B **206–207**, 795 (1995)
13. S. Nimori, H. Aoki, T. Terashima, G. Kido and T. Suzuki: J. Phys. Soc. Jpn. **65**, 2728 (1996)
14. K. Sugiyama, T. Inoue, Y. Haga, T. Suzuki and M. Date: J. Magn. Magn. Mater. **140–144**, 1255 (1995)
15. A. Kido, S. Nimori, G. Kido, Y. Nakagawa, Y. Haga and T. Suzuki: Physica B **186–188**, 185 (1993)
16. R.A.B. Devine: Phys. Rev. B **18**, 5877 (1978)
17. D.X. Li, Y. Haga, H. Shida, Y.S. Kwon and T. Suzuki: Phys. Rev. B **54**, 10483 (1996); D.X. Li, PhD Thesis, Tohoku University (1995)
18. J.M. Mao, M.E. Zudov, R.R. Du, P.P. Lee, L.P. Sadwick and R.J. Hwu: J. Appl. Phys. **87**, 5170 (2000)

19. I.M. Lifshitz and R.M. Kosevich; Zh. Eksp. Theor. Fiz. **29**, 730 (1955) [Sov. Phys. JETP **2**, 636 (1956)]
20. A. Chadpadhyay, P. Burlet, J. Rossat-Mignot, H. Bartholin, C. Vietter and O. Vogt: J. Magn. Magn. Mater. **63–64**, 52 (1987)
21. Y. Nakanishi, T. Sakon, Y. Takahashi, A. Uesawa, M. Kubota, T. Suzuki and M. Motokawa: Physica B **281–282**, 742 (2000)
22. G. Busch, P. Junod, O. Vogt and F. Hullinger: Phys. Lett. **6**, 79 (1963)
23. H.R. Child, M.K. Wilkinson, J.W. Cable, W.C. Koehler and E.O. Wollan: Phys. Rev. **131**, 922 (1963); G. Busch and O. Vogt: J. Appl. Phys. **39**, 1334 (1968)
24. H. Kitazawa, T. Suzuki, M. Seda, I. Oguro, A. Yanase, A. Hasegawa and T. Kasuya: J. Magn. Magn. Mater. **31–34**, 421 (1983)
25. H. Aoki, G.W. Crabtree, W. Joss and F. Hullinger: J. Magn. Magn. Mater. **52**, 389 (1985)
26. A. Kido, S. Nimori, G. Kido, Y. Haga and T. Suzuki: Physica B **186–188**, 187 (1993)
27. S. Nimori, G. Kido, H. Matsui, T. Goto, D.X. Li and T. Suzuki: Physica B **201**, 111 (1994)
28. R. Settai: PhD Thesis, Tohoku University (1994)
29. S. Nimori: PhD Thesis, Tohoku University (1994)
30. O. Vogt and K. Mattenberger: *Handbook on the physics and chemistry of the rare earth*, Vol. **17** (Elsevier, Amsterdam, 1993) p. 308
31. B.R. Cooper and O. Vogt: Phys. Rev. B **1**, 1218 (1970)
32. A. Buschbeck, C. Chojnovski, J. Kotzler, R. Sonder and G. Thummes: J. Magn. Magn. Mater. **69**, 171 (1987)
33. M. Wulff: PhD Thesis, Cambridge Univ. (1985)
34. M. Wulff, G.G. Lonzarich, D. Fort and H.L. Skriver: Europhys. Lett. **7**, 629 (1988)

10 High-Field Magnetization Process and Crystalline Electric Field Interaction in Rare-Earth Permanent-Magnet Materials

H. Kato, T. Miyazaki and M. Motokawa

During the last two decades the performance of permanent magnets has been greatly improved by introducing rare-earth (R) elements to their constituents [1]. It is doubtless that the high coercivity of these magnets comes from the large magnetic anisotropy originated by the crystalline electric field (CEF) acting on R ions with large orbital angular momentum. Magnetization measurements up to the high-field region, where the hard-axis magnetization saturates, are indispensable in order to obtain the basic insight into the magnetic anisotropy. Since 1985, we have been investigating systematically the high-field magnetization in a series of $Nd_2Fe_{14}B$-type compounds using mainly single crystal samples [2]. On the other hand, we have developed a method of analyzing these magnetization curves which consists of a simplified Hamiltonian taking the exchange and crystal field at the R ions into account, with the Fe sublattice being treated phenomenologically [3]. The essential feature in this model is the coupling of the two different types of sublattices. One is the R sublattice, which gives a large magnetic anisotropy owing to the CEF interaction. Another is the Fe sublattice, which determines the large magnetization and high Curie temperature as a result of strong Fe-Fe exchange interactions. This method has proven to be applicable not only to the $R_2M_{14}B$ system, with $M = $ Fe or Co, but also to the pseudo-ternary system $(R_{1-x}R'_x)_2Fe_{14}B$ [4] or other R-Fe-X systems such as $R_2Fe_{17}N_x$ [5]. In these systems an interplay among the R-Fe exchange interaction, CEF potential acting on R ions and a large magnetic moment of the Fe sublattice leads to a variety of magnetic properties such as a first-order magnetization process (FOMP) and spin reorientation (SR) transitions. In general, such SR transitions will be accompanied by a considerable lattice deformation, since there is a large orbital contribution to the R magnetic moments, resulting in a strong coupling between the spin and lattice systems. It is therefore of interest to investigate the magnetoelastic properties of these materials.

On the other hand, an extension of the idea above to nanoscale two-phase magnets, is the exchange-spring magnets [6] or nanocomposite magnets. That is, high-coercivity and high-magnetization permanent magnets can be realized by combining the nanoscale particles of the "hard" phase, which has a large coercivity, and the "soft" phase with high magnetization.

In this paper we present the general method of calculating the magnetization and magnetostriction in the R_hFe_kX system, where R is a pure or mixed rare-earth element and X is a non-magnetic element such as B or N.

Experimental data on $(Er_{1-x}Tb_x)_2Fe_{14}B$ and $SmFe_{12}/\alpha$-Fe nanocomposite systems are reviewed.

10.1 Exchange and Crystal Field Model for the $(R_{1-x}R'_x)_h Fe_k X$ System

We deal with the magnetic properties of a system containing h $(R_{1-x}R'_x)$ and k Fe atoms as magnetic elements. In the case of the $R_2Fe_{17}N_3$ system, $x = 0$, $h = 2$ and $k = 17$, whereas for $(R_{1-x}R'_x)_2Fe_{14}B$, $h = 4$ and $k = 28$, since the f and g sites for the R or R' ions must be subdivided magnetically into f_1, f_2, g_1 and g_2 sites. The total free energy of the system at temperature T in an external field H is assumed to be simply given by

$$F(\mathbf{H},T) = (1-x)F_R + xF_{R'} + E_{Fe} + C(\Delta V/V_0)^2 , \qquad (10.1)$$

where F_R $(F_{R'})$ is the free energy for the R (R') sublattice given by

$$F_R(\mathbf{H},T) = -k_B T \sum_{i=1}^{h} \ln \sum_s \exp(-E_s(i)/k_B T) , \qquad (10.2)$$

in which $E_s(i)$ is the sth eigenvalue of the Hamiltonian for the ith R ion,

$$\mathcal{H}_R(i) = \lambda \mathbf{L} \cdot \mathbf{S} + \mathcal{H}_{CEF}(i) + 2\mu_B \mathbf{S} \cdot \mathbf{H}_m(i) + \mu_B(\mathbf{L} + 2\mathbf{S}) \cdot \mathbf{H} , \qquad (10.3)$$

where $\mathbf{H}_m(i)$ is the molecular field and $\mathcal{H}_{CEF}(i)$ is the CEF Hamiltonian, which can be written as

$$\mathcal{H}_{CEF}(i) = \sum_{n,m} A_n^m(i) \sum_j V_n^m(j) , \qquad (10.4)$$

in which $A_n^m(i)$ is the CEF coefficient and $\sum_j V_n^m(j)$ is the unnormalized tesseral function for the jth $4f$ electron. The non-zero terms in (10.4) are determined depending on the point symmetry of the R or R' ion site and the choice of quantization axes. The matrix elements of $\mathcal{H}_{CEF}(i)$ can be calculated using the tensor-operator technique which has already been described in detail [3]. The third term in (10.1) expresses the anisotropy and Zeeman energy of the Fe sublattice,

$$E_{Fe} = k\{K_{Fe} \sin^2 \theta - \mathbf{m}_{Fe} \cdot \mathbf{H}\} , \qquad (10.5)$$

where K_{Fe} is the uniaxial anisotropy constant per Fe atom and θ is the angle between the z-axis and the direction of the Fe moment $\mathbf{m}_{Fe}$.

The last term in (10.1) denotes the elastic energy for the volume change, in which C is an elastic constant. As for the CEF Hamiltonian, for the R and R' sites we simply assume that (i) only the axial CEF coefficients A_2^0, A_4^0, and A_6^0 are modified by the volume change, and (ii) the volume dependence of

A_n^0 is determined by the point-charge approximation. The latter assumption directly leads to the following relations:

$$
\begin{aligned}
A_2^0(V) &= A_2^0(V_0)G_2^0\{1 - \Delta V/V_0\}, \\
A_4^0(V) &= A_4^0(V_0)G_4^0\{1 - (5/3)\Delta V/V_0\}, \\
A_6^0(V) &= A_6^0(V_0)G_6^0\{1 - (7/3)\Delta V/V_0\},
\end{aligned}
\tag{10.6}
$$

where $V = V_0 + \Delta V$ and G_n^0 should be equal to unity if the expansion is isotropic. On condition that the total free energy of $Eq.$ (10.1) takes a minimum, we can simultaneously calculate $\Delta V/V$ ($\approx \Delta V/V_0$) and the magnitude and direction of the magnetic moments in each sublattice for various temperature and field values.

10.2   High-Field Magnetization, Spin Reorientation and Magnetostriction in $(Er_{1-x}Tb_x)_2Fe_{14}B$

The magnetic properties of $R_2Fe_{14}B$ have been extensively studied in connection with high-performance Nd-Fe-B permanent magnets [1]. In the case of $Er_2Fe_{14}B$, an abrupt SR transition at $T_{SR} = 323\,K$ was observed [7], above which the direction of the magnetic moments changes to tetragonal [001] from [100]. The partial replacement of Tb for Er is known to cause a rapid decrease of T_{SR} and an appearance of an intermediate phase with a tilted easy axis [8], as shown in Fig. 10.1. Such a behavior can be understood qualitatively by the competing magnetic anisotropy arising from the Er, Tb, and Fe sublattices. That is, Er favors the easy [100] direction, while Tb and Fe tend to align along [001]. In this section, we demonstrate how the model calculation described in the preceding section can reproduce, or sometimes predict, the experimental data in the $(Er_{1-x}Tb_x)_2Fe_{14}B$ system.

Polycrystalline samples of $(Er_{1-x}Tb_x)_2Fe_{14}B$ with $0 \leq x \leq 0.6$ were prepared by melting under an argon atmosphere using an induction furnace. Magnetically aligned samples were prepared by orienting the crushed powders in a field of 10 kOe using epoxy resin. The temperature dependence of the magnetization was measured using a vibrating-sample magnetometer with a field of 1 kOe applied perpendicular to the aligned direction. High-field magnetization measurements at 4.2 K were carried out by using a sample-extraction magnetometer in fields up to 270 kOe generated by a hybrid magnet.

The solid lines in Fig. 10.1 are the phase boundaries calculated from the simple-mixture model described in the previous section with the CEF and molecular field parameters for the end-member compounds [9]. These calculations are in good agreement with the experiments, although no adjustable parameters are assumed. Calculated magnetization curves for $(Er_{1-x}Tb_x)_2Fe_{14}B$ with $x = 0.4$ are shown in Fig. 10.2a, which exhibit abrupt increases in magnetization at 80 kOe, 130 kOe, and 290 kOe with the field along the [100], [110],

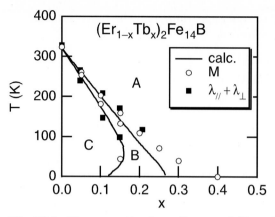

Fig. 10.1. The magnetic phase diagram of $(Er_{1-x}Tb_x)_2Fe_{14}B$. The solid lines are the results of calculations, while the open and solid symbols are the experimental spin-reorientation temperatures determined from magnetization and magnetostriction measurements, respectively. The easy-axis direction lies parallel to [001] and [100] in the A and C regions, respectively, and is tilted in region B

and [001] directions, respectively. Such a FOMP is a result of discontinuous rotations of Er, Tb, and Fe moments caused by the double-minimum behavior of the CEF potentials [3,10]. In order to confirm this prediction, we measured the high-field magnetization curves. Figure 10.2b shows the experimental magnetization curves for the aligned polycrystal of $(Er_{0.6}Tb_{0.4})_2Fe_{14}B$. Although somewhat smeared owing mainly to incomplete alignment, anomalous increases in magnetization were observed around 70 kOe and 240 kOe with the field perpendicular and parallel to the aligned direction, respectively. Therefore, it has been confirmed that not only the zero-field properties such as SR, but also the high-field magnetization behavior in these mixed compounds can be predicted successfully by using the simple-mixture model.

Next, we measured the magnetostriction in the $(Er_{1-x}Tb_x)_2Fe_{14}B$ system in order to investigate the lattice deformation at the SR [11]. Ribbon samples of $(Er_{1-x}Tb_x)_2Fe_{14}B$ with $0 \leq x \leq 0.45$ were prepared by a single-roller rapid-quenching method in an Ar atmosphere. The surface velocity of the copper wheel was varied between $V_s = 4 \sim 31$ m/s, the ejection pressure of Ar gas was $P_{Ar} = 0.6 \sim 1.4$ kg/cm^2, and the orifice diameter of the quartz crucible was fixed to 0.5 mm. Compositions were checked by an inductively coupled plasma (ICP) analysis. Magnetostriction measurements were performed by a capacitance method apparatus [12], by which we observed $\Delta l/l$ parallel ($\lambda_{//}$) and perpendicular ($\lambda_\perp$) to the field direction within the ribbon plane.

X-ray diffraction measurements of these ribbons showed that the c-axis of the tetragonal cell in each crystallite nicely aligns perpendicular to the ribbon surface when $V_s = 15.7$ m/s and $P_{Ar} = 0.6$ kg/cm^2. The degree of alignment

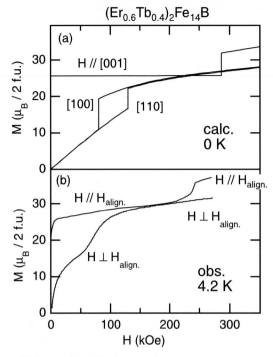

Fig. 10.2. (a) The calculated and (b) the observed magnetization curves for $(\mathrm{Er}_{0.6}\mathrm{Tb}_{0.4})_2\mathrm{Fe}_{14}\mathrm{B}$

was found not to be affected by the Tb replacement; the full width at half maximum of the c-axis orientation distribution was $(12 \pm 5)°$. Figure 10.3 shows the field dependence of the longitudinal and transverse magnetostriction constants $\lambda_{//}$ and $\lambda_\perp$ in $\mathrm{Er}_2\mathrm{Fe}_{14}\mathrm{B}$, in which the field was applied along the ribbon surface, namely, perpendicular to the c-axis. Both $\lambda_{//}$ and $\lambda_\perp$ increase monotonically with increasing field even in a field of $20\,\mathrm{kOe}$. Such a behavior is consistent with the report by Algarabel [13] for non-aligned polycrystals and suggests that the contribution from the volume magnetostriction is due mainly to the Fe $3d$ band. At $331\,\mathrm{K}$, which is very near to the T_{SR} of this compound, both $\lambda_{//}$ and $\lambda_\perp$ are larger than those at $294\,\mathrm{K}$. We plotted the areal magnetostriction, i.e. the sum of the longitudinal and transverse magnetostriction $\lambda_{//} + \lambda_\perp$ against temperature, as shown in Fig. 10.4. The most significant feature in Fig. 10.4a is a maximum of $\lambda_{//} + \lambda_\perp$ around $325\,\mathrm{K}$. This temperature of the maximum slightly increases with increasing field. It must be noted that the anisotropic magnetostriction $\lambda_{//} - \lambda_\perp$ does not change significantly in this temperature and field range. Since the observed peak temperature of $\lambda_{//} + \lambda_\perp$ almost coincides with T_{SR}, we can interpret the phenomena as the areal increase of the c-plane at the SR transition.

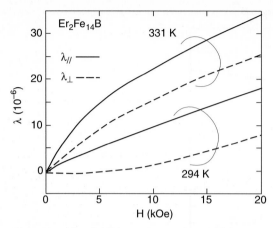

Fig. 10.3. The field dependence of the longitudinal and transverse magnetostriction $\lambda_{//}$ and $\lambda_{\perp}$ in $Er_2Fe_{14}B$ at 294 K and 331 K

The $x = 0.106$ sample, on the other hand, exhibits a double peak at $T_1 \sim 210\,\mathrm{K}$ and $T_2 \sim 145\,\mathrm{K}$, as shown in Fig. 10.4b. Lim et al. [9] reported such successive transitions in $(Er_{1-x}Tb_x)_2Fe_{14}B$ with $0.05 \leq x \leq 0.15$ in a similar temperature range and explained the higher and lower transitions as the SRs from axial to tilted and tilted to planar phases. The observed double peak, therefore, appears to correspond to the areal increase of the c-plane at the beginning and closing of the easy-axis rotation, that is, the two SR transitions. The temperatures at which $\lambda_{//}+\lambda_{\perp}$ takes a maximum for samples with $0 \leq x \leq 0.21$ are plotted in Fig. 10.1 and are in good agreement with the transition temperatures determined by magnetization measurements.

In order to analyze the experimental data above, we have made a model calculation of the volume magnetostriction arising from the CEF potential of the Er and Tb sites, as was described in Sect. 10.2. The same values of the CEF parameters A_n^m and molecular field H_m were used as those given in [3], except that A_2^0 for $Er_2Fe_{14}B$ was reduced by 7% ($283\,\mathrm{K}\ a_0^{-2}$) so as to best fit the observed T_{SR} ($x = 0$) in the present sample. As for the G_n^0 parameters defined in (10.7), we adopted the set $G_2^0 = 1$, $G_4^0 = -1$, and $G_6^0 = 1$. Figure 10.5 shows the calculated easy-axis direction θ and the volume change $\Delta V/V$ as a function of temperature. For $x = 0$, as shown in Fig. 10.5a, a discontinuous change of $\Delta V/V$ is seen at T_{SR} for $H = 0$. Upon increasing the strength of the field applied along the [100] direction, T_{SR} increases and the anomalous change in $\Delta V/V$ around T_{SR} becomes smaller. This result is in qualitative agreement with the experimental data given in Fig. 10.4a, although the observed $\lambda_{//} + \lambda_{\perp}$ does not show a discontinuous change but instead exhibits just a maximum. In the case of $x = 0.106$, the calculated values of $\Delta V/V$ exhibit a double peak at T_1 and T_2. It should be noted that, with increasing field, the peak at T_2 just shifts to a higher value,

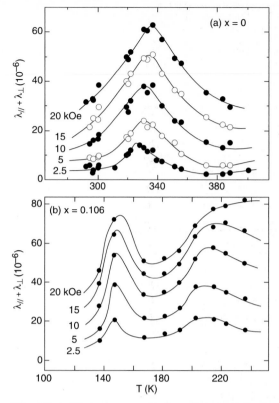

Fig. 10.4. The temperature dependence of the areal magnetostriction $\lambda_{//} + \lambda_{\perp}$ in $(\mathrm{Er}_{1-x}\mathrm{Tb}_x)_2\mathrm{Fe}_{14}\mathrm{B}$ with (**a**) $x = 0$ and (**b**) $x = 0.106$. The field was applied within the ribbon plane. The solid lines are guides for the eye

whereas the T_1 peak becomes smeared. This behavior around T_2 is similar to the result for $x = 0$. Such calculated results appear to correspond to the observations (Fig. 10.4b), although the field dependence of the observed T_1 peak is not obvious for the lack of data points. It was found that the effect of the magnetoelastic interaction on the magnitude of T_{SR} is negligibly small in the present system.

10.3   Magnetic Properties of c-Axis Oriented SmFe$_{12}$: α-Fe Nanocomposite Thin Films

A study of the film-type exchange-coupled hard/soft magnets has recently been the one of the main topics in the study of hard-magnetic materials [6,14]. Shindo et al. [15,16] have succeeded in fabricating $\mathrm{Nd}_2\mathrm{Fe}_{14}\mathrm{B}/\alpha$-Fe multilayers by a sputtering technique. Cadieu et al. [17] have reported that it is possible, by using rf sputtering, to synthesize binary SmFe$_{12}$ films of

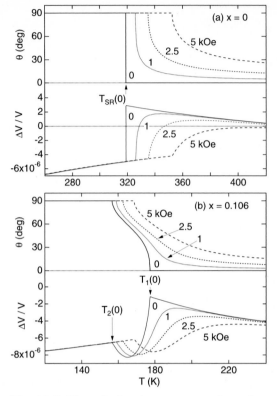

Fig. 10.5. The calculated temperature dependence of the easy-axis direction θ and the volume change $\Delta V/V$ in $(Er_{1-x}Tb_x)_2Fe_{14}B$ with (**a**) $x = 0$, and (**b**) $x = 0.106$. The field is applied in the [100] direction ($\theta = 90°$)

the ThMn$_{12}$ structure without a third element. We have synthesized thin-film nanocomposite magnets, composed of α-Fe and c-axis oriented SmFe$_{12}$ phases by using rf-magnetron sputtering onto heated glass substrates. The hard SmFe$_{12}$ and soft α-Fe phases are not placed in a multilayer geometry but are instead randomly dispersed in the films. We report here the results of x-ray and magnetization measurements [18].

The SmFe$_{12}$/α-Fe nanocomposite films were deposited by rf magnetron sputtering in an Ar atmosphere. The volume fraction of the α-Fe, V_{Fe}, was adjusted systematically by putting a different number of Sm tips on the disk-shaped Fe target. The films have the forms of glass/Sm-Fe(1 μm)/Ti(50 nm), in which the Ti layer was deposited to keep the magnetic layer from oxidizing. Magnetization curves were measured by two kinds of vibrating sample magnetometer systems, with maximum applied fields of 16 kOe and 140 kOe. When the field is applied perpendicular to the film plane, a demagnetizing field correction was made by using the coefficient 4π. The final values of V_{Fe} in the deposited films were estimated by combining the results of Rietveld

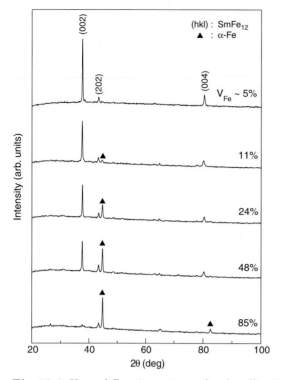

Fig. 10.6. X-ray diffraction patterns for glass/Sm-Fe/Ti films with different volume fractions of the α-Fe phase

analysis [19] of x-ray diffraction patterns and the temperature dependence of the magnetization at high temperatures.

In order to obtain the optimum condition for synthesizing highly-aligned single-phase $SmFe_{12}$ films with the $ThMn_{12}$ structure, 25 kinds of films were deposited systematically by adjusting the number of Sm tips from 25 to 37 and the substrate temperature T_s from 300°C to 600°C. According to the X ray diffraction (XRD) and magnetization measurements of these samples, the best value of T_s for synthesizing highly-aligned single-phase $SmFe_{12}$ films is 550 °C. Next we tried to synthesize the aligned $SmFe_{12}/\alpha$-Fe nanocomposite films by using the sputtering parameters obtained above and gradually reducing the number of Sm tips.

Typical x-ray diffraction patterns for films with different values of V_{Fe} are shown in Fig. 10.6. Although, with increasing V_{Fe}, the intensity of the α-Fe peak grows, the intensity of the (002) peak for the $SmFe_{12}$ phase maintains a large value up to $V_{Fe} = 48\%$, exhibiting the conservation of the (001) texture. We then estimated the average grain size of both phases by using the full-width at half maximum (FWHM) of the diffraction peaks and Scherrer's formula. The average size of the α-Fe grains increases with V_{Fe} for $V_{Fe} < 40\%$

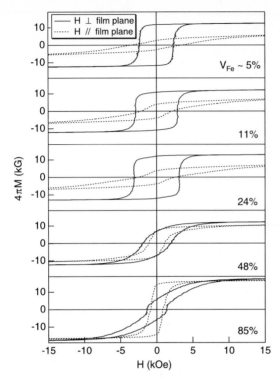

Fig. 10.7. The magnetization curves for glass/Sm-Fe/Ti films for which x-ray data were shown in Fig. 10.6. Fields of up to 140 kOe were applied parallel (dotted curves) and perpendicular (solid curves) to the film plane

and becomes roughly constant at about 30 nm for $V_{Fe} > 40\%$. In the case of SmFe$_{12}$ grains, the estimated values do not depend on V_{Fe} very much and fall into the range of $30 \sim 40$ nm.

The magnetization curves for films given in Fig. 10.6 are shown in Fig. 10.7. For films with $V_{Fe} < 30\%$, demagnetization curves along the easy axis keep a good squareness and the hard-axis curves exhibit low magnetization values. Both of these indicate a firm exchange coupling between the SmFe$_{12}$ and α-Fe phases and that the magnetic moments of α-Fe are forced to be almost parallel to the SmFe$_{12}$ moments via the exchange coupling. It should be noted that, for the larger V_{Fe} films, the difference between the magnetization curves for the parallel and perpendicular directions becomes small and the coercivity H_c is also reduced.

Magnetization curves were also measured at low temperatures. Figure 10.8 shows the differential magnetization curves dM/dH at 5 K with fields applied along the film plane, that is, along the hard direction of SmFe$_{12}$. These dM/dH curves exhibit a peak behavior around 100 kOe, which means an anomalous increase in the magnetization. The existence of such an anomaly

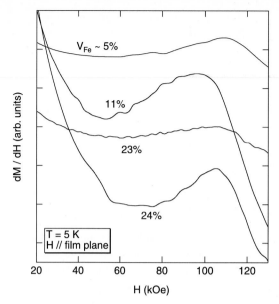

Fig. 10.8. The differential magnetization dM/dH at 5 K plotted against the field H applied along the film plane for glass/Sm-Fe/Ti films

in $SmFe_{11}Ti$ was first reported by Hu et al. [20] and then by several groups [21,22]. The origin of this phenomenon is the combined effect of the CEF at the Sm site and the Sm-Fe exchange interactions [3]. It should be noted that the peak field H_p of the dM/dH curves is smallest for $V_{Fe} = 11\%$ films and largest for $V_{Fe} = 5\%$. We have plotted in Fig. 10.9 the H_p values as a function of the average grain size of $SmFe_{12}$. It appears that a linear correlation exists between H_p and the $SmFe_{12}$ grain size, although the number of data points is small.

In order to interpret these results, we have calculated the magnetization curves for $SmFe_{12}$ by using the exchange and crystal field Hamiltonian described in Sect. 10.2. We have included the first excited $J = 7/2$ and second excited $J = 9/2$ multiplets in addition to the ground state $J = 5/2$ multiplet, since the mixing of excited multiplets is known to be very important in the Sm^{3+} system, especially for the occurrence of an anomalous increase in magnetization [3,21].

Initial parameters for the calculations are those reported by Kaneko et al. [21], that is, $K_{Fe} = 2.27$ K, $m_{Fe} = 1.86 \mu_B$, $H_m = 240$ K, $A_2^0 = -143$ Ka_0^{-2}, $A_4^0 = 4.0 Ka_0^{-4}$, and $A_6^0 = 6.7 Ka_0^{-6}$. We have confirmed that the calculated magnetization curves using these parameters agree with those given by Kaneko et al. [21], when the prefactor k in (10.5) is 11 instead of 12, corresponding to the composition of $SmFe_{11}Ti$. We then changed the magnitudes of H_m and A_2^0 systematically in order to investigate the variation of H_p. Although the dependence of H_p on H_m was found to be small, H_p

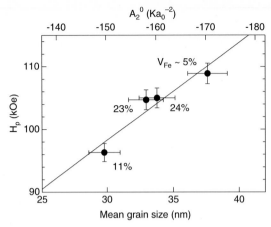

Fig. 10.9. The peak field H_p in dM/dH curves given in Fig. 10.8 as a function of the average grain size of SmFe$_{12}$. The solid line denotes the calculated dependence of H_p on A_2^0 plotted by using the top scale. Other parameters used in the calculation are given in text

changed significantly with A_2^0. We have found that H_p increases almost linearly with increasing A_2^0, as shown by the solid line in Fig. 10.9. This line is in accordance with the experimental data, which thus suggests that, in smaller-sized grains, the magnitude of A_2^0, namely the uniaxial anisotropy, is reduced. Intrinsic magnetic properties such as the CEF parameters are in general independent of the microstructure. However, when the grain size is reduced to the nanometer level, a contribution from the surface, at which the CEF interaction may be significantly modified, will be non-negligible, resulting in the change of the effective CEF parameters. Dahlgren et al. [23] reported a similar reduction in the spin reorientation temperature in smaller-sized Nd$_2$Fe$_{14}$B grains prepared by melt spinning.

10.4 Conclusion

High-field magnetic properties in rare-earth iron intermetallics are reported. A general method of calculating magnetization curves in the ternary system R-Fe-X is given, where R is a rare-earth element and X a non-magnetic element such as B. In this method the ferromagnetism of the Fe sublattice is treated phenomenologically, while the exchange and crystal fields acting on the R ions are exactly taken into account. Experimental data on $(Er_{1-x}Tb_x)_2Fe_{14}B$ and SmFe$_{12}/\alpha$-Fe nanocomposite systems are reviewed in comparison with the calculations.

Acknowledgments

The authors would like to thank Dr. D.W. Lim, T. Ishizaki and T. Nomura for their collaboration. We are grateful to Professor M. Yamada for his fruitful discussion about the crystal-field calculations. Thanks are also due to Professor Y. Nakagawa, Professor G. Kido and staff members of the High Field Laboratory for Superconducting Materials. This work was partly supported by the Murata Science Foundation, by a Grant-in-Aid for Scientific Research (No. 09650002) from the Ministry of Education, Science, Sports and Culture, by the Mazda Foundation and by the Iketani Science and Technology Foundation.

References

1. J.F. Herbst: Rev. Mod. Phys. **63**, 819 (1991)
2. Y. Nakagawa, H. Kato, D. W. Lim, G. Kido and M. Yamada: *Proc. 6th. Int. Symposium on Magnetic Anisotropy and Coercivity in Rare Earth-Transition Metal Alloys*, ed. by S. G. Sanker, Pittsburgh, 25 October 1990 (Carnegie Mellon University, Pittsburgh, 1990) p. 12.
3. M. Yamada, H. Kato, H. Yamamoto and Y. Nakagawa: Phys. Rev. B **38**, 620 (1988)
4. D.W. Lim, H. Kato, M. Yamada, G. Kido and Y. Nakagawa: Phys. Rev. B **44**, 10014 (1991)
5. H. Kato, M. Yamada, G. Kido, Y. Nakagawa, T. Iriyama and K. Kobayashi: J. Appl. Phys. **73**, 6931 (1993)
6. E.F. Kneller and R. Hawig: IEEE Trans. Magn. **27**, 3588 (1991)
7. S. Hirosawa and M. Sagawa: Solid State Commun. **54**, 335 (1985)
8. R.T. Obermyer and F. Pourarian: J. Appl. Phys. **69**, 5559 (1991)
9. D.W. Lim, H. Kato, M. Yamada, G. Kido and Y. Nakagawa: J. Magn. Magn. Mater. **104–107**, 1429 (1992)
10. H. Kato, D.W. Lim, M. Yamada, Y. Nakagawa, H. Aruga Katori and T. Goto: Physica B **211**, 105 (1995)
11. H. Kato, T. Ishizaki, F. Sato and T. Miyazaki: J. Magn. Soc. Jpn. **23** 495 (1999)
12. S. Ishio and F. Sato: J. Magn. Soc. Jpn., **12**, 259 (1988) (*in Japanese*)
13. P.A. Algarabel, A. Del Moral, M.R. Ibara and C. Marquina: J. Magn. Magn. Mater. **114**, 161 (1992)
14. R. Skomski and J. M. D. Coey: Phys. Rev. B **48**, 15812 (1993)
15. M. Shindo, M. Ishizone, H. Kato, T. Miyazaki and A. Sakuma: J. Magn. Magn. Mater. **161**, L1 (1996)
16. M. Shindo, M. Ishizone, A. Sakuma, H. Kato and T. Miyazaki: J. Appl. Phys. **81**, 4444 (1997)
17. F.J. Cadieu, H. Hegde, A. Navarathna, R. Rani and K. Chen: Appl. Phys. Lett. **59**, 875 (1991)
18. H. Kato, T. Nomura, M. Ishizone, H. Kubota, T. Miyazaki and M. Motokawa: J. Appl. Phys. **87**, 6125 (2000)
19. F. Izumi: *The Rietveld Method*, ed. by R.A. Young (Oxford Univ. Press, Oxford, 1993) Chap. 13

20. B.-P. Hu, H.-S. Li, J.P. Gavigan and J.M.D. Coey: J. Phys. Condense. Matter **1**, 755 (1989)
21. T. Kaneko, M. Yamada, K. Ohashi, T. Tawara, R. Osugi, H. Yoshida, G. Kido and Y. Nakagawa: *Proc. 10th Int. Workshop on Rare Earth Magnets and Their Applications*, Kyoto, 16-19 May 1989 (The Society of Nontraditional Metallurgy, Tokyo 1989) p.191.
22. X.C. Kou, T.S. Zhao, R. Grössinger and H.R. Kirchmayr: Phys. Rev. B **47**, 3231 (1993)
23. M. Dahlgren, X.C. Kou, R. Grössinger, J.F. Liu, I. Ahmad, H.A. Davis and K. Yamada: IEEE Trans. Magn. **33**, 2366 (1997)

11 Study of Covalent Spin Interactions in $Cd_{1-x}Mn_xSe$ by Cryobaric Magnetophotoluminescence

N. Kuroda, Y.H. Matsuda, G. Kido, I. Mogi, J.R. Anderson and W. Giriat

Hydrostatic pressure can compress a solid at a rate as high as of the order of 1% per 1 GPa in volume. Since a volume change of over 1% causes a large modification of the electron structure, various electronic properties of a material are strongly influenced by a high pressure and the study of the effects provides a wealth of information on the nature of the underlying physical processes. There have been a number of studies on the pressure dependence of the magnetism of magnetic materials. Among them the substances belonging to diluted magnetic semiconductors (DMSs) are unique in the sense that in the absence of a magnetic field they behave as ordinary semiconductors, while if a magnetic field is applied they exhibit unusual, striking galvanomagnetic and magneto-optical properties. In DMSs transition-metal elements are substituted for a fraction of 0.1–10% of cations of host semiconductors. Their unique properties arise from the strong interactions of electrons and/or holes of the host semiconductors with magnetic ions. The $s/p - d$ hybridization is also an important ingredient. The pressure effects on those properties have been studied extensively [1].

The magnetism of magnetic ions in a DMS depends on the electronic structure of the host semiconductor. For instance, p-type $Pb_{1-x-y}Sn_yMn_xTe$ of $x = 0.03$ with a hole concentration of $5 \sim 10 \times 10^{20}$ cm^{-3} is a degenerate semiconductor but shows ferromagnetic order due to Mn^{2+} spins below the Curie temperature of 2–4 K, which is determined by the hole concentration. Exploiting the property of this compound that the hole concentration can be tuned by pressure, Suski et al. [2] have shown the clear evidence that the ferromagnetic order arises from the Ruderman-Kittel-Kasuya-Yoshida (RKKY) interaction mediated by the degenerate holes. Pressure-induced switchover of the $d - d$ interaction from an antiferromagnetic to ferromagnetic regime has been observed by Chudinov et al. [3] in the Shubnikov-de Haas experiment in n-type HgMnSe system. Since this material is a zero-gap DMS, under low pressures the Bloembergen-Roland and/or kinetic superexchange mechanism dominates the $d - d$ interactions [4]. According to Chudinov et al., pressure induces an energy gap, and as a result, as in the case of $Pb_{1-x-y}Sn_yMn_xTe$, the RKKY mechanism tends to dominate the $d - d$ interactions under high pressures.

In the present work we are concerned with the exchange mechanisms in the CdMnSe system, which is one of the wide-gap and nondegenerate II-VI DMSs. If the Mn^{2+} ions are sufficiently dilute, they are scattered throughout the network of the semiconducting sp^3 covalent bonds. Turning our attention to the cation sublattice, we note that the majority of the Mn^{2+} ions are isolated and the rest form small clusters such as pairs, triads, or quartets [5]. For pairs in $Cd_{1-x}Mn_xSe$, it has recently been confirmed from cryobaric studies [6] of the exciton magnetophotoluminescence that the interaction between the two Mn^{2+} spins of a pair can be described well in terms of the kinetic superexchange theory based on the three-level model of Larson et al. [7], the model being comprised of the upper and lower Hubbard states of the d electrons and the valence band of anion p orbitals of the host semiconductor. To date, however, the mechanism of interactions among the small clusters, including singles, has been controversial. Theoretically, by adapting the perturbation theory to their three-level model, Larson et al. [7] have argued that the kinetic (antiferromagnetic) exchanges mediated by the extended anion p states are responsible also for the second-neighbor and more distant interactions and that the exchange constant decreases rapidly with the radial distance R between two magnetic ions as

$$ J = J_0 \exp(-4.89r^2) \,, \quad r = \frac{R}{a} \,, \tag{11.1} $$

where a is $\sqrt{2}$ times the nearest-neighbor distance R_1 and J_0 is a constant giving $J = J_1$ for $r = r_1 \equiv 1/\sqrt{2}$, where J_1 is the nearest-neighbor exchange constant. In addition to the work of Larson et al., several different relationships, that is, $J = J_1(r/r_1)^{-6.8}$, $J = J_1(r/r_1)^{-8.5}$ and $J_1 = J_2/(4\gamma) = J_3/(2\gamma) = J_4/\gamma = J_5/(4\gamma^2)$, have been proposed by Twardowski et al. [8], Rusin [9], Bruno and Lascaray [10] and Shen et al. [11], respectively, where the subscript n denotes nth-neighbor exchanges and the parameter γ is a numerical factor of the order of 0.04. The power laws are derived empirically [8] from the x dependence of the freezing temperature of a spin-glass state, or theoretically [9] within the framework of the three-level model. In contrast, the last two relationships are based on a different notion, named the independent-exchange-path (IEP) model by Shen et al., that the strength of the interaction between a pair of localized spins of transition-metal ions is determined by the number of cation-anion bonds connecting the relevant ions. Bednarski et al. [12] have made a theoretical analysis of the magnetization profile in $Zn_{1-x}Mn_xTe$, $Zn_{1-x}Mn_xSe$ and $Cd_{1-x}Mn_xTe$ at low temperatures to compare with available experimental data. Viewed from their results, the experimental magnetization data of the three Mn-based systems appear to favor the power law.

The purpose of the present study is to investigate which notion is valid in $Cd_{1-x}Mn_xSe$. Measurement of the pressure dependence of the magnetization profile may give the decisive information on our issue, since pressure can change J_n over a wide range without changing the constituent elements. In particular, the information on the pressure dependence of the relative magnitudes of J_n ($n \geq 2$) to J_1 would be crucial. It is known that the magneto-optical spectroscopy of the band-gap exciton enables us to probe the magnetization of magnetic ions via the interaction of the exciton with the localized spins [13]. In fact, it has been established from the aforementioned cryobaric magnetophotoluminescence experiment by Kuroda and Matsuda [6] that J_1 in $Cd_{1-x}Mn_xSe$ of $x = 0.05$ is enlarged prominently by pressure. In the present study we examined the pressure dependence of J_n for $n \geq 2$ in the CdMnSe system by using the same technique.

There are several ways to evaluate J_n for $n \geq 2$ in a DMS. One is to observe, as demonstrated by Vu et al. [14] with respect to $Zn_{1-x}Co_xTe$, the magnetization steps due to second- and third-nearest-neighbor pairs. In Mn-based II-VI DMSs, however, the measurement of the J_2- and J_3-steps is difficult, because as discussed later J_2 and J_3 are so small that individual clusters behave as isolated ones under magnetic field. The interactions among clusters produce an internal magnetic field upon the clusters themselves. Consequently, the spin temperature is altered by a specific temperature T_0, depending on x, from the lattice temperature [15]. In the present study we have investigated the composition (x) and pressure dependencies of T_0. We discuss the result in comparison with the information on J_1, which is concurrently obtained from observation of the J_1-steps.

11.1   Experiment

Figure 11.1 shows the experimental setup for cryobaric photoluminescence spectroscopy in a strong magnetic field. An optical system [16] consisting of a diamond-anvil cell (DAC) of the clamp type and fiber optics is used to measure the near-gap photoluminescence radiation. The optical system is shown in Fig. 11.2. This system is designed to be suitable for combination with a hybrid magnet. The light beam from an Ar-ion laser, which is used as the light source to excite photoluminescence, is introduced into the optical chamber through a cable of glass fiber of core diameter 100 μm. The beam is focused on the sample in the DAC by a lens (L1) and two prisms.

To observe the near-gap photoluminescence the light radiated backward from the sample is collected with another lens (L2). The signal light is fed into a silica fiber of diameter 1.0 mm and then is guided to a multichannel spectrometer through a cable of bundled silica fibers. Here the laser beam is incident at an oblique angle on the diamond anvil, so that the flux of the beam reflected by the diamond and sample goes outside the aperture of the light-collecting lens L2. Because of this optical geometry the fluorescence of the fibers themselves is suppressed significantly. This is essential for observing a weak photoluminescence signal from the sample.

The diamonds used as anvils have a culet of 0.6 mm diameter, a girdle of 3.2 mm and a table of 2.0 mm. The gasket is a stainless-steel plate of thickness 0.2 mm. Pressure is generated in a 0.3 mm diameter hole in the gasket. Condensed argon is employed as the pressure-transmitting medium. Argon is liquefied and loaded into the DAC under atmospheric pressure by using

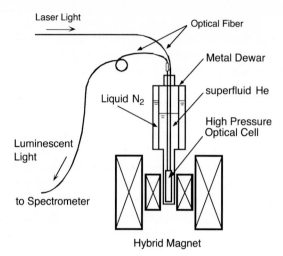

Fig. 11.1. Experimental setup for cryobaric photoluminescence spectroscopy in a strong magnetic field

an apparatus that refrigerates the DAC together with the anvil-clamping jig with liquid nitrogen [17]. When the sample cell is filled with liquid argon, the cell is closed by clamping the diamond anvils. The DAC and jig are taken out of the apparatus when the whole system is warmed up to room temperature. Afterwards, the pressure is raised to an appropriate value.

Since the substances studied here undergo a structural phase transition to a dark rock-salt phase just above 2 GPa, the pressure range is limited to 0–2 GPa in the present work. The optical system is immersed in liquid He at 4.2 K or in pumped superfluid He at 1.4 K. A static magnetic field of up to 23 T generated by the hybrid magnet is applied parallel to the c-axis of the wurtzite crystal structure of the sample. The 514.5 nm line of an argon-ion laser is used as the light source to excite photoluminescence.

The value of pressure is deduced from the pressure-induced energy shift of the exciton photoluminescence band of the sample itself at zero magnetic field on the basis of the pressure versus energy gap relationship obtained from the absorption measurement at room temperature. In this absorption measurement a microscope-spectrometer system was used and the ruby fluorescence method was employed to calibrate the pressure. An example of the pressure dependence of the fundamental absorption band is shown in Fig. 11.3. The observed shift of the absorption edge is represented well by

$$E_G = E_{G0} + cP + dP^2 , \tag{11.2}$$

where P denotes the pressure. The values of the coefficients c and d in the substances examined in this study are listed in Table 11.1.

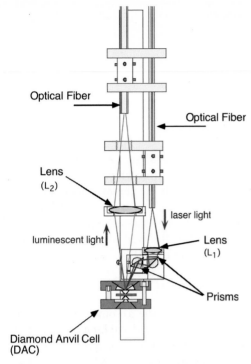

Fig. 11.2. Arrangement of a DAC and optics in the optical chamber

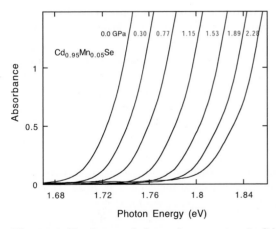

Fig. 11.3. Fundamental absorption spectrum in Cd$_{0.95}$Mn$_{0.05}$Se under various pressures at room temperature

11.2   Experimental Results

Figures 11.4, 11.5 and 11.6 show the photoluminescence energy of the A-exciton in Cd$_{1-x}$Mn$_x$Se of $x = 0.01$, 0.05 and 0.10, respectively, as a function

Table 11.1. Values of the pressure coefficients c and d in $Cd_{1-x}Mn_xSe$

x	c (meV/GPa)	d (meV/GPa^2)
0.01	59±1	−4.1±0.6
0.05	55±3	−2.1±1.3
0.10	54±1	−3.8±0.4
0.25	44±3	−1.1±1.3

of magnetic field under various pressures at $1.4\,K$ and/or $4.2\,K$. The observed state is the lower magnetic sublevel of the A-exciton which is originally of spin singlet. The large red shift is induced mainly by the exchange interactions of the exciton with Mn^{2+} ions magnetized by the magnetic field. If the distribution of the Mn^{2+} ions in the Cd sublattice is random, since the external magnetic field H is parallel to the c-axis, the observed photoluminescence energy is written in the mean field approximation as [17]

$$E_A = E_0 - \frac{1}{2}N_0(\alpha - \beta)x\langle S_z\rangle - \frac{1}{2}(g_e - g_h)\mu_B H + \sigma H^2, \qquad (11.3)$$

where E_0 is the exciton energy at $H = 0$, $\langle S_z\rangle$ is the absolute, mean value of S_z of the Mn^{2+} spins, N_0 is the density of the Cd sublattice, α and β are the $s - d$ and $p - d$ exchange constants, respectively, g_e and g_h are the g-parameters of the conduction electrons and holes, respectively, and σ is the coefficient of the diamagnetic shift of the A-exciton.

The Mn^{2+} clusters of which the ground states have nonzero total spins are magnetized continuously in the same manner as singles. Pairs, on the other hand, have a singlet ground state because of an antiferromagnetic coupling of the nearest-neighbor spins, so that they undergo a staircase-like magnetization. Let the probability that a Mn^{2+} ion can be regarded magnetically as a single be p_1^* and the probability that a Mn^{2+} ion forms a pair be p_2. Then, $\langle S_z\rangle$ at temperature T can be expressed to a good approximation by [18–21]

$$\langle S_z\rangle = p_1^* S B_S\left\{\frac{2S\mu_B H}{k(T + T_0)}\right\} + \frac{p_2}{2}\langle S_z\rangle_p, \qquad (11.4)$$

where $S = 5/2$, B_S is the normalized Brillouin function, μ_B is the Bohr magneton, T_0 is the specific spin temperature mentioned in Sect. 11.1 and $\langle S_z\rangle_p$ is the absolute, mean value of S_z of a pair. We have

$$\langle S_z\rangle_p = -\frac{\displaystyle\sum_{S_T=0}^{2S}\sum_{m=-S_T}^{S_T} m\exp\left\{-\frac{E_{p,m}}{k(T+T_p)}\right\}}{\displaystyle\sum_{S_T=0}^{2S}\sum_{m=-S_T}^{S_T}\exp\left\{-\frac{E_{p,m}}{k(T+T_p)}\right\}}, \qquad (11.5)$$

with

$$E_{p,m} = -J_1\left\{S_T(S_T+1) - 2S(S+1)\right\} + 2\mu_B m(H - H_d), \qquad (11.6)$$

where $E_{p,m}$ represents the energy of a pair with a total spin S_T and a magnetic quantum number m; H_d is an internal field due to clusters surrounding the pair. The quantity T_p in (11.5) is a temperature parameter which is introduced to represent the broadening of the magnetization steps occurring at $H = H_j = -jJ_1/\mu_B + H_d,\ j = 1, 2, \ldots 2S$ [22].

Theoretical curves of E_A are calculated with E_0, $N_0(\alpha - \beta)xp_1^*$, $N_0(\alpha - \beta)xp_2$, T_0, J_1 and T_p taken as adjustable parameters. The linear Zeeman energy $(g_e - g_h)\mu_B H$, the diamagnetic shift σH^2 and the shift $-2\mu_B H_d$ are not negligible but are very small compared with the total shift of E_A. Therefore, $g_e - g_h$, σ and H_d are assumed to be independent of pressure. Referring to the literature, their values are taken to be $1.7, 6.5 \times 10^{-6}\,\text{eV}/\text{T}^2$ and $0.7\,\text{T}$, respectively.

For $x = 0.01$, p_2 is so small compared to p_1^* that we may neglect the energy part arising from pairs:

$$E_s = -\frac{p_2}{4}N_0(\alpha - \beta)x\langle S_z\rangle_p .\tag{11.7}$$

The calculated curves of $E_A - E_s$ are shown by the dotted lines in Fig. 11.4. The contribution of E_s becomes significant for $x = 0.05$ and 0.10. The calculated curves of $E_A - E_s$ and E_A for the compounds of $x = 0.05$ and 0.10 are shown by dotted and solid lines, respectively, in Figs. 11.5 and 11.6 along with the experimental data.

The spin-temperature parameter T_0 emerges from this analysis to be $0.7 \pm 0.3\,\text{K}$ regardless of pressure in $Cd_{0.99}Mn_{0.01}Se$. The change in the value of T_0 due to pressure in this substance seems to be comparable to the experimental errors of $\pm 0.3\,\text{K}$ at most. As x increases, T_0 grows significantly. In the compounds of $x = 0.05$ and 0.10, T_0 amounts to $2.0 \pm 0.2\,\text{K}$ and $2.6 \pm 0.4\,\text{K}$, respectively, at 1 atm. Accordingly they show an appreciable pressure dependence as shown in Fig. 11.7. The positive sign of T_0 means that the distant-neighbor interactions are antiferromagnetic. We see from Fig. 11.7 that the interactions are strengthened by pressure.

Figure 11.8 shows experimental and theoretical values of the contribution from pairs, $-E_s$, as a function of magnetic field in $Cd_{0.95}Mn_{0.05}Se$ at several pressures. The experimental values are obtained by subtracting the theoretical values of $E_A - E_s$ from experimental values of E_A. Although the steps are broadened, the first and second steps can still be identified distinctly. At 1 atm, they are located at $H_1 = 11.5$ and $H_2 = 22.1\,\text{T}$, respectively. The positions of these steps yield $J_1/k = -7.2 \pm 0.3\,\text{K}$ at 1 atm. Similarly, for $Cd_{0.90}Mn_{0.10}Se$ we obtain $J_1/k = -7.6 \pm 0.3\,\text{K}$ at 1 atm. The average of these values of J_1/k is $-7.4\,\text{K}$, which agrees within experimental errors with the value $-7.6 \pm 0.2\,\text{K}$ obtained by Foner et al. [21]. As the pressure increases, the steps are shifted towards higher magnetic fields because of the interplay of the enhancement of the $p-d$ hybridization and the weakening of the onsite and intersite Coulomb energies U and V of the Mn d electrons [6]. The values of $-J_1/k$ at various pressures are plotted in Fig. 11.9.

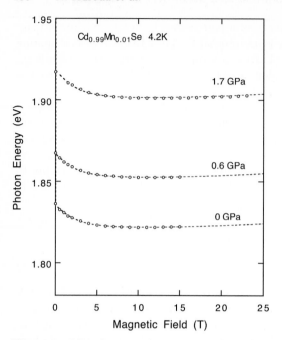

Fig. 11.4. Photoluminescence energy of excitons in $Cd_{0.99}Mn_{0.01}Se$ in a magnetic field at 4.2 K at several pressures. The dotted lines are the theoretical curves of $E_A - E_s$

Table 11.2. Experimental values of J_1, T_0 and their pressure coefficient in $Cd_{1-x}Mn_xSe$

x	J_1/k (K) at 1 atm	$d\ln|J_1|/dP$ (GPa^{-1})	T_0 at 1 atm	$d\ln T_0/dP$ (GPa^{-1})
0.01	-	-	0.7 ± 0.3	-
0.05	-7.2 ± 0.3	0.25 ± 0.05	2.0 ± 0.2	0.24 ± 0.1
0.10	-7.6 ± 0.2	0.25 ± 0.05	2.6 ± 0.4	0.40 ± 0.1

The exciton-Mn^{2+} exchange constant $N_0(\alpha - \beta)$ also changes with pressure, as reported elsewhere [6,17]. In addition, the value of T_p is found to be in a range between 0.5 and 4.5 K and between 0.8 and 5.4 K for our samples of $x = 0.05$ and 0.10, respectively, showing a tendency to increase with pressure. However, the larger part of the observed changes of T_p is likely to be produced by pressure-induced strains of crystals, because the changes are almost irreversible upon releasing the pressure.

The experimental results presented above are summarized in Table 11.2.

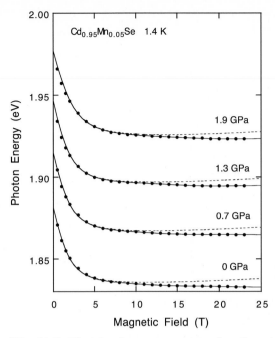

Fig. 11.5. Photoluminescence energy of excitons in Cd$_{0.95}$Mn$_{0.05}$Se in a magnetic field at 1.4 K at several pressures. The solid and dotted lines are the theoretical curves of E_A and $E_A - E_s$, respectively

11.3   Discussion

11.3.1   Mean-Field Approximation

Barilero et al. [15] have discussed the mechanism of T_0 in zinc blende DMSs in terms of the mean-field approximation. They have argued that for a random distribution of Mn^{2+} ions T_0 is directly related to the internal exchange field produced by interactions among small clusters. In fact, as shown in Fig. 11.10, T_0 exhibits a systematic dependence on x. The internal field arises mainly from Mn-Mn interactions associated with the covalent bond pathways of the atomic sequence Mn-Se-Cd-Se-Mn: The difference in the spatial configuration of atoms in the crystal gives rise to the difference in the radial distance between the two terminating Mn sites [10]. Also shown in Fig. 11.10 are the experimental values of T_0 available for Cd$_{1-x}$Mn$_x$Se [13], Cd$_{1-x}$Mn$_x$S [18], Cd$_{1-x}$Mn$_x$Te [18,23] Zn$_{1-x}$Mn$_x$Se [23,24] and Zn$_{1-x}$Mn$_x$Te [15]. It is apparent from Fig. 11.8 that Mn-based II–VI DMSs have similar properties of spin interactions.

The cation sites of a wurtzite crystal form a hcp sublattice. There are 6 second-nearest-neighbor sites at $R = a$, 2 third-nearest-neighbor sites at $R = \sqrt{4/3}a$, 18 fourth-nearest-neighbor sites at $R = \sqrt{3/2}a$ and 12 fifth-

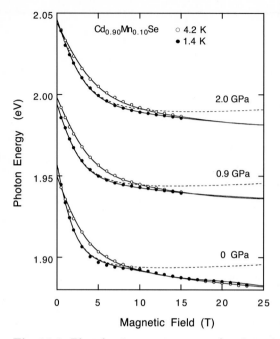

Fig. 11.6. Photoluminescence energy of excitons in $Cd_{0.90}Mn_{0.10}Se$ in a magnetic field at 1.4 K and 4.2 K at several pressures. The solid and dotted lines are the theoretical curves of E_A and $E_A - E_s$, respectively

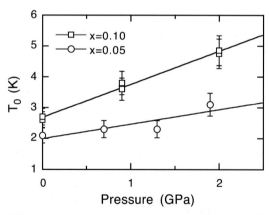

Fig. 11.7. Pressure dependence of T_0 in $Cd_{0.95}Mn_{0.05}Se$ and $Cd_{0.90}Mn_{0.10}Se$

nearest-neighbor sites at $R = \sqrt{11/6}a$ [19]. The third- and fourth-nearest-neighbor sites correspond to the third-nearest-neighbor sites at $R = \sqrt{3/2}a$ of the fcc sublattice of zinc blende DMSs and the fifth-nearest-neighbor sites correspond to the fourth-nearest-neighbor sites at $R = \sqrt{2}a$ of the fcc sublattice. Approximating the third-nearest-neighbor exchange constant J_3' to

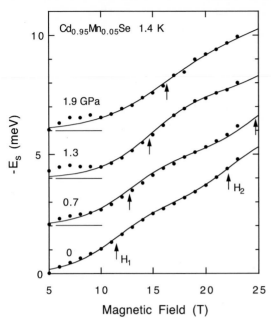

Fig. 11.8. $-E_s$ versus magnetic field at $1.4\,$K in Cd$_{0.95}$Mn$_{0.05}$Se at several pressures. The solid lines are the theoretical curves. The vertical arrows show the positions of H_1 and H_2

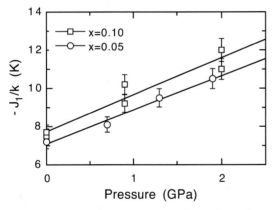

Fig. 11.9. Pressure dependence of $-J_1/k$ in Cd$_{0.95}$Mn$_{0.05}$Se and Cd$_{0.90}$Mn$_{0.10}$Se

be equal to the fourth-nearest-neighbor constant J_3 and reading the fifth-nearest-neighbor exchanges as the fourth-nearest-neighbor ones to make an argument parallel to Barilero et al., we obtain the relationship

$$kT_0 = -4xp_1^*S(S+1)J^*\,, \tag{11.8}$$

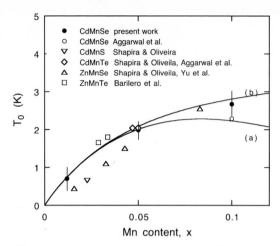

Fig. 11.10. The x dependence of T_0 in $Cd_{1-x}Mn_xSe$ at 1 atm. Reported experimental data for $Cd_{1-x}Mn_xSe$ [13], $Cd_{1-x}Mn_xS$ [23], $Cd_{1-x}Mn_xTe$ [18,23], $Zn_{1-x}Mn_xSe$ [23,24] and $Zn_{1-x}Mn_xTe$ [15] are also shown for comparison. The solid lines (**a**) and (**b**) are the theoretical curves calculated from (11.8) with p_1^* given by (11.10) and (11.11), respectively

with an effective internal exchange constant J^* of

$$J^* = J_2 + \frac{10}{3}J_3 + 2J_4 . \tag{11.9}$$

In a zinc blende DMS, J^* is given by $J_2 + 4J_3 + 2J_4$. The contribution from pairs to this internal field is neglected in deriving (11.8), because the continuous magnetization of singles and single-like clusters, which is described by $B_S(H; T + T_0)$, is sensitive to T_0 particularly in the low-field region, but pairs are in the singlet ground state magnetic field for up to about 10 T, as we have seen in the preceding section.

Taking account of the internal fields due to clusters up to triads, p_1^* can be written by the probabilities p_1, p_3, p_4 of singles, open triads and closed triads, respectively, as [25]

$$p_1^* = p_1 + \frac{p_3}{3} + \frac{p_4}{15} . \tag{11.10}$$

If the contribution from clusters greater than triads is taken into account, the upper bound of p_1^* is given by [23]

$$p_1^* = p_1 + \frac{p_3}{3} + \frac{p_4}{15} + \frac{1 - p_1 - p_2 - p_3 - p_4}{5} . \tag{11.11}$$

If the distribution of Mn^{2+} ions in the hcp sublattice is random, the probabilities of finding respective clusters are known to be [26]

$$p_1 = (1 - x)^{12} ,$$

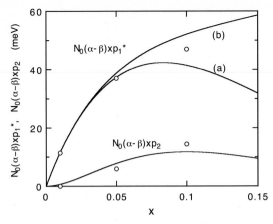

Fig. 11.11. The x dependence of of $N_0(\alpha-\beta)xp_1^*$ and $N_0(\alpha-\beta)xp_2$. The solid lines are the theoretical curves. The difference between lines (**a**) and (**b**) comes from the use of (11.10) and (11.11), respectively, for p_1^*

$$p_2 = 12x(1-x)^{18},$$
$$p_3 = 18x^2(1-x)^{23}(7-5x),$$
$$p_4 = 3x^2(1-x)^{21}\{7-6x+(1-x)^2\}. \tag{11.12}$$

For a fcc lattice, p_1, p_2 and p_3 are identical with those given by (11.12) but p_4 is replaced by $24x^2(1-x)^{22}$. Numerically, however, the difference in p_1^* between hcp and fcc lattices is very small.

11.3.2   Analysis of the Experimental Data

To begin with let us look at the x dependence of $N_0(\alpha - \beta)xp_1^*$ and $N_0(\alpha - \beta)xp_2$. Figure 11.11 shows their experimental values at 1 atm as a function of x. The best-fit theoretical curves are also shown in Fig. 11.11 along with the experimental data. These theoretical curves are calculated from (11.10), (11.11) and (11.12) with a common parameter of $N_0(\alpha - \beta) = 1.30$ eV.

The theoretical curves explain the experimental data of $N_0(\alpha-\beta)xp_1^*$ and $N_0(\alpha - \beta)xp_2$ well. The value calculated from (11.10), which takes only the clusters smaller than quartets into account, disagrees to some extent with the experimental data for $N_0(\alpha-\beta)xp_1^*$ for $x = 0.1$, but in view of the theoretical upper bound given by (11.11) the disagreement is rather reasonable. It turns out from these data that Mn^{2+} ions are indeed distributed throughout the hcp sublattice at random.

The result shown in Fig. 11.11 assures also that the exchange interaction between an exciton and a Mn^{2+} ion is almost independent of x. In addition, as we have seen in Sect. 11.2, our experimental values of J_1/k for $x = 0.05$ and 0.10 agree with each other within experimental errors. These findings suggest that Mn-Mn interactions are almost independent of x for the compounds

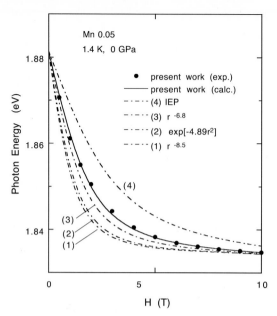

Fig. 11.12. Comparison of theoretical curves with the experimental values of the exciton energy E_A in $Cd_{0.95}Mn_{0.05}Se$ in a magnetic field at 1.4 K and 1 atm

of $x \leq 0.1$. Consequently, the observed x dependence of T_0 permits us to evaluate J^* at a given pressure. In Fig. 11.10 the theoretical curves of T_0 calculated by putting (11.10), (11.11) and $J^*/k = -2.0$ K into (11.8) are compared with experimental values at 1 atm. The mean field theory is found to explain the experimental data very well. Taking the experimental errors into account, the present data show J^*/k to be -2.0 ± 0.4 K at 1 atm.

We now proceed to the comparison of models for the spatial variation of J. Here we look at the case of $x = 0.05$ as a typical example. With the value of $J_1/k = -7.4$ K at 1 atm, the Gaussian form, $\exp(-4.89r^2)$, and power laws $r^{-8.5}$ and $r^{-6.8}$ predict $T_0 = 0.89$, 0.71 and 1.54 K, respectively, whereas the independent-exchange-path model predicts $T_0 = 4.3$ K if Shen's γ of 0.044 is adapted. In Fig. 11.12 the photon energies $E_A - E_s$ calculated by putting these values of T_0 and $N_0(\alpha - \beta)xp_1^* = 37.0$ meV into (11.3) are compared with our experimental data for E_A in $Cd_{0.95}Mn_{0.05}Se$ at 1.4 K and 1 atm.

The calculated curves show significant differences from each other in the low-field region. This is because the initial slope scales with $T + T_0$ as $(T + T_0)^{-1}$. Twardowski's power law $r^{-6.8}$ matches the experimental data rather well, but the other three laws mismatch the data. If the exponent q of the power law r^{-q} is adjusted, our experimental data of $J_1/k = -7.4$ K and $J^*/k = -2.0$ K yield $q = 6.1$ with $J_2/k = -0.89$ K, $J_3/k = -0.26$ K and $J_4/k = -0.14$ K. The theoretical curve is shown by the solid line in Fig. 11.12.

Although Larson's three-level model appears to underestimate J_n for $n \geq 2$, it puts forward the notion that the spatial variation of covalent spin interactions in II–VI DMSs can be expressed as $J = J_1 f(r)$, with a volume-independent function $f(r)$ of $f(1/\sqrt{2}) = 1$. This notion gives a physical basis for the power laws [9]. The volume-independence is common to indirect interactions mediated by extended electronic states. In fact, a similar property is also seen in the RKKY interaction . Apart from volume-dependent prefactors, including the squared $s - d$ exchange constant, the spatial dependence of the RKKY interaction is a function of the product $k_F R$ of the Fermi momentum k_F and R, and thus is invariable as long as the number of electrons forming the Fermi sphere is conserved. It is worthwhile to see how T_0 in $Cd_{1-x}Mn_xSe$ varies with pressure relative to J_1. As seen in Fig. 11.7, the normalized pressure coefficient of T_0, that is, $d \ln T_0/dP$, which is equal to $d \ln |J^*|/dP$, is $0.24 \pm 0.1\,GPa^{-1}$ in $Cd_{0.95}Mn_{0.05}Se$. This result agrees well with $d \ln |J_1|/dP = 0.25 \pm 0.05\,GPa^{-1}$ indicated by the data of $Cd_{0.95}Mn_{0.05}Se$ and $Cd_{0.90}Mn_{0.10}Se$ in Fig. 11.9. Although $d \ln T_0/dP$ of $0.4 \pm 0.1\,GPa^{-1}$ at $x = 0.10$ is a little larger than the value at $x = 0.05$, it is still comparable within experimental errors to $d \ln |J_1|/dP$.

The present result suggests that the power law r^{-q} holds even under high pressures, while retaining the exponent at $q \approx 6$. This finding supports Larson's covalent spin interaction picture that the exchange interactions between dilute Mn^{2+} spins in the covalent bond network of a II–VI semiconductor are determined by kinetic exchanges mediated by the extended p orbitals constituting the valence band of the host semiconductor. Shen et al. have dealt with $Zn_{1-x}Mn_xTe$ of $x = 0.938$, in which the Zn sites are mostly replaced by Mn, to deduce the IEP model. In such a dense magnetic alloy, the d electrons could have a significant itinerancy, and thus the dominant exchange mechanism could be different from that in the usual DMSs.

11.4   Conclusion

The pressure dependence of the exchange interactions among small Mn clusters scattered throughout the network of sp^3 covalent bonds in the diluted magnetic semiconductor $Cd_{1-x}Mn_xSe$ has been studied by a cryobaric measurement of the exciton magnetophotoluminescence. A pressure of up to $2\,GPa$ is generated with a diamond anvil cell, being subjected at low temperatures to the static magnetic field up to $23\,T$ generated with a hybrid magnet. The observed specific spin temperature T_0 of the clusters shows the effective internal exchange constant $J^*/k \equiv J_2/k + (10/3)J_3/k + 2J_4/k$ to be $-2.0 \pm 0.4\,K$ at $1\,atm$, where J_n denotes the nth-neighbor exchange constant. The nearest-neighbor interaction constant is estimated to be $J_1/k = -7.4 \pm 0.4\,K$ at $1\,atm$ from an analysis of the effect of the stepwise magnetization of Mn pairs. J^*, as well as J_1, increases rapidly with increasing pressure. The pressure coefficient $d \ln |J^*|/dP = 0.2 \sim 0.4\,GPa^{-1}$ agrees with

$d \ln |J_1|/dP = 0.25 \pm 0.05 \, \text{GPa}^{-1}$ within experimental error. This result supports Larson's covalent spin interaction picture that the exchange interactions between the scattered, localized spins are determined by kinetic exchanges mediated by the extended p orbitals constituting the valence band of the host II-VI semiconductor.

Acknowledgment

The experiments were performed at the High Field Laboratory for Superconducting Materials, Tohoku University. The authors (N.K. and Y.M.) are grateful to K. Sai and Y. Ishikawa for operation of the high field magnets and helpful assistance throughout the whole course of the experiments.

References

1. N. Kuroda: *Semiconductors and Semimetals*, ed. by R. K. Willardson and R. Weber, Vol. 54, *High Pressure in Semiconductor Physics I*, ed. by T. Suzuki and W. Paul (Academic Press, San Diego, 1998) Chap. 6
2. T. Suski, J. Igalson and T. Story: J. Magn. Magn. Mater. **66**, 325 (1987)
3. S.M. Chudinov, D. Tu. Rodichev, G. Manici and S. Sizza: Phys. Status Solidi B, **175**, 213 (1993)
4. N.B. Brandt, V.V. Moshchalkov, A.O. Orlov, L. Skrbek, I.M. Tsidil'kovski, and S.M. Chudinov: Sov. Phys. JETP **57**, 614 (1983)
5. S. Oseroff and P. H. Keesom: *Semiconductors and Semimetals*, ed. by R. K. Willardson and A. C. Beer, Vol. 25, *Diluted Magnetic Semiconductors*, ed. by J. K. Furdyna and J. Kossut (Academic Press, San Diego, 1988) Chap. 3
6. N. Kuroda and Y. Matsuda: Phys. Rev. Lett. **77**, 1111 (1996)
7. B.E. Larson, K.C. Hass, H. Ehrenreich and A.E. Carlsson: Phys. Rev. B **37**, 4137 (1988)
8. A. Twardowski, H.J. Swagten, W.J.M. de Jonge and M. Demianiuk: Phys. Rev. B **36**, 7013 (1987)
9. T.M. Rusin: Phys. Rev. B **53**, 12577 (1996)
10. A. Bruno and J.P. Lascaray: Phys. Rev. B **38**, 9168 (1988)
11. Q. Shen, H. Luo and J.K. Furdyna: Phys. Rev. Lett. **75**, 2590 (1995)
12. H. Bednarski, J. Cisowski and J.C. Portal: Phys. Rev. B **55**, 15762 (1997)
13. R.L. Aggarwal, S.N. Jasperson, J. Stankiewicz, Y. Shapira, S. Forner, B. Khazai and A. Wold: Phys. Rev. B **28**, 6907 (1983)
14. T.Q. Vu, V. Bindilatti, Y. Shapira, E.J. McNiff Jr., C.C. Agosta, J. Papp, R. Kershaw, K. Dwight and A. Wold: Phys. Rev. B **46**, 11617 (1992)
15. G. Barilero, C. Rigaux, N.H. Hau, J.C. Picoche and W. Giriat: Solid State Commun. **62**, 345 (1987)
16. Y. Matsuda, N. Kuroda and Y. Nishina: Rev. Sci. Instrum. **63**, 5764 (1992)
17. Y. Matsuda and N. Kuroda: Phys. Rev. B **53**, 4471 (1996)
18. R.L. Aggarwal, S.N. Jasperson, P. Becla and R.R. Galazka: Phys. Rev. B **32**, 5132 (1985)
19. B.E. Larson, K.C. Hass and R.L. Aggarwal: Phys. Rev. B **33**, 1789 (1986)
20. J.R. Anderson: Physica B **164**, 67 (1990)

21. S. Foner, Y. Shapira, D. Heiman, P. Becla, R. Kershaw, K. Dwight and A. Wold: Phys. Rev. B **39**, 11793 (1989)
22. Y. G. Rubo, M. F. Thorpe and N. Mousseau: Phys. Rev. B **56**, 13094 (1997)
23. Y. Shapira and N.F. Oliveira Jr.: Phys. Rev. B **35**, 6888 (1987)
24. Y.W.Y. Yu, A. Twardowski, L.P. Fu, A. Petrou and B.T. Jonker: Phys. Rev. B **51**, 9722 (1995)
25. Y. Shapira, S. Forner, D.H. Ridgley, K. Dwight and A. Wold: Phys. Rev. B **30**, 4021 (1984)
26. R.E. Behringer: J. Chem. Phys. **29**, 573 (1958); M.M. Kreitman and D.L. Barnett: ibid. **43**, 364 (1965)

Part V

Other High Field Physical Properties

12   Magnetic Properties
of III–V Ferromagnetic Semiconductor
(Ga,Mn)As

F. Matsukura, T. Dietl, T. Omiya, N. Akiba, D. Chiba, E. Abe, H.
Hashidume, K. Takamura, Y. Ohno, T. Sakon, M. Motokawa and H. Ohno

Modern information technology utilizes the charge degree of freedom of electrons to process information in semiconductors and the spin degree of freedom to store information in magnetic materials. The next step is to explore the combination of the two degrees of freedom to look into the possibility of realizing new functionalities.

An approach compatible with present-day electronic materials is to make non-magnetic III–V semiconductors magnetic, and even ferromagnetic, by introducing a high concentration of magnetic ions. III–V compound semiconductors and heterostructures are widely used for high-speed electronic devices as well as for optoelectronic devices, especially those based on the GaAs/(Al,Ga)As system, which has been the test bench for new physics and new device concepts. The effort to grow such III–V–based diluted magnetic semiconductors (DMSs) was rewarded with successful molecular beam epitaxial (MBE) growth of uniform (In,Mn)As films on GaAs substrates [1]. Subsequent discovery of hole-induced ferromagnetic order in p-type (In,Mn)As [2] encouraged researchers to investigate GaAs based systems [3] and led to the successful growth of ferromagnetic (Ga,Mn)As [4]. Currently a number of groups are working on the MBE growth of III–V–based DMSs to advance the understanding of this new material as well as heterostructures based on it [2,5–9].

12.1   Preparation of (Ga,Mn)As
by Molecular Beam Epitaxy and its Lattice Properties

In order to observe magnetic cooperative effects, one needs to introduce a sizable amount of magnetic elements (a few percent or more). Although non-equilibrium epitaxial growth methods such as MBE could offer doping in excess of the thermodynamic solubility limit, segregation of impurities during MBE growth was an obstacle in obtaining a high concentration of magnetic elements [10]. This obstacle has been overcome by low-temperature MBE.

Typical MBE growth of (Ga,Mn)As is carried out under As-stabilized conditions by using solid source MBE with elemental sources Ga, Mn, In, Al and As. Mn provides both localized spins and carriers (holes) due to its acceptor nature. Mn content x in the $(Ga_{1-x}Mn_x)As$ epitaxial films on GaAs (001) substrates can be determined by X-ray diffraction (XRD) measurements, once it is calibrated by other means such as electron probe microanalysis (EPMA). Normally, either a GaAs buffer layer or a (Al,Ga)As buffer layer is grown before growth of (Ga,Mn)As. For the control of strain in the film, strain-relaxed (In,Ga)As buffer layers with lattice constants a greater than the subsequent (Ga,Mn)As layer can be used. The (Ga,Mn)As growth can be started by simply commencing with the Mn beam during the low-temperature GaAs growth and keeping the substrate temperature T_S constant at 250 °C. Although the properties of grown (Ga,Mn)As do depend on growth parameters such as the As overpressure and T_S [11,12], as long as the established growth procedure is followed, the properties of the (Ga,Mn)As films are reproducible. Reflection high-energy electron diffraction (RHEED) patterns are used to monitor the surface reconstruction during growth. The surface reconstruction of (Ga,Mn)As is (1×2) during and after growth. When the Mn flux or T_S or both are too high, a RHEED pattern indicative of the appearance of the MnAs (NiAs-structure) second phase on the surface emerges. The maximum x so far obtained is about 0.07, above which surface segregation occurs even at low growth temperature. Clear RHEED oscillations are observed at the initial growth stage (even without Mn), which indicate the growth mode being two-dimensional [12,13].

The lattice parameter a of strain-free (Ga,Mn)As determined by XRD increases linearly with x up to 0.071, following Vegard's law. A high-quality interface between GaAs and (Ga,Mn)As was confirmed by XRD of GaAs/(Ga, Mn)As superlattice structures [14]. Extended X-ray absorption fine structure (EXAFS) measurements of (Ga,Mn)As ($x = 0.005$ and 0.074) by Shioda et al. showed that Mn was substitutionally incorporated into the Ga sublattice [15].

12.2   Magnetic and Magnetotransport Properties

12.2.1   Magnetic Properties

Magnetization measurements revealed the presence of ferromagnetic order in the (Ga,Mn)As films at low temperatures [4]. Sharp, square hysteresis loops, indicating a well-ordered ferromagnetic structure, appeared in the magnetization M versus magnetic field B curves when B was applied in the plane of the film. When B was applied perpendicular to the sample surface, an elongated magnetization curve with little hysteresis was obtained, indicating that the easy-axis for magnetization is in the plane. This difference may be explained by the magnetoelastic effect; a perpendicular easy-axis can be realized in (Ga,Mn)As with the use of biaxial tensile stress, which can be built in using a (In,Ga)As buffer layer [16]. The highest ferromagnetic transition temperature T_c so far obtained is 110 K [17]. The relationship between x and T_c was found to be approximately $T_c/x = 2000$ K up to $x = 0.05$. The low temperature saturation magnetization, M_S, of the (Ga,Mn)As film is consistent

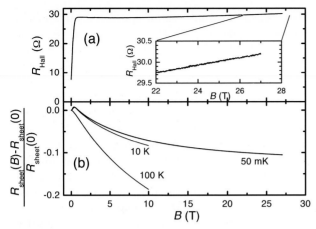

Fig. 12.1. The magnetotransport properties of $(Ga_{0.947}Mn_{0.053})As$: (**a**) the Hall resistance at $50\,mK$ in the magnetic field up to $27\,T$ (the inset shows the Hall resistance in the high-field region); (**b**) the sheet magnetoresistance ratio at several temperatures

with the spin of Mn, $S = 5/2$. The temperature, T, dependence of the susceptibility χ of (Ga,Mn)As films follows the Curie-Weiss form $\chi = C/(T-\theta)$, where C is the Curie constant and θ the paramagnetic Curie temperature; θ appears to fall close to T_c. In the absence of carriers, however, θ becomes negative $(-2\,K)$ indicating that the direct exchange among Mn is antiferromagnetic and the ferromagnetic interaction observed in these films is carrier (hole) induced [18].

12.2.2   Magnetotransport Properties

Depending on the Mn content x, the temperature dependence of (Ga,Mn)As shows a metallic or insulating behavior. Below, we focus on the behavior of metallic 200-nm thick (Ga,Mn)As with a Mn content $x = 0.053$ on a $(Al_{0.9}Ga_{0.1})As$ buffer layer in order to avoid complications arising from localization [19].

Magnetotransport measurements were performed with B perpendicular to the sample plane. The Hall resistance, R_{Hall}, and the sheet resistance, R_{sheet}, were measured simultaneously using a Hall bar geometry. The measurement at temperature $T = 50\,mK$ with B up to $27\,T$ was carried out in a hybrid-magnet system equipped with a dilution refrigerator. The T dependence of R_{Hall} and R_{sheet} in the range from 1.5 to $300\,K$ with B up to $10\,T$ was measured using a standard setup.

Figure 12.1 shows the results of the magnetotransport measurements at $50\,mK$. R_{Hall} of magnetic materials can be expressed by the sum of the ordi-

nary Hall resistance and the anomalous Hall resistance, which is proportional
to the perpendicular component of M:

$$R_{\text{Hall}} = \frac{R_0}{d} B + \frac{R_S}{d} M \,, \tag{12.1}$$

where R_0 is the ordinary Hall coefficient, R_S the anomalous Hall coefficient,
and d the thickness of the conducting layer. R_S is proportional to either
R_{sheet} or R_{sheet}^2, depending on the mechanism of the anomalous Hall effect
– skew or side-jump scattering, respectively. As shown in Fig. 12.1a, R_{Hall}
shows a steep rise in the low-field region ($< 0.5\,\text{T}$) due to the rotation of M
from the in-plane orientation (magnetic easy plane) to the direction of B,
then drops slightly ($< 8\,\text{T}$) and finally increases monotonically in the high-
field region ($> 8\,\text{T}$). In the highest-B regime, the anomalous Hall resistance
saturates because of the saturation of both M and magnetoresistance (MR)
(Fig. 12.1b). The remaining linear slope reflects the ordinary Hall term, from
which one can determine the hole concentration p unambiguously. We ob-
tained $p = 3.5 \times 10^{20}\,\text{cm}^{-3}$ from the slope of R_{Hall} versus B curve above
22 T (the inset to Fig. 12.1). The present value of p corresponds to 30% of
the nominal Mn content ($\approx 1.2 \times 10^{21}\,\text{cm}^{-3}$), which suggests that there is a
compensation of Mn acceptors by deep donors, most probably As antisites,
a high concentration of which is known to be present in GaAs grown at low
temperature.

We hereafter use p determined from the measurements shown in Fig. 12.1
in order to analyze the T and B dependence of the magnetotransport properties in the high-temperature paramagnetic region. In this temperature range,
we can neglect the strong spin correlation, magnetic anisotropy, and localization effects seen at low temperatures. In the present analysis, we assumed
that skew-scattering controls the magnitude of the anomalous Hall effect, but
the choice of skew scattering or side-jump scattering gives only a small effect
(20% error at most) on the final results.

The T dependence of the inverse of χ, obtained from the magnetotransport properties around $B = 0$ using (12.1) is well described by the Curie-
Weiss law with $\theta = 105\,\text{K}$. The result demonstrates that one can determine
χ for (Ga,Mn)As from the transport data alone.

The T dependence of the resistivity ($\rho = dR_{\text{sheet}}$) of metallic (Ga,Mn)As
shows a peak around T_{C}. Negative MR is observed in the same temperature
range (ρ decreases by 20% from 0 to 7 T at 100 K). The magnitude of the
negative MR and the T dependence of ρ suggests that spin-disorder scattering
by thermodynamic fluctuations of the magnetic spins is involved. A peak
around T_{C} can be interpreted as critical scattering. Within this framework
the negative MR is caused by the reduction of scattering associated with
spin-alignment. The corresponding contribution to the ρ is given by [20]

$$\rho = 2\pi^2 \frac{k_{\text{F}}}{pe^2} \frac{m^2 \beta^2}{h^3} \frac{k_{\text{B}} T^2}{g^2 \mu_{\text{B}}^2} [2\chi_\perp(T, B) + 2\chi_{//}(T, B)] \,, \tag{12.2}$$

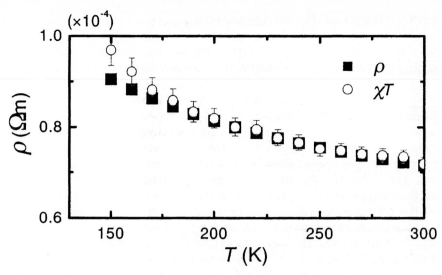

Fig. 12.2. Temperature dependence of the resistivity in the high-temperature paramagnetic region. Solid squares and open circles in this figure show the experimental
data and the fit, respectively

where k_F is the Fermi wave vector, which is determined from p assuming a
spherical Fermi surface, m is the hole effective mass, taken here as $0.5m_0$
(m_0: the mass of free electron), β is the exchange integral of the interaction between the holes and the magnetic spins (p-d exchange), k_B is the
Boltzmann constant, h is the Planck constant, and $\chi_\perp$ and $\chi_{//}$ are the transverse and longitudinal magnetic susceptibilities, respectively, which are determined from the magnetotransport data by the use of (12.1); $\chi_\perp = M/B$,
and $\chi_{//} = \partial M/\partial B$.

As shown in Fig. 12.2, the temperature dependence of ρ at $B = 0$ far above
T_C can be well reproduced by (12.2) using β and a background resistivity as
fitting parameters. The deviation near T_C may be explained by noting that
the long-range nature of the carrier-mediated magnetic interactions reduces
the magnitude of χ for nonzero wave vectors, an effect disregarded in (12.2).
From the fit we obtained β, which is usually expressed in terms of $|\,N_0\beta\,|$
(N_0 is the density of the cation sites) as 1.7 eV. The B dependence of ρ at
high temperatures is also well reproduced by scattering from the fluctuations
of the magnetic spins. Again, we used (12.2) to fit the data at $T \geq 200\,\mathrm{K}$,
and obtained the value of $|\,N_0\beta\,|$ as 1.3 eV.

Typically $|\,N_0\beta\,|$ is around 1 eV for II–VI DMSs. According to photoe-
mission studies on (Ga,Mn)As, $N_0\beta = -1.2\,\mathrm{eV}$ [21]. In order to determine
$N_0\beta$ from transport measurements, one has to take into account the localization effect which enhances the exchange between carrier and magnetic spins
through the carrier-carrier interaction [22]. The present value obtained for
$|\,N_0\beta\,|$ may be influenced by such an effect.

12.3   Origin of Ferromagnetism

Since the magnetic interaction between the Mn ions has been shown to be antiferromagnetic in semi-insulating fully compensated (Ga,Mn)As using Sn as a donor [18], the ferromagnetic interaction in magnetic III–VDMSs is most likely hole induced. One approach to the understanding of the carrier-induced ferromagnetic interaction starts from approximating the interaction as being between the conduction carrier Fermi sea and the localized magnetic moments. This approach has been employed for understanding various optical and magnetic phenomena observed in II–VI-based paramagnetic semiconductors. The nature of exchange is expressed in terms of exchange integrals $N_0\alpha$ for the conduction band and $N_0\beta$ for the valence band. A mean-field theory has been developed taking into account the feedback mechanism between the magnetization polarization and the carrier polarization [23]. This mean-field theory results in the same expression of T_C as that of the well-known Ruderman-Kittel-Kasuya-Yosida (RKKY) interaction. The RKKY interaction is carrier-induced and sufficiently long-range to account for the magnetic interaction in dilute systems. It has been put forward to explain the carrier (hole)-induced ferromagnetism as well as the spin-glass phase observed in a IV–VI compound (Pb,Sn,Mn)Te [24].

T_C is calculated from the hole-free energy in the framework of the mean-field theory by solving a 6×6 Luttinger Hamiltonian taking into account the presence of the p-d interaction. We obtain $T_C = 110\,\mathrm{K}$ for $p = 3.5 \times 10^{20}\,\mathrm{cm}^{-3}$, $\mid N_0\beta \mid = 1.3\,\mathrm{eV}$, in rather good agreement with the range of $N_0\beta$ and p inferred from photoemission and from transport measurements [25]. The mean-field approach can explain the absence of ferromagnetism in n-type materials; the small effective mass together with the small $N_0\alpha$ (about $0.2\,\mathrm{eV}$) makes it difficult for the ferromagnetic interaction to overcome the direct antiferromagnetic coupling among the Mn ions. The mean-field picture was also used to explain the hole-induced ferromagnetism observed in II–VI (Zn,Mn)Te [26].

Another approach emphasizes the role of the d-band formed by transition metal impurities. Based on a first-principles calculation of the ground state energy of (In,Mn)As, Akai stressed the partial d-character of the holes and invoked a double-exchange picture for the observed ferromagnetism [27]. The energy gain of the ferromagnetic phase is proportional to the carrier concentration and therefore the calculation showed that compensating the hole results in destabilization of the ferromagnetic state.

Whether the two approaches are different or not probably depends on the physical properties under discussion, because the partial d-character of the holes is the source of enhancement of the hole-localized spin exchange $\mid N_0\beta \mid$ over the electron-localized spin exchange $N_0\alpha$.

12.4   Heterostructures

New physics such as the fractional quantum Hall effect have emerged from non-magnetic III–V semiconductor heterostructures. They have also provided a test bench for a number of new device concepts, among which are quantum well lasers and high electron mobility transistors. Ferromagnetic III-VDMSs can add a new dimension to the III–V heterostructure systems because they can introduce magnetic cooperative phenomena not present in conventional III–V materials.

12.4.1   Trilayers

Although a number of studies on giant magnetoresistance (GMR) effects and interlayer exchange coupling between ferromagnetic layers in metallic magnetic multilayer systems have been reported, only very little has been known about semiconducting counterparts. We therefore investigated MR effects and interlayer coupling of magnetic trilayer structures using (Ga,Mn)As [28].

On a $1\,\mu$m-thick (In,Ga)As buffer, we prepared trilayer structures of $30\,$nm $(Ga_{0.95}Mn_{0.05})As/2.8\,$nm $(Al_yGa_{1-y})As/30\,$nm $(Ga_{0.97}Mn_{0.03})As$, which gives the direction of easy axis perpendicular to the plane by introducing tensile strain. Samples with various Al contents y (0.14, 0.30, 0.42 and 1.00) were prepared to change the barrier height ($0.55y\,$eV) in the valence band.

R_{sheet} and R_{Hall} were measured with B perpendicular to the sample plane. These results showed that the MR ratio $(R_{\text{sheet}}\text{-}R_0)/R_0$ of the trilayer increases when M of the two (Ga,Mn)As layers are aligned antiparallel (Fig. 12.3), indicating the presence of spin-dependent scattering, which was supported by the minor loop measurements (open symbols in Fig. 12.3).

From the magnetic field shift of the minor loops with respect to $B = 0$, the interlayer coupling strength J was calculated. The MR ratio as well as J decreases with the increase of T and/or the increase of y. The magnitudes were much smaller than those of metallic systems, and the coupling was always ferromagnetic.

The vertical magnetotransport of a $20 \times 20\,\mu$m^2 trilayer with an AlAs barrier showed a resistance increase between $8\,$mT and $16\,$mT, where M of the two (Ga,Mn)As layers are aligned antiparallel as shown in Fig. 12.4. The MR ratio was about 5.5% at $20\,$K, which is much higher than the MR effect due to the spin-dependent scattering, indicating that this is a tunneling MR effect.

12.4.2   Resonant-Tunneling Structures

When a semiconductor becomes ferromagnetic, spin splitting of the conduction as well as the valence bands is expected to occur due to s-d and p-d

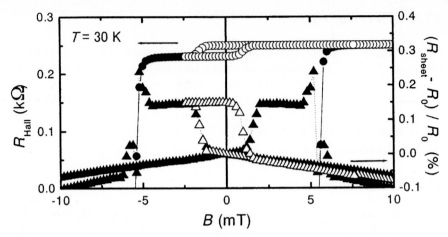

Fig. 12.3. The $(Ga_{0.95}Mn_{0.05})As/(Al_{0.14}Ga_{0.86})As/(Ga_{0.97}Mn_{0.03})As$ trilayer structure magnetotransport properties at $T = 30\,K$. Circles and triangles correspond to the Hall resistance and magnetoresistance ratio, respectively. Closed and open symbols show the major and minor loops, respectively. The minor loop of R_{Hall} is skewed by the presence of ferromagnetic coupling between the two $(Ga,Mn)As$ layers. The coupling strength J can be calculated by this shift

exchange interactions. We fabricated resonant-tunneling diodes (RTD) to investigate the spontaneous band splitting in $(Ga,Mn)As$. The T and B dependence of the current-voltage (I–V) characteristics of p-type AlAs/GaAs/AlAs double-barrier RTDs having $(Ga,Mn)As$ as the emitter material were studied. Spontaneous splitting of resonant peaks in the absence of B was observed upon lowering the temperature below T_C, which was interpreted as the spin splitting of the valence band of ferromagnetic $(Ga,Mn)As$ observed in tunneling spectra [29].

12.4.3 Electrical Spin Injection in Ferromagnetic Semiconductor Heterostructures

It has recently been demonstrated, by the use of a light-emitting diode structure, that spin injection is indeed possible using $(Ga,Mn)As$ [30]. Holes injected from the p-type $(Ga,Mn)As$ layer through an intrinsic GaAs layer into an $(In,Ga)As$ quantum well in p-i-n structures retain their spin imbalance information over distances greater than 200 nm and give rise to electroluminescence polarization in the absence of an external B. The demonstration of electrical spin injection in all semiconductor light-emitting structures is believed to open a variety of new possibilities.

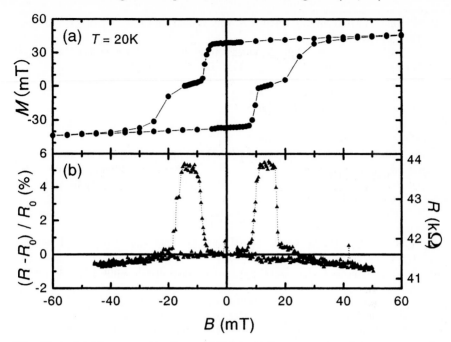

Fig. 12.4. (a) The magnetization and (b) tunneling magnetoresistance curve of a $(Ga_{0.95}Mn_{0.05})As/AlAs/(Ga_{0.97}Mn_{0.03})As$ tunnel junction at $20\,K$

12.5   Conclusion

This review covers the experimental and theoretical results on the III–V ferromagnetic semiconductor (Ga,Mn)As accumulated to date. Since ferromagnetic (Ga,Mn)As can readily be incorporated into the existing semiconductor heterostructure systems, it allows us to explore the physics and applications of previously unavailable combinations of quantum structures and ferromagnetism in semiconductors.

Acknowledgement

The work at RIEC, Tohoku University was supported by the "Research for the Future" Program (# JSPS-RFTF97P00202) from the Japan Society for the Promotion of Science and by Grant-in-Aid on Priority Area "Spin Controlled Semiconductor Nanostructures" (# 09244103) from the Ministry of Education, Japan.

References

1. H. Munekata, H. Ohno, S. von Molnar, A. Segmuller, L. L. Chang and L. Esaki: Phys. Rev. Lett. **63**, 1849 (1989)

2. H.Ohno, H.Munekata, T. Penny. S. von Molnar and L. L. Chang: Phys. Rev. Lett. **68**, 2664 (1992)
3. J. De Boeck, R. Ocstcrholt, A. Van Esch, H. Bender, C. Bruynscracde, C. Van Hoof and G. Borghs: Appl. Phys. Lett. **68**, 2744 (1996)
4. H. Ohno, A. Shen, F. Matsukura, A. Oiwa, A. Endo, S. Katsumoto and Y. Iye: Appl. Phys. Lett **69**, 363 (1996)
5. H. Ohno: Science **281**, 951 (1998)
6. T. Hayashi, M. Tanaka, T. Nishinaga, H. Shimada, H. Tsuchiya and Y. Otsuka: J. Cryst. Growth **175/176**, 1063 (1997)
7. Y. Nishikawa, Y. Satoh and J. Yoshino: Extended Abstracts of Second Symp. on Physics and Application of Spin-Related Phenomena in Semiconductors, H. Ohno, Y. Oka and J. Yoshino, eds., 27-28 Jan. 1997, Sendai, Japan. p. 122.
8. A. Van Esch, L. Van Bockstal, J. De Boeck, G. Verbanck, A. S. van Steenbergen, P. J. Wellmann, B. Grietens, R. Bogaerts, F. Herlach and G. Borghs: Phys. Rev. B **56**, 13103 (1997)
9. J. Sadowski, J. Domagala, J. Bak-Misiuk, M. Sqialek, J. Kanski, L. Ilver and H. Oscarsson: Acta Physica Polonica A **94**, 509 (1998)
10. D. DeSimone, C. E. C. Wood and C. A. Evans Jr.: J. Appl. Phys. **53**, 4938 (1982)
11. F. Matsukura, A. Shen, Y. Sugawara, T. Omiya, Y. Ohno and H. Ohno: *Proc. of 25th Int. Symp. Compound Semiconductors*, Institute of Physics Conf. Series, H. Sakaki, N. Yokoyama and Y. Hirayama eds., 12-16 Oct. 1998, Nara, Japan (IOP Publishing Ltd., Bristol, 1999) p. 547
12. H. Shimizu, T. Hayashi, T. Nishinaga, M. Tanaka: Appl. Phys. Lett., **74**, 1540 (1990)
13. A. Shen, H. Ohno, F. Matsukura, Y. Sugawara, N. Akiba, T. Kuroiwa, A. Oiwa, A. Endo, S. Katsumoto and Y. Iye: J. Cryst. Growth **175/176**, 1069 (1997)
14. A. Shen, H. Ohno, F. Matsukura, Y. Sugawara, Y. Ohno, N. Akiba and T. Kuroiwa: J. Jpn. Appl. Phys. **36**, L73 (1997)
15. R. Shioda, K. Ando, T. Hayashi and M. Tanaka, Phys. Rev. B **58**, 1100 (1998)
16. H. Ohno, F. Matsukura, A. Shen, Y. Sugawara, A. Oiwa, A. Endo, S. Katsumoto and Y. Iye: *Proc. of 23rd Int. Conf. Physics of Semiconductors*, M. Scheffler and R. Zimmermann eds., Berlin, 21-26 July, 1996 (World Scientific, Singapore, 1996) p. 405
17. F. Matsukura, H. Ohno, A. Shen and Y. Sugawara, Phys. Rev. B **57**, R2037 (1998)
18. Y. Satoh, N. Inoue, Y. Nishikawa and J. Yoshino: *Extended Abstracts of Third Symp. on Physics and Application of Spin-Related Phenomena in Semiconductors*, H. Ohno, Y. Oka and J. Yoshino, eds., Sendai, 17-18 Nov. 1997, p. 23
19. S. Katsumoto, A. Oiwa, Y. Iye, H. Ohno, F. Matsukura, A. Shen and Y. Sugawara: Phys. Status Solidi B **205**, 115 (1998)
20. T. Dietl: *Handbook on Semiconductors* Vol. **36**, T. S. Moss ed. (North-Holland, Amsterdam, 1994) pp. 1282-1287
21. J. Okabayashi, A. Kimura, O. Rader, T. Mizokawa, A. Fujimori, T. Hayashi and M. Tanaka: Phys. Rev. B **58**, R4211 (1998)
22. B. L. Altshuler and A G. Aronov: Zh. Eksp. Theo. Fiz. 38, 128 (1983): JETP Lett., **38**, 153 (1983)

23. T. Dietl, A. Haury and Y. MerledŘAubigne: Phys. Rev. B **55**, R3347 (1997)
24. T. Story, R. R. Galazka, R. B. Frankel and P. A. Wolff: Phys. Rev. Lett. **56**, 777 (1986)
25. T. Dietl: H. Ohno, F. Matsukura, J. Cibert and D. Ferrand: Science **287**, 1019 (2000)
26. D. Ferrand, J. Cibert, A. Wasiela, C. Bourgognon, S. Tatarenko, G. Fishman, S. Kolesnik, J. Jaroszynski, T. Dietl, B. Barbara and D. Dufeu: J. Appl. Phys. **87**, 6451 (2000)
27. H. Akai: Phys. Rev. Lett. **81**, 3002 (1998)
28. N. Akiba, D. Chiba, K. Nakata, F. Matsukura, Y. Ohno and H. Ohno: J. Appl. Phys. **87**, 6436 (2000)
29. H. Ohno, N. Akiba, F. Matsukura, A. Shen, K. Ohtani and Y. Ohno: Appl. Phys. Lett., **73**, 363 (1998)
30. Y. Ohno, D. K. Young, B. Beschoten, F. Matsukura, H. Ohno and D. D. Awschalom: Nature **402**, 790 (1999)

13   Transport Properties
of the Half-Filled Landau Level
in GaAs/AlGaAs Heterostructures:
Temperature Dependence
of Electrical Conductivity
and Magnetoresistance of Composite Fermions

R. Jahana, S. Kawaji, T. Okamoto, T. Fukase, T.Sakon and M. Motokawa

The fractional quantum Hall effect (FQHE) was discovered at the Landau level filling factor of $\nu = 1/3$ in two-dimensional electron systems of GaAs/AlGaAs heterostructures in 1982 by Tsui, Stormer and Gossard [1]. For FQHE at $\nu = 1/(2m + 1)$, where m is an integer, Laughlin proposed trial states describing highly correlated quantum liquids, which provide a good description of FQHE at these filling factors [2]. Later on, higher quality heterostructures have experimentally made it possible to observe FQHE at $\nu = n/(2mn + 1)$ for integer n. Jain proposed an elegant explanation for FQHE at the filling factor $\nu = n/(2mn + 1)$ by introducing new quasi-particles, called composite fermions (CFs) [3]. A CF is a composite particle comprising an electron and an even number, $2m$, of flux quanta $\phi_0 = h/e$. In this picture, the integer quantum Hall effect (IQHE) of the CFs with $2m$ flux quanta corresponds to FQHE at $\nu = n/(2mn+1)$ as follows. The inverse of the filling factor $\nu^{-1} = 2m + n^{-1}$ is given by the number of available flux quanta per electron. However, $2m$ flux quanta are already a part of the CFs. Therefore, in the CF picture, the number of available flux quanta, excluding the gauge flux which is a part of the CFs, is n^{-1} per CF. Thus the CF state at the filling factor n is equivalent, in a mean field sense, to the electron state at the filling factor $\nu = n/(2mn+1)$. Following a similar procedure, Halperin, Lee and Read [4] developed a theory for the state at exactly half filling. At $\nu = 1/2$, the CF system behaves like a system of fermions in the absence of a magnetic field, with a well-defined Fermi surface at wave vector $k_{\mathrm{F}} = 1/\ell_0$ with $\ell_0 = (\hbar/eB)^{1/2}$.

The quantum correction to the classical Drude conductivity of the CFs and the magnetic field effect on the quantum correction to the conductivity is a subject to be studied in comparison with the results developed in two-dimensional electron systems with disorder in the early 1980s [5]. In particular, it is interesting that the impurities cause not only the usual potential scattering in the transport of the CFs but also scattering through random fluctuations of the Chern-Simons gauge field. Halperin, Lee and Read [4] discussed the effect of the random gauge fields on the transport of the CFs. Kalmeyer and Zhang [6] argued that the localization induced by the potential

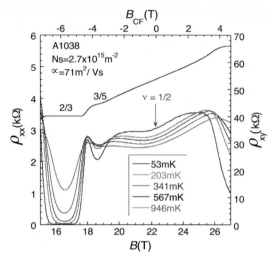

Fig. 13.1. Magnetic field dependence of ρ_{xy} at 50 mK and ρ_{xx} at four temperatures up to 946 mK in the A1038 sample

scattering is suppressed by the random gauge fields and the metallic state is realized at $\nu = 1/2$. Therefore, a significant difference may be observed between the CFs and normal fermions in the low-temperature behavior of their transport properties. Recently Nagaosa and Fukuyama formulated the quantum correction to the conductivity of the CFs at $\nu = 1/2$ and the magnetoresistance of the CFs around $\nu = 1/2$ [7].

We measured the diagonal resistance and Hall resistance of samples with Hall bar geometries made from three GaAs/AlGaAs heterostructure wafers. We obtained the conductivity of the CFs from the reciprocal of the resistivity of the samples in the magnetic field for $\nu = 1/2$. First, we compared the conductivity in the absence of a magnetic field to that of the CFs at the lowest temperature for each sample and obtained a good proportionality relation. Then we fitted the temperature dependence of the quantum correction term to the classical Drude conductivity of the CFs to the theory of Nagaosa and Fukuyama [7] and obtained the ratio between the relaxation time of electron scattering and that of the CFs and the coupling constant of the quantum correction. Magnetoresistance in the CFs was also studied based on Nagaosa and Fukuyama's theory and it was found that the experimental results were well explained by the theory.

13.1   Experiments

13.1.1   Samples and Measurement Procedures

We used three Hall bar samples prepared from GaAs/AlGaAs heterostructure wafers. The thickness of the spacer layer, $d_{\rm s}$, electron concentration, $N_{\rm s}$, and

Table 13.1. Characteristics of samples. N_s: electron concentration, μ: mobility, d_s: thickness of the spacer layer

Sample	$N_s(10^{15}\mathrm{m}^2)$	$\mu(\mathrm{m}^2/\mathrm{Vs})$	$d_s(\mathrm{nm})$	Laboratory
A1038	$2.4 - 3.2$	$61 - 75$	40	IMR
R115-b5	$1.3 - 1.8$	$101 - 155$	50	GU
R483-FET	$1.0 - 1.8$	$57 - 117$	40	GU

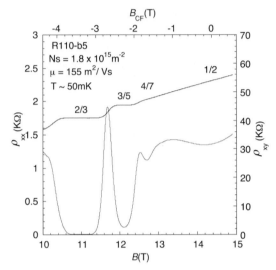

Fig. 13.2. Magnetic field dependence of ρ_{xy} and ρ_{xx} at $50\,\mathrm{mK}$ in the R110-b5 sample

electron mobility, μ, of the samples are summarized in the Table 13.1. The electron concentration was controlled by illumination with light by means of an LED or the application of a front-gate voltage. The length between the source- and drain-electrodes is $2100\,\mu\mathrm{m}$ in the A1038 sample and $600\,\mu\mathrm{m}$ in the R110-b5 and R138-FET samples. The diagonal resistance and Hall resistance were measured by passing a low frequency ($5.7-19.3\,\mathrm{Hz}$) ac current of $10\,\mathrm{nA}$ through the samples cooled by dilution refrigerators. Experiments on the A1038 sample in magnetic fields up to $27\,\mathrm{T}$ were made in the High Field Laboratory for Superconducting Materials in the Institute for Materials Research, Tohoku University. Experiments on the R110-b5 and R483-FET samples in fields up to $15.6\,\mathrm{T}$ were made at the Gakushuin University.

13.1.2   Conductivity of the CFs

Comparison Between the Conductivity of the CFs $\sigma^{\mathrm{CF}} = 1/\rho_{xx}(\nu = 1/2)$ and the Conductivity of the Electrons σ_0 in the Absence of the Magnetic Field. The magnetic field dependence of ρ_{xx} at different temperatures between $53\,\mathrm{mK}$ and $946\,\mathrm{mK}$ and the magnetic field dependence of ρ_{xy} at $50\,\mathrm{mK}$ are shown in Fig. 13.1. Here B_{CF} is the effective magnetic field on the CFs where $B_{\mathrm{CF}} = 0$ at $\nu = 1/2$. Note that the temperature dependence of ρ_{xx} near $\nu = 1/2$ is opposite to that at $\nu = 3/2$. Figure 13.2 shows the magnetic field dependence of ρ_{xx} and ρ_{xy} at $50\,\mathrm{mK}$ in the sample R110-b5.

We obtained the conductivity of the CFs simply defined by $\sigma^{\mathrm{CF}} = 1/\rho_{xx}$ ($\nu = 1/2$) in all samples. Here the contribution of $\rho_{xy}(\nu = 1/2)$ is ignored because the Hall voltage is induced due to the motion of magnetic flux in a magnetic field and is not related to the Hall resistivity of the CFs. We can obtain the conductivity of the CFs at absolute zero $\sigma^{\mathrm{CF}}_{T=0}$ by simple extrapolation of $\sigma^{\mathrm{CF}}(T)$ to $T = 0$. We obtained that $\sigma^{\mathrm{CF}}_{T=0}$ is roughly proportional to the conductivity of the two-dimensional fermions at low temperature in the absence of the magnetic field.

Temperature Dependence of the Conductivity σ^{CF}: Analysis Based on Nagaosa-Fukuyama Theory. The theoretical result for the conductivity of the CFs at $\nu = 1/2$ derived by Nagaosa and Fukuyama [7] is summarized as follows. The conductivity σ^{CF} is given by sum of the classical Drude conductivity σ^{CF}_0 and the quantum correction $\sigma^{\mathrm{CF}}_{\mathrm{I}}$ due to the CF-CF interaction as

$$\sigma^{\mathrm{CF}} = \sigma^{\mathrm{CF}}_0 + \sigma^{\mathrm{CF}}_{\mathrm{I}}, \tag{13.1}$$

where the interaction term is given by

$$\sigma^{\mathrm{CF}}_{\mathrm{I}} = -\frac{e^2}{2\pi^2\hbar}g^{\mathrm{CF}}[\Psi(1 + \frac{\hbar}{2\pi k_{\mathrm{B}}T\tau^{\mathrm{CF}}_0}) - \Psi(1)]. \tag{13.2}$$

Here $\Psi(x)$ is the Digamma function and the coupling constant g^{CF} is given by the contribution of the Chern-Simon gauge field as

$$g^{\mathrm{CF}} = \frac{A}{1-A}\ln A; \quad \text{with} \quad A = 8\left[\frac{\sigma^{\mathrm{CF}}_0}{e^2/h}\right]^2. \tag{13.3}$$

We can describe the interaction term

$$-\sigma^{\mathrm{CF}}_{\mathrm{I}}(T) = \sigma^{\mathrm{CF}}_0 - \sigma^{\mathrm{CF}}(T) \tag{13.4}$$

by introducing an effective mass ratio m^*/m^{CF} and the scattering time ratio $\tau^{\mathrm{CF}}_0/\tau_0$ as

$$-\sigma^{\mathrm{CF}}_{\mathrm{I}}(T) = \sigma_0(m^*/m^{\mathrm{CF}})(\tau^{\mathrm{CF}}_0/\tau_0) - \sigma^{\mathrm{CF}}(T)$$

$$= \frac{e^2}{2\pi^2\hbar}g^{\mathrm{CF}}[\Psi(1 + \frac{\hbar}{2\pi k_{\mathrm{B}}T\tau^{\mathrm{CF}}_0}) - \Psi(1)]. \tag{13.5}$$

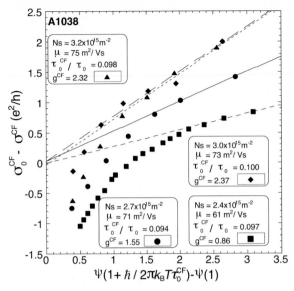

Fig. 13.3. Temperature dependence of the quantum correction to the Drude conductivity of the CFs in the A1038 sample

We can measure σ_0 and $\sigma^{CF}(T)$. Then, when we assume the effective mass ratio m^*/m^{CF}, we can determine the scattering time ratio τ_0^{CF}/τ_0 as an unknown parameter by fitting the temperature dependence data at low temperatures to the quantum correction given by (13.5), which passes through the origin of the plot at the high-temperature limit. The slope of this fit gives the coupling constant g^{CF} as given by (13.3).

In the following fit we use the effective mass of the CFs, $m^{CF} = 0.51m_e$, obtained by the temperature dependence of the Schubnikov-de Haas oscillations by Leadley et al. [8]. The results are shown in Fig. 13.3 for the A1038 sample.

The scattering times of the CFs in the absence of the effective magnetic field at low temperature τ_0^{CF} in all the samples are summarized in a plot versus the scattering time of the electrons in the absence of magnetic field, τ_0, in Fig. 13.4. The plot shows that τ_0^{CF} is proportional to τ_0 as approximately expressed by $\tau_0^{CF} = 0.11\tau_0$.

The scattering time of the CFs, τ_0^{CF}, determined by the temperature dependence of the quantum correction to the classical Drude conductivity at low temperature, extrapolated to the high-temperature limit using Nagaosa and Fukuyama's theory of the interaction effect of the CFs, can be used to calculate the conductivity of the CFs at $\nu = 1/2$, which can be compared to the conductivity of the CFs obtained by simple extrapolation of the low-temperature conductivity data to absolute zero $\sigma^{CF}(T = 0)$. The con-

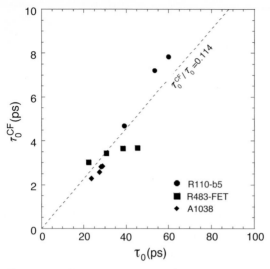

Fig. 13.4. Comparison between the scattering time of the CFs and that of the 2D fermions in various samples

ductivity calculated by $\sigma_0^{CF} = \sigma_0(m^*/m^{CF})(\tau_0^{CF}/\tau_0)$, with $m^{CF} = 0.51m_e$, when compared to σ_0, is in good agreement with the simple results.

Coupling Constant g^{CF} in the Interaction Term in the Conductivity. The coupling constant g^{CF} evaluated from the slope of the plot of the quantum correction to the conductivity versus the theoretically expected temperature dependence of the interaction term derived by Nagaosa and Fukuyama's theory is compared with the theoretically evaluated g^{CF} as a function of σ_0^{CF} given by (13.3). The coupling constants obtained experimentally are about 0.3 times the theoretical values. Both τ_0^{CF}/τ_0 and g^{CF} determined by fitting the theory to the experimental interaction term of the conductivity depend on the fitting parameter m^{CF}. The scattering time ratio τ_0^{CF}/τ_0 is approximately proportional to m^{CF}/m_e, but g^{CF} depends on this parameter very weakly.

13.1.3 Magnetoresistance of the CFs

Fluctuations of the gauge field are expected to cause a large magnetoresistance of the CFs from the effective magnetic field $B_{CF} = B - B_{\nu=1/2}$. Nagaosa and Fukuyama formulated the magnetoresistance as

$$\sigma^{CF}(B_{CF}, T) - \sigma^{CF}(T) = -\alpha(\frac{B_{CF}}{B_{\nu=1/2}})^2 , \tag{13.6}$$

$$\alpha = \sigma_0^{CF}\left[\frac{2(\sigma_0^{CF})^2}{k_F d_S}\right]^2 \quad \propto \quad (\sigma_0^{CF})^5 , \tag{13.7}$$

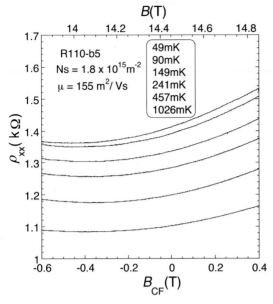

Fig. 13.5. Changes in the diagonal resistivity ρ_{xx} versus the effective magnetic field B_{CF} at several temperatures T measured in the R110-b5 sample

where k_{F} is the Fermi wave number of the CFs and d_{s} is the thickness of the spacer layer in the GaAs/AlGaAs heterostructure. Here the conductivities are given in unit of e^2/h. The conductivity of the CFs is calculated from the measured diagonal and Hall resistivities by

$$\sigma^{\mathrm{CF}}(B_{\mathrm{CF}}, T) = \frac{\rho_{xx}^{\mathrm{CF}}}{(\rho_{xx}^{\mathrm{CF}})^2 + (\rho_{xy}^{\mathrm{CF}})^2}, \tag{13.8}$$

where

$$\rho_{xx}^{\mathrm{CF}} = \rho_{xx}, \tag{13.9}$$

$$\rho_{xy}^{\mathrm{CF}} = \rho_{xy} - \rho_{xy}(\nu = 1/2). \tag{13.10}$$

Figure 13.5 shows the changes in the diagonal resistivity ρ_{xx} versus the effective magnetic field B_{CF} at several temperatures T measured in the R110-b5 sample. Changes in the magnetoconductance calculated by (13.8) at four temperatures between 49 mK and 1029 mK are shown in Fig. 13.6. The coefficient α in (13.6), which reproduces the experimental magnetoconductance curve, is shown as a function of the range of magnetic field in Fig. 13.7. Then we determine α by extrapolating the effective field B_{CF} to zero. The coefficient α obtained is about one tenth of the theoretical value calculated from (13.7).

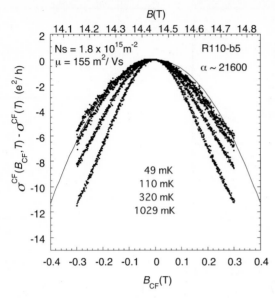

Fig. 13.6. Magnetoconductance of the CFs versus the effective magnetic field B_{CF} at several temperatures T measured in the R110-b5 sample

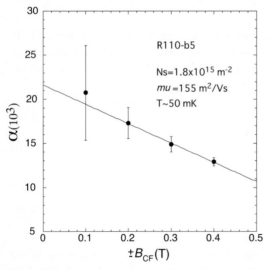

Fig. 13.7. The dependence of the coefficient α on the magnetic field range used in the curve fitting

13.2   Discussion

The experimental evidence for the CFs studied by the end of 1996 has been extensively discussed by Willett [9], including dc transport experiments. However, the dc transport measurements so far reported do not explain the logarithmic temperature dependence for the following two reasons. Measurements of the diagonal conductivity σ_{xx} using Corbino discs with a large ratio of the diameters of the inner- and outer-electrodes [10] are quantitatively insufficient to avoid the effects of the electric field dependence of the conductivity. When the mobility of the CFs is very high, the quantum correction to the Drude conductivity can be masked by the usual strongly temperature dependent scattering processes. Therefore, we have made careful measurements of the temperature-dependent dc transport and magnetoresistance of the CFs in various low mobility samples.

The scattering times of the CFs obtained by fitting the experimental temperature dependence of the electrical conductivity of the CFs at $\nu = 1/2$ to Nagaosa and Fukuyama's theory in three samples whose electron concentrations were controlled are proportional to the scattering times of fermions in the absence of the magnetic field, as shown in Fig. 13.4. In Fig. 13.4, the scattering time in the R483-FET sample does not show a clear linear relation. This fact probably shows that the change in the electron concentration produced by the application of a gate voltage is not homogeneous or that there may exist a small leakage current in the gate structure.

The analysis of the magnetoresistance shows that the coefficient α in (13.6) is one-tenth of the value calculated by (13.7). As shown in (13.7), the coefficient α depends strongly on the conductivity of the CFs, σ_0^{CF}. Therefore, we calculate the conductivity σ_0^{CF} from (13.7) for the experimentally obtained coefficient α, and compare the conductivity σ_0^{CF} with the conductivity σ^{CF} (experiment) $= \sigma_0(m^*/m^{CF})(\tau_0^{CF}/\tau_0)$, where $m^{CF} = 0.51m_e$ and τ_0^{CF}/τ_0 is the value obtained in Fig. 13.4. The results are shown in Fig. 13.8. The figure shows that a small error in the experimental conductivity of the CFs, σ_0^{CF} (experiment), leads to a large difference between the coefficient α calculated from (13.7) and the α experimentally determined from magnetoresistance measurements. Therefore, we conclude that the theory explains well our experimental magnetoresistance.

In conclusion, the experimental results observed for the temperature dependence of the conductivity of the two-dimensional electron systems at the half-filled Landau level and for their magnetoresistance versus small changes in the field are generally explained by Nagaosa and Fukuyama's theory based on the composite fermion picture with fluctuations of the gauge field.

13.3   Conclusion

The temperature dependence of the electrical conductivity of composite fermions at the Landau level filling factor $\nu = 1/2$ and the magnetoresistance of the composite fermions around $\nu = 1/2$ were measured for GaAs/AlGaAs heterostructures in magnetic fields up to 27 T. The experimental results are

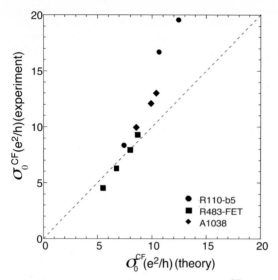

Fig. 13.8. Conductivity of the CFs, σ_0^{CF} (experiment), caluculated as $\sigma_0 \times (m^*/m^{\mathrm{CF}}) \times (\tau_0^{\mathrm{CF}}/\tau_0)$ versus the conductivity, σ_0^{CF} (theory), calculated from α given by (13.7)

compared with the theoretical results of Nagaosa and Fukuyama. The theory can generally explain the experimental results.

Acknowledgment

We are grateful to Professors H. Fukuyama and N. Nagaosa for their helpful discussions. We thank also Professors H. Sakaki and K. Hirakawa for providing us with the R110-b5 and R483-FET samples used in this work.

References

1. D.C. Tsui, H.L. Stormer and A.C. Gossard: Phys. Rev. Lett. **48**, 1559 (1982)
2. R.B. Laughlin: Phys. Rev. Lett. **50**, 1395 (1983)
3. J.K. Jain: Phys. Rev. Lett. **63**, 199 (1989)
4. B.I. Halperin, P.A. Lee and N. Read: Phys. Rev. B **47**, 7312 (1993)
5. S. Kawaji: Surface Sci. **299/300**, 563 (1994)
6. V. Kalmeyer and S.C. Zhang: Phys. Rev. B **46**, 9889 (1992)
7. N. Nagaosa and H. Fukuyama: J. Phys. Soc. Jpn. **67**, 3353 (1998)
8. D.R. Leadley, R.J. Nicholas and J.J. Harris: Phys. Rev. Lett. **72**, 1906 (1994)
9. R.L. Willett: Adv. Phys. **46**, 447 (1997)
10. R.P. Lokhinson, B. Su and V.J. Goldman: Phys. Rev. B **52**, R11588 (1995)

14   Novel Electronic States
in Low-Dimensional Organic Conductors

N. Toyota, T. Sasaki, T. Fukase, H. Yoshino and K. Murata

There has been a great interest in chemistry/physics joint research to create a conductor based on an organic molecule. After the breakthrough in synthesizing a TTF (tetrathiafulvalene) radical donor, its salt with a counter-anion TCNQ (tetracyanoquinodimethane) was found in the early 1970s to show quasi one-dimensional (Q1D) metallic conduction with high electrical conductivity as good as a conventional metal. During the last three decades there have been a number of important discoveries of novel phenomena resulting from low-dimensional electronic states. (There are available several review articles [1–4].)

Magnetic fields as well as pressure have routinely played an important role in probing and/or perturbing the electronic states and their physical properties. One of the first successful examples found in the early 1980s was the discovery of a so-called field-induced spin-density-wave state in TMTSF (tetramethyl-tetraselenafulvalene) salts followed by exotic high-field states which are still an extensively studied subject at high field facilities. In 1987, Shubnikov-de Haas oscillations were observed for the first time in quasi two-dimensional (Q2D) organic salts such as κ-(BEDT-TTF)$_2$Cu(NCS)$_2$, β-(BEDT-TTF)$_2$I$_3$, and β-(BEDT-TTF)$_2$IBr$_2$ [5–7]. These studies were soon followed by independent observations (in Sendai [8] and in Grenoble [9]) of the magnetic-breakdown phenomena in κ-(BEDT-TTF)$_2$Cu(NCS)$_2$, which is re-addressed here. This so-called Fermiology has greatly confirmed that these low-dimensional electronic states are well understood in terms of a rather simple tight-binding approximation based on Debye-Hückel molecular orbitals. Furthermore, it has led to discussions, based on these reliable electronic states, of a variety of electronic condensates such as low-dimensional superconductivity, metal-to-insulator transitions, charge- or spin-density-waves, band magnetism, and so on. At present, these high field studies are being carried out in parallel with low temperature and/or pressure experiments in water-cooled, hybrid and long-pulse magnets at major facilities all over the world.

The present article covers some of important results brought about by the groups at Tohoku University, Electrotechnical Laboratory, and Osaka City University, all of which have been intensive users of the High Magnetic Field Center at IMR. The subjects are:

1. the magnetic breakdown in κ-(BEDT-TTF)$_2$Cu(NCS)$_2$;

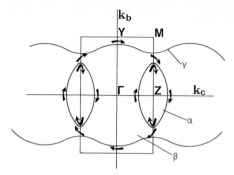

Fig. 14.1. The calculated Fermi surface in the k_b-k_c plane of κ-(BEDT-TTF)$_2$Cu(NCS)$_2$

2. the H_{c2} study of organic superconductor β-(BEDT-TTF)$_2$I$_3$ under pressure;
3. the magnetoresistance symmetry of the two-dimensional organic τ-conductors; and
4. the direct observation of reconstructed Fermi surfaces of (TMTSF)$_2$ClO$_4$ utilizing the third angular effect of magnetoresistance.

14.1   Magnetic Breakdown in κ-(BEDT-TTF)$_2$Cu(NCS)$_2$

Just after the discovery of the new class organic superconductor κ-(BEDT-TTF)$_2$Cu(NCS)$_2$ ($T_c = 10.4$ K) by Urayama et al. [10], the Fermi surface was studied by Oshima et al. [5]. Shubnikov-de Haas (SdH) oscillations were clearly observed in the magnetoresistance below 1 K and above 8 T. They observed the α orbit which has a SdH frequency of 600 T in a magnetic field perpendicular to the conducting b-c plane. The orientation dependence of the frequency assured the existence of a cylindrical Fermi surface, which is expected for a quasi two-dimensional conductor such as the present organic salt. The band structure calculations by the two-dimensional tight binding approximation based on the extended Hückel method predicted the existence of two bands at the Fermi level; one is a hole-like closed Fermi surface (α) centered at Z and the other a pair of electron-like open sheets (γ) along the Γ-Z direction (Fig. 14.1) The observed α orbit corresponds to the closed Fermi surface at Z point.

The calculated Fermi surface implies the possibility of a magnetic breakdown effect at the points on Z-M where the two bands come close to each other. The topology of the Fermi surface is the same as the textbook model for magnetic breakdown [11]. Such magnetic breakdown orbits of the present salt have been observed for the magnetoresistance (SdH effect) [8,9,12,13] and magnetization (de Haas – van Alphen (dHvA) effect) [14] in higher magnetic fields.

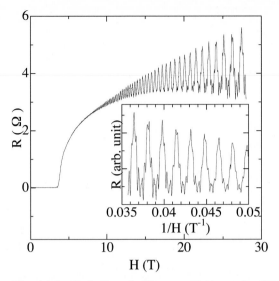

Fig. 14.2. Shubnikov-de Haas oscillations of κ-(BEDT-TTF)$_2$Cu(NCS)$_2$. SdH oscillations with a low frequency of 600 T appear in low magnetic fields and rapid oscillations at a frequency of 3800 T (magnetic breakdown oscillations) are superimposed on the slow oscillations in high fields. The inset shows the oscillatory part in high magnetic fields plotted versus the inverse of the magnetic field

Figure 14.2 shows SdH oscillations with a frequency of 3800 T in the high magnetic field region above 20 T which are superimposed on the slow SdH oscillations due to the closed α orbit. This high frequency just corresponds to the extremal cross sectional Fermi surface area of 3.63×10^{15} cm^{-2}, which is 100% of the first Brillouin zone in the k_b-k_c plane [8]. From the band structure calculations, no closed orbit corresponding to that area is expected. In a sufficiently strong magnetic field, carriers orbiting on the Fermi surface can, by the Lorentz force, tunnel through the potential barrier, traveling from one orbit to another. In the present case, a high-frequency β orbit could be realized by tunneling the gap four times in a circle. The inset (b) in Fig. 14.3 shows the schematic trajectory of the magnetic breakdown β orbit. The tunneling probability P at the gap is approximately described as $P = \exp(-AE_g^2/\hbar\omega_c E_F) \equiv \exp(-H_{MB}/H)$, where $A \simeq 1$, E_g, E_F and $\hbar\omega_c$ are the energy gap, Fermi energy and cyclotron energy, respectively. H_{MB} is the characteristic magnetic breakdown field. With the combination of the semiclassical tunneling probability and the coupled network model [15–17], the possible orbits are α, 2α, 3α, β, $\beta + \alpha$, $\beta + 2\alpha$, $2\beta - \alpha$, ...

Fourier transform spectra of the SdH oscillations in Fig. 14.3, however, demonstrate that not only the possible orbits from the network model but also the "forbidden orbits", the $\beta - \alpha$ "orbit", for example, which is shown in the inset (c) of Fig. 14.3, are discernible in the spectra. The "forbidden orbits"

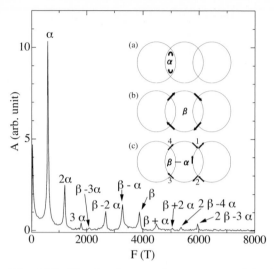

Fig. 14.3. The Fourier transform spectra of the SdH oscillations. The insets are the schematic figures of (**a**) the α orbit, (**b**) the magnetic breakdown β orbit and (**c**) the "forbidden $\beta - \alpha$ orbit" in the coupled network model

require a sudden reversal of the carrier orbiting on the path along its semiclassical trajectory. These "forbidden orbits" ($\beta - \alpha, \beta - 2\alpha, \beta - 3\alpha, 2\beta - 3\alpha, 2\beta - 4\alpha, \ldots$) are not explained in the semiclassical picture. The origin could be ascribed to the Chambers-Shiba-Fukuyama-Stark (CSFS) quantum interference effect[1] [18–20]. The "forbidden" frequency has been mostly observed in the magnetotransport measurements (SdH oscillations). This quantum interference effect occurs for the electrons passing through open orbits connecting two magnetic breakdown junctions. In the present $\beta - \alpha$ case, as shown in the inset (c) of Fig. 14.3, two orbits (2–3(MB)–4(MB)–1) and (2–1) interfere at the two junctions 1 and 2. Such interference provides the magneto-transport oscillations with the same frequency corresponding to the "$\beta - \alpha$ orbit" in the SdH oscillation.

It is well known that the CSFS quantum interference effect should not contribute to the free energy of the system and so the relevant frequencies should not be detected in the magnetization. However, the observation of such "forbidden orbits" in the dHvA effect of κ-(BEDT-TTF)$_2$Cu(NCS)$_2$ by Meyer et al. [14], suggests that the $\beta - \alpha$ frequency comes from a real oscillation of the thermodynamic functions of state such as the free energy. But the amplitude of the $\beta - \alpha$ frequency in the dHvA effect is rather weaker than that observed

---

[1] This quantum interference effect has been given the name "Stark quantum interference effect" in the literature. However, it should be named as in the text, because Chambers [18] and Shiba and Fukuyama [19] had independently predicted in their theoretical papers a few years before Stark and his co-worker [20] observed the effect in Mg.

in the SdH effect. Several attempts to explain the $\beta - \alpha$ frequency have been made. Machida, Kishigi and Hori [21] demonstrated the magnetization oscillation with the $\beta - \alpha$ frequency by full-quantum mechanical calculations with the condition of fixed electron number. After their work, Harrison et al. [13] and Nakano [22] obtained the the same results of the $\beta - \alpha$ frequency of the magnetization due to the chemical potential oscillation [24] using the density of state derived from the Pippard network model and the multi independent band model, respectively. Full-quantum calculations by Kishigi [23] confirmed that the $\beta - \alpha$ frequency in the magnetization appears under the condition of fixed electron number (i.e. chemical potential oscillation) and does not appear when the chemical potential is fixed. Therefore, the origin of the "forbidden" frequencies in the dHvA effect (thermodynamic phenomenon) is the chemical potential oscillations. The enhancement of the amplitude of the "forbidden" frequency in SdH effect (transport phenomenon) may be due to the combination of the CSFS interference effect and the chemical potential oscillations.

14.2 H_{c2} Study of Organic Superconductor β-(BEDT-TTF)$_2$I$_3$ Under Pressure

Prior to 1985, the superconducting transition temperature, T_c, of known organic superconductors stayed around 1 K, irrespective of whether they were one- or two-dimensional conductors, except for the 2 K of (BEDT-TTF)$_4$(ReO$_4$)$_2$ [25] and (TMTSF)$_2$FeO$_3$ [26]. It has been argued whether the superconductivity is s-, p- or d- like. Meanwhile, there was the discovery of a T_c increase from around 1–8 K in β-(BEDT-TTF)$_2$I$_3$ [27,28] by application of a pressure of 1.3 kbar (0.13 GPa) early in 1985. Since this material was known to have a T_c of 1 K at ambient pressure, one of the possibilities was the existence of two kinds of superconductivity next to each other in the thermodynamic phase diagram. If that was the case, it was direct proof of non s-paired superconductivity, since s-like superconductivity does not have internal degrees of freedom. Similar interests in superconductivity have been discussed for the heavy-Fermion system of Th-doped UBe$_{13}$[29]. It was made clear later that the apparent two-T_cs of β-(BEDT-TTF)$_2$I$_3$ originated from a structural transition that takes place at 175 K and the pressure boundary for the high T_c of 8 K was 0.4 kbar instead of 1.3 kbar. However, it was still open whether the superconductivity is conventional or unconventional.

The type of the superconductivity can be examined in various ways, e.g. by studying the temperature dependence of T_1 in nuclear magnetic resonance experiments, the temperature dependence of the London penetration depth, the gap anisotropy, and the temperature dependence of the upper critical field, H_{c2}. For this purpose, the study of H_{c2} was planned in the high field facility at Tohoku University.

What was known for β-(BEDT-TTF)$_2$I$_3$ was that the Fermi surface and, hence, the electronic properties of β-(BEDT-TTF)$_2$I$_3$, is two-dimensional (2-D). Therefore, H_{c2} takes its maximum with the magnetic field parallel to the 2D plane. Since T_c of β-(BEDT-TTF)$_2$I$_3$ was known to vary (decrease) with pressure monotonically, T_c was adjustable below 8 K with pressure. With the benefit of this property, the temperature dependence of H_{c2} as a function of T_c was obtained before using the high field facility of Tohoku University. Since, our interest was in finding the maximum values, the 2D plane of the samples was always aligned perpendicular to the magnetic field by using a rotating magnetic field at the Electrotechnical Laboratory in Tsukuba. When T_c was below 4 K, H_{c2} varied linearly with T. If this feature continued for $T_c > 4$ K, it would be expected that H_{c2} at low temperature would exceed the Clogston-Pauli limit ($H_{c2} = 1.82\,T_c$, in units of tesla and kelvin, respectively). With these data at hand, it was planned to study H_{c2} for $T_c > 4$ K in the high field facility at Tohoku University.

The experiment was quite challenging in many respects:(a) since the sample was inevitably tilted in the high-pressure bomb, the pressure bomb made of BeCu was rotated in the high field to compensate the tilt of the sample in the bomb; (b) remote control was designed for the regulation of the facility; and (c) the measurement in the sub-microvolt range had never been done there before. To achieve (a) and (b), a stepping motor was adopted, which was located 1.5 m above the center of the magnetic field. The length of the electrical leads for measurement and control of the stepping motors was about 50 m. By overcoming these technical problems, the data acquisition was successfully achieved. It turned out that H_{c2}(T) was very saturated by lowering the temperature when T_c was around 6 K. The saturation value of H_{c2} corresponded exactly to the Clogston-Pauli limit. The saturation tendency was never observed when $T_c < 4$ K. This result strongly suggests that the superconductivity was of s-type paired origin.

Subsequent H_{c2} studies have produced the following results. In the 2D organic superconductor (BEDT-TTF)$_4$Hg$_{2.89}$Br$_8$ there appeared the phenomenon that H_{c2} greatly exceeded the Clogston-Pauli limit [30]. In (TMTSF)$_2$ PF$_6$, a 1D organic superconductor, an extremely high H_{c2} was produced by adjusting the accurate field angle with respect to the sample axis, suggesting a p-type superconductor [31].

14.3   Magnetoresistance Symmetry of Two-Dimensional Organic τ-Conductors

The quasi-two-dimensional (Q2D) organic conducting crystal τ-(EDO-S,S-DMEDT-TTF)$_2$(AuBr$_2$)(AuBr$_2$)$_y$, where $y \sim 0.75$, consists of an unsymmetrical organic donor, EDO-S, S-DMEDT-TTF (ethylenedioxy-S,S-dimethyl-ethylenedithio-tetrathiafulvalene, as shown in Fig. 14.4 and an inorganic linear anion, AuBr$_2^-$. The crystal structure is tetragonal, with space group I4$_1$22,

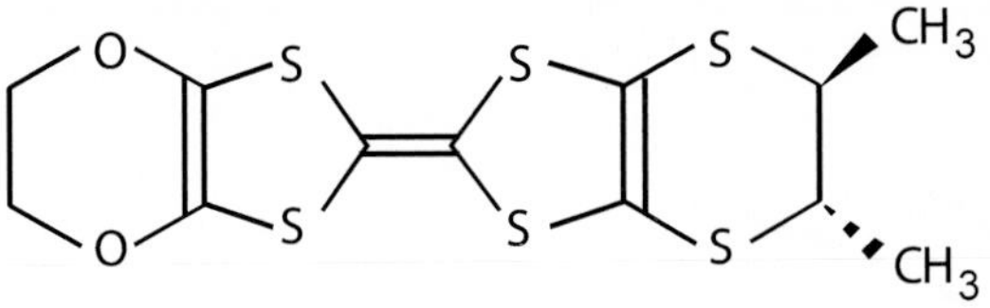

EDO-*S, S*-DMEDT-TTF

Fig. 14.4. Molecular structure of EDO-*S, S*-DMEDT-TTF

and the packing of the donor molecules within the conducting layers ($// ab$) has fourfold symmetry [32]. Each conducting layer is made of a 2:1 mixed packing of donors and the linear anions are sandwiched between insulating layers of the same linear anions whose locations are randomly and partially occupied. Then, with $y \cong 0.75$, a cross-shaped 2D Fermi surface is realized [32]. This is consistent with the metallic electrical resistivity [32].

Murata et al. [33,34], reported the change in periodicity of the angular dependence of the c-axis magnetoresistance (MR), $\rho_c(\phi)$, for a magnetic field within the $a - b$ plane by varying the magnitude of the magnetic field and/or temperature for the AuBr$_2$ salt [34]. Namely, $\rho_c(\phi)$ has the period of $180°$ at 1.1 K and 1.5 T, while the period is $90°$ at higher temperature and/or higher magnetic field, as is expected from the fourfold symmetry of the crystal structure. If the lowering of the symmetry is accompanied by a phase transition (i.e. the metal-insulator (M-I) transition at about 50 K), it is difficult to observe the twofold symmetry by measuring macroscopic properties because, in general, the lowering of the crystal symmetry produces a mixed-domain structure.

Therefore, in this work $\rho_c(\phi)$ was studied to clarify whether there is a twofold symmetry phase by measuring the MR. The MR below 5 T was studied at Osaka City University (OCU), and above 5 T the 23- and 27-T hybrid magnets were used at the Institute for Materials Research (IMR) at Tohoku University. The crystals were obtained by the standard electrochemical oxidation method described elsewhere [32]. The crystals were black square platelets and they were thinnest along the most resistive c-axis. The size of the crystals was $0.88 \times 0.63 \times 0.038$ mm^2 and $0.75 \times 0.75 \times 0.18$ mm^3 for samples 1 and 2, respectively. The c-axis MR was measured using a dc four-probe method. Two pairs of annealed Au wires (10 μm in diameter) were attached, using Au paste, to the opposite crystal surfaces of the ab plane. The magnetic field was applied and rotated within the ab plane. Thus, the magnetic field was kept perpendicular to the current path and the angular dependence of the MR is considered to represent the anisotropy of the electronic structure within the ab plane.

Figure 14.5 shows the temperature dependence of the c-axis electrical resistivity, ρ_c of the two samples. The magnitude of σ_c ($= 1/\rho_c$) is about $(3.1–3.9) \times 10^{-3}$ S·cm^{-1} at 300 K. The temperature dependence of ρ_c is metallic

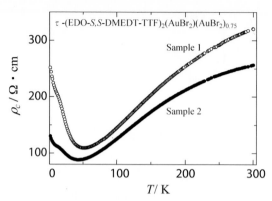

Fig. 14.5. The temperature dependence of the c-axis resistivity of τ-(EDO-S, S-DMEDT-TTF)$_2$(AuBr$_2$)(AuBr$_2$)$_y$, ($y \sim 0.75$, samples 1 and 2)

and turns to semiconducting at about 50 K, which is reproducible among samples. The overall temperature dependence of ρ_c in Fig. 14.5 is similar to that of ρ_{ab} in [32]. Thus, ρ_c is considered to be governed by the motion of carriers within the most conducting ab plane in spite of the strong 2D character of the electronic system.

The MR measurement at and below 5.0 T was carried out at 1.3 K in the semiconducting regime, as shown in Fig. 14.6. The normalized angular dependence of the MR, $\Delta\rho_c(\phi)/\rho_{c,B=0}$ ($\rho_{c,B=0} = 2.5 \times 10^2$ Ω·cm) at 5.0 T is shown in Fig. 14.6a and that at 0.5 T in Fig. 14.6b, respectively. These data are always in the negative magnetoresistance region.

At 5.0 T, $\Delta\rho_c(\phi)/\rho_{c,B=0}$ shows no hysteresis behavior and takes maxima at 0° and 90° periodically, which seems to be consistent with the crystal symmetry within the ab plane at room temperature. The curve #1 in Fig. 14.6b shows the virgin data obtained before the field sweep above 0.5 T. The curves #2-4 in Fig. 14.6b were obtained after the field sweep from 5.0 to 0.5 T at 0°, 90° and 45°, respectively. The curves #2 and #3 have a period of 180°, though the phase between them is 90°. The curve #4 is almost flat and very similar to #1. No distinct hysteresis for the angular sweep was observed for the curves in Fig. 14.6.

The 180° periodicity of the curves #2 and #3 in Fig. 14.6b implies that the electronic structure in the magnetic phase has twofold symmetry around the c-axis, while the crystal structure at room temperature has fourfold symmetry. In general, a single crystal of a material becomes multidomain if the crystal symmetry is broken when a phase transition occurs and a number of orientations of the low symmetry domains may appear. A similar situation in AuBr$_2$ salt is expected, since at least two orientations are possible for the lower symmetry domain. The hysteresis behavior of the MR suggests that the phase transition is likely and that, further, the lower symmetry domains are magnetic.

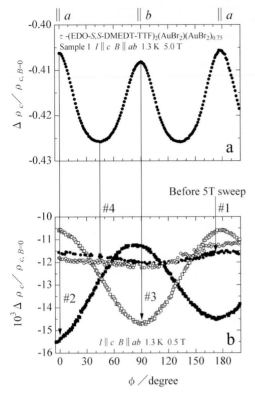

Fig. 14.6. The angular dependence of the magnetoresistance of τ-(EDO-S,S-DMEDT-TTF)$_2$(AuBr$_2$)(AuBr$_2$)$_y$, ($y \sim 0.75$, sample 1) at 1.3 K for I // c and B // ab at 5.0 T (**a**) and at 0.5 T (**b**) before the field sweep above 0.5 T (#1) and after the field sweep from 5.0 T at 0° (#2), 90° (#3) and 45° (#4), respectively. The magnetoresistance is normalized by the resistivity (2.5×10^2 $\Omega \cdot$ cm) at $B = 0$ T

When the crystal of AuBr$_2$ salt is cooled below the possible phase transition temperature without any magnetic field, the multidomain structure is considered to appear. The MR measured for the multidomain sample with low enough field is expected to show no angular dependence as is observed for curve #1 in Fig. 14.6b, implying that the domain walls stay unchanged. On the other hand, the 180° period of curves #2 and #3 suggests that the magnetic domains, each of which has twofold symmetry, are aligned along the a- and b-axes, respectively, by applying the magnetic field of 5.0 T. The 90° period of $\Delta \rho_c(\phi)/\rho_{c,B=0}$ at 5.0 T in Fig. 14.6a is evidence that the magnetic field of 5.0 T is strong enough to align the magnetic domains along the crystallographically equivalent directions $+a$, $+b$, $-a$ or $-b$. Once the domains are aligned along these directions, they cannot be realigned by a low magnetic field such as 0.5 T. In this sense, the 90° symmetry at 5 T at low temperature is apparent and 180° symmetry is always realized, but the 5 T field forces the domains to follow its direction. This explains the 180° symmetry of the curves #2 and #3 without hysteresis behavior and suggests that the a- and b-axes are equivalently the easy axes for the magnetization within the ab plane. If this is the case, the same amount of the two kinds of domains, whose spontaneous magnetization is parallel to the a- and b-axes, respectively, are considered to exist in the sample after the magnetic field of 5.0 T is applied along the $a+b$ direction (45° in Fig. 14.6b). This multidomain structure for #4 is artificially achieved, while that for #1 naturally appears. Both multidomain structures give almost the same flat angular dependence of the MR, because two kinds of MR components whose phases are shifted by 90° from each other are considered to contribute equally.

Figure 14.7 shows $\Delta\rho_c(\phi)/\rho_{c,B=0}$ ($\rho_{c,B=0} = 2.6 \times 10^2$ Ω·cm) of the AuBr$_2$ salt above 5 T measured for sample 1 at 0.5 K at the Institute for Material Research. The 90° period at 5.0 T and 1.3 K is still observed. The negative MR is almost independent of the magnitude of the magnetic field at 0° (// a) and 90° (// b), while, at angles off the crystal axes, the MR continues to decrease with increasing magnetic field. This is clearly shown as the field dependence of the MR in Fig. 14.8 obtained for sample 1, but during another run, at the Institute for Material Research. From 0 T, the MR rapidly decreases with increasing magnetic field up to about 5 T at 0° (// a) and 45° (// $a + b$), respectively. Above 5 T, the MR is almost independent of the magnetic field at 0°, while it gradually decreases at 45°. Finally, above 25 T the MR shows a slight upturn both at 0° and 45°.

With decreasing magnetic field, the MR shows hysteresis behavior below 5 T as in the inset in Fig. 14.6. However, the zero-field value of ρ_c is recovered

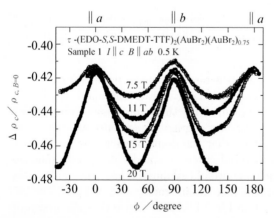

Fig. 14.7. The angular dependence of the magnetoresistance of τ-(EDO-S,S-DMEDT-TTF)$_2$(AuBr$_2$)(AuBr$_2$2)$_y$, ($y \sim 0.75$, sample 1) for I // c and B // ab at 0.5 K and at 7.5 (open circles), 11 (closed circles), 15 (open squares) and 20 T (closed squares), respectively. The magnetoresistance is normalized by the resistivity (3.8 × 10^2 Ω·cm) at $B = 0$ T

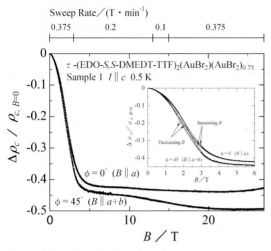

Fig. 14.8. The field dependence of the magnetoresistance of τ-(EDO-S, S-DMEDT-TTF)$_2$(AuBr$_2$)(AuBr$_2$)$_y$, ($y \sim 0.75$, sample 1) for $I \parallel c$ at 0.5 K with increasing and decreasing magnetic field. The direction of the magnetic field is $0°$ ($// a$) and $45°$ ($// a+b$), respectively. The inset shows the same data on an enlarged scale to display the hysteresis clearly. The magnetoresistance is normalized by the resistivity (3.8×10^2 $\Omega\cdot$cm) at $B = 0$ T

when the field is zero, which is reproducible for all samples. This is not instrumental since it is observed at IMR with the sweep rate of 0.375 T/min and also at OCU at 0.167 T/min. The domain model cannot explain the hysteresis behavior in the field sweep.

Our present idea about the low-temperature state for AuBr$_2$ salt is that it is weakly ferromagnetic. The negative MR is probably due to the aligning of thermally fluctuating spins in the strong magnetic field as is often observed for ferromagnets. It is worth noting, however, that two kinds of disorder, concerning the position of a vacancy at the anion site and the orientation of the ethylenedioxy group of the donor molecule, are recognized at room temperature for the AuBr$_2$ salt [32]. Though the relation between these disorders and the negative MR as well as the semiconducting behavior of the resistivity is not clear, Papavassiliou et al. [35], pointed out the possibility of weak localization in the AuBr$_2$ salt based on the similarity of the field dependence of the MR ($B // c$ and $I // ab$), which has negative slope at least up to 15 T at 2.1 K, to those of other organic compounds showing negative MR.

In summary, we studied the electronic states at two temperatures of the quasi-two-dimensional (Q2D) organic conducting crystal τ-(EDO-S,S-DMEDT-TTF)$_2$(AuBr$_2$)(AuBr$_2$)$_y$, ($y \sim 0.75$). Combining the low-field work below 5 T at Osaka City University and the high-field work up to 27 T at the Institute for Materials Research at Tohoku University, we could demonstrate the presence of a specific field 5 T at which the magnetoresistance behavior changes. Also, depending on the angle of the field, negative magnetoresistance does or does not saturate. But the hysteresis behavior upon field sweeping remains unsolved.

14.4   Direct Observation of Reconstructed Fermi Surfaces of (TMTSF)$_2$ClO$_4$ Utilizing the Third Angular Effect of Magnetoresistance

The first ambient-pressure organic conductor, (TMTSF)$_2$ClO$_4$ (TMTSF = tetramethyltetraselenafulvalene, $T_c = 1.2$ K) [36], is known to exhibit a cascade of field-induced spin-density-waves (FI-SDW) accompanied by the quantum Hall effect [37], rapid oscillation of magnetoresistance, etc. The variety of electronic properties is, in part, the consequence of the quasi-one-dimensional (Q1D) Fermi surface with a quarter-filled band, which is associated with the anisotropic tight-binding band parameters ($t_a : t_b : t_c = 300 : 30 : 1 - 0.1$) through the interaction between the TMTSF molecules. However, the information on the band parameters has not been sufficient for describing the various electronic properties mentioned above, so the unified understanding of the temperature-field phase diagram between (TMTSF)$_2$ClO$_4$ and (TMTSF)$_2$PF$_6$ has not been established yet. Furthermore, in (TMTSF)$_2$ClO$_4$, it is known that the non-centrosymmetric anion ClO$_4^-$, which takes two orientations, becomes ordered at $(0, 1/2, 0)$ in the k-representation below 24 K by slow cooling (dT/d$t < 15$ K/h) [38]. The slowly cooled state, or the so-called relaxed state (R-state), generates a band folding at $(0, 1/2, 0)$, which makes two pairs of 1 D Fermi surfaces. Although this band folding is well known, no direct demonstration of the reconstructed Fermi surface has been achieved.

Besides the estimation of the band parameters t_a, t_b and t_c, based on the crystallographic lattice parameters, Yoshino et al. [39], proposed a more direct method to estimate the ratio of the transfer integrals, t_a/t_b, from experiment by "the third angular effect" (TAE) of magnetoresistance, which is already widely accepted. The TAE is an anomalous hump that is observed in the angular dependence of ρ_{c^*}, with the magnetic field parallel to the conducting ab plane. This magnetoresistance exhibits a broad hump around $B//a$ (or b) and (mostly sharp) dips on both sides of the hump for the field rotation within the most conducting ab planes of the metallic quasi-one-dimensional (Q1D) DMET [39,40] and TMTSF salts [41–43]. The angular positions (ϕ_c) of the dips and the width of the hump anomaly ($\Delta\phi$) are independent of the magnitude of the magnetic field, indicating that the TAE is a geometric phenomenon, where $\Delta\phi$ is defined as the angle between the two local minima (dips) at $\phi_c \sim \pm(10 - 20°)$ of the magnetoresistance. Recent studies [41–46] have revealed the basic understanding of the TAE and that the phenomenon is related to the carrier motion on the Q1D Fermi surface and ϕ_c is close to the inflection angle of the Q1D Fermi surface. Although the detailed mechanism is presented elsewhere [46], an important consequence for TAE is that the experimentally obtained $\Delta\phi$ is almost proportional to the ratio of t_b/t_a, with the proportionality factor being the ratio of the lattice constants [39]. It should be noted that the dip structure becomes much clearer either by lowering the temperature or increasing the magnetic field.

From the systematic phase diagram of the TMTSF salts, $(TMTSF)_2PF_6$ is thought to be more one-dimensional than $(TMTSF)_2ClO_4$, from which a larger $\Delta\phi$ was expected for $(TMTSF)_2ClO_4$. Contrary to this expectation, for $(TMTSF)_2PF_6$ $\Delta\phi$ is 38° at 8.5 kbar [42,43] and for $(TMTSF)_2ClO_4$ $\Delta\phi$ is roughly 20° [41]. For comparison with other Q1D organic conductors, for $(DMET)_2I_3$ $\Delta\phi$ is 28° [40] and for $(DMET)_2AuBr_2$ it is 27° [39]. The exceptionally small $\Delta\phi$ in $(TMTSF)_2ClO_4$ would have been explained in terms of the band folding mentioned above, but there had been no experimental evidence. The $\Delta\phi$ of $(TMTSF)_2ClO_4$ strongly suggests the existence of two kinds of Q1D Fermi surfaces because the reconstruction of the Brillouin zone due to the AO (anion ordering) makes the Q1D Fermi surface flatter than that without the anion potential [39]. Probably previous reports with just one hump at 1.7 K and up to 12 T by Osada et al. [41], were the result of not using a large enough field or low enough temperature to resolve the structure of the two pairs of Fermi surface sheets. The experiment in the high field facility of the Institute for Materials Research at Tohoku University was motivated to demonstrate directly the two sheets of Fermi surface of $(TMTSF)_2ClO_4$.

The experiment was performed at $T = 0.5$ K and in fields up to 20 T. The samples were slowly cooled to realize the R-state. The effect of the folding of the Fermi surface sheet in $(TMTSF)_2ClO_4$ was *first directly observed* in this work. Moreover, with this result, we could estimate the strength of anion ordering potential.

Figure 14.9a shows the angular dependence of the magnetoresistance of $(TMTSF)_2ClO_4$ (#1, R-state) at $T = 0.5$ K and fields up to 20 T. The TAE anomaly is appreciable above 7.5 T between $\phi_c = \pm 12°$. The $\Delta\phi$, whose definition is shown in Fig. 14.9a, is 24° and 20% larger than that reported by Osada et al. [41]. This is probably due to the slight misalignment of the sample to the magnetic field in [41], though one cannot discard the possibility of the rather rapid cooling of the sample across TAO in the present study. The much smaller $\Delta\phi$ for $(TMTSF)_2ClO_4$ (24°) in the R-state than that for a similar system of $(TMTSF)_2PF_6$ (38° at 8.5 kbar) [42,43] is due to the AO in the former. Thus, a more incomplete AO may give a wider $\Delta\phi$.

The result for the other sample (#2) is shown in Fig. 14.9b. Although the TAE is observed at 7.5 T, the magnetoresistance curve for #2 is apparently different from that for #1. The oscillatory structure recognized between $-20°$ and $20°$ above 11 T in Fig. 14.9b is probably due to the off-angle effect reported for $(TMTSF)_2PF_6$ under pressure [42,43] and caused by the slight misalignment of the sample #2 to the magnetic field. On the other hand, the lack of oscillation in the data in Fig. 14.9a guarantees that the magnetic filed was kept parallel to the ab plane of #1.

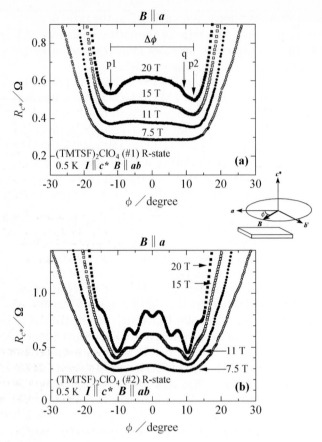

Fig. 14.9. The angular dependence of the c^*-axis magnetoresistance of $(TMTSF)_2ClO_4$ at $0.5\,K$ in the R-state for samples (**a**) #1 and (**b**) #2. The magnetic field was rotated within the ab plane. The inset between (**a**) and (**b**) shows the definition of ϕ with respect to the crystal axes. The third angular effect anomaly with width $\Delta\phi$ is recognized in (**a**). The oscillations seen in (**b**) are probably due to the off-angle effect of the magnetic field from the ab plane

In Fig. 14.9a, a step is recognized at $9°$, as is indicated by the arrow q, in addition to the TAE minima (p1 and p2) at $\phi_c = \pm 12°$. In the previous paper [39], a simple method to estimate the anisotropy of Q1D metals by utilizing the TAE was reported and extended to estimate the anion gap, E_g of $(TMTSF)_2ClO_4$ in the R-state. The dispersion relations (14.1) from perturbation theory are used:

$$E = -2t_z \cos(k_x x) \pm \sqrt{4t_y^2 \cos^2(k_y y) + E_g^2}, \qquad (14.1)$$

where the small dimerization of the TMTSF molecules along the a-axis is ignored [47].

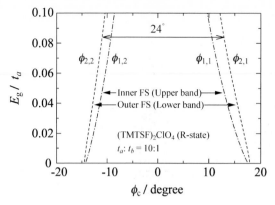

Fig. 14.10. The dependence of the angular positions ($\phi_{1,1}$, $\phi_{1,2}$, $\phi_{2,1}$ and $\phi_{2,2}$) of the third angular effect anomaly on the anion gap (E_g) determined from the inflection angle of the Fermi surface by using (14.1) assuming $t_x/t_y = 10$. The broken and chain curves are for the outer and inner Fermi surfaces, respectively

Two kinds of Q1D Fermi surface of $(TMTSF)_2ClO_4$ give different values of $\Delta\phi$ in general, namely, the inner Fermi surface from the lower band gives a narrower $\Delta\phi$. Thus, the anomalies p1 and p2 in Fig. 14.9a are probably from the outer Fermi surface and q is from the inner one.

Following [39], one can calculate ϕ_c as the inflection angle of the Q1D Fermi surface for various combinations of the parameters t_x/t_y and E_g. The E_g-dependence of the four inflection angles for the inner ($\phi_{1,1}$, $\phi_{1,2}$) and outer ($\phi_{2,1}$ and $\phi_{2,2}$) Fermi surfaces, respectively, are obtained by assuming $t_x/t_y = 10$, as shown in Fig. 14.10. From the present results, $\phi_{2,1} - \phi_{2,2} = \Delta\phi = 24°$, E_g is estimated to be $0.083t_x$ (25 meV for $t_x = 300$ meV). It is noteworthy that $E_g = 25$ meV is the same order of that determined from the amplitude of the rapid oscillation of the magnetoresistance by Uji et al. [48,49] They obtained $E_g \sim 5$ meV under the assumption that $E_F = 100$ meV, where the latter is about one-fourth of the present energy scale. Figure 14.11 shows the first Brillouin zone and Fermi surfaces of $(TMTSF)_2ClO_4$ determined experimentally by using (14.1) and the lattice parameters at 1.7 K [50], where $x = a/2 = 3.534°$, $y = b = 7.638°$ and $g = 68.92°$. The solid and broken lines/curves in Fig. 14.11 are for the R- and Q-states, respectively.

The large E_g, however, suggests that (14.1) based on perturbation theory is insufficient to describe the electronic state of $(TMTSF)_2ClO_4$ in the R-state. Although we have attempted to estimate E_g in the present study, an improved treatment should be developed to understand the physical properties of $(TMTSF)_2ClO_4$. Anyway, it must be stressed that $E_g = 0.083t_x$ is rather large and comparable to $t_y = t_x/10$. One can also introduce the effect of the AO by the splitting of t_y into t_{y1} and t_{y2} ($t_{y1} \neq t_{y2}$) as a weak dimer-

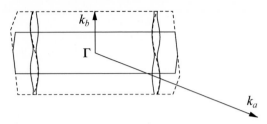

Fig. 14.11. The first Brillouin zone and Fermi surfaces of $(TMTSF)_2ClO_4$ in the R-(solid lines/curves) and Q- (broken ones) states respectively. The lattice constants are from [50]. The anisotropy of the transfer integral (t_x/t_y) is assumed to be 10 and the anion gap, E_g is determined to be $0.083t_x$ experimentally

ization of the TMTSF molecules along the $b(y)$-axis. Then the dispersion relations

$$E = -2t_z \cos(k_x x) \pm \sqrt{t_{y1}^2 + t_{y2}^2 + 2t_{y1}t_{y2} \cos(2k_y y)} \qquad (14.2)$$

are also used to describe the R-state of $(TMTSF)_2ClO_4$. It is found that anomalously large dimerization $(t_{y1}/t_{y2} = 1.9)$ is needed to get $\Delta\phi = 24°$ by assuming $(t_{y1} + t_{y2})/2 = t_x/10$. The results of fitting by using (14.1) and (14.2) show that the effect of the AO in reconstructing the band structure is enormous and the Fermi surfaces of $(TMTSF)_2ClO_4$ are much flatter than expected from the anisotropy of $t_x/t_y = 10$.

14.5   Conclusion

This article presents selected topics which have come out of high magnetic field studies of the novel electronic states in quasi-one- or two-dimensional organic conductors, including:

1. magnetic breakdown in κ-$(BEDT\text{-}TTF)_2Cu(NCS)_2$;
2. H_{c2} study of the organic superconductor β-$(BEDT\text{-}TTF)_2I_3$ under pressure;
3. magnetoresistance symmetry of the two-dimensional organic τ-conductors; and
4. direct observation of reconstructed Fermi surfaces of $(TMTSF)_2ClO_4$ utilizing the third angular effect of magnetoresistance.

Acknowledgment

The authors would like to thank the former directors of the High Magnetic Field Center, Emeritus Professors Y. Muto and Y. Nakagawa, as well as the present director Professor M. Motokawa. The academic and technical staffs are also greatly appreciated. The authors (N.T., T.S., T.F.) of Tohoku University appreciate a Grant-in-Aid for Scientific Research from the Ministry of Education, Science and Culture of Japan. The works headed by the authors (K.M., H.Y) were carried out as a part of the "Research for the Future Project" supported by the Japan Society for the Promotion of Science and as a part of the JISEDAI Project from the Agency of Industrial Science and Technology and the Ministry of International Trade and Industry.

References

1. J.M. Williams, J.R. Ferraro, R.J. Thorn, K.D. Carlson, U. Geiser, H.H. Wang, A.M. Kini and M.-H. Whangbo: *Organic Superconductors: Synthesis, Structure, Properties and Theory* (Prentice Hall, Englewood Cliffs, 1992)
2. J. Wosnitza: *Fermi Surfaces of Low-Dimensional Organic Metals and Superconductors* (Springer-Verlag, Berlin,1996)
3. T. Ishiguro, K. Yamaji and G. Saito: *Organic Superconductors* (Springer-Verlag, 1998)
4. M. Lang: Superconductivity Review **2**, 1 (1996)
5. K. Oshima, T. Mori, H. Inokuchi, H. Urayama, H. Yamochi and G. Saito: Phys. Rev. B **38**, 938 (1988)
6. K. Murata, N. Toyota, Y. Honda, T. Sasaki, M. Tokumoto, H. Bando, H. Anzai, Y. Muto and T. Ishiguro: J. Phys. Soc. Jpn. **57**, 1540 (1988)
7. N. Toyota, T. Sasaki, K. Murata, Y. Honda, M. Tokumoto, H. Bando, H. Anzai, T. Ishiguro and Y. Muto: J. Phys. Soc. Jpn. **57**, 2616 (1988)
8. T. Sasaki, H. Sato and N. Toyota: Solid State Commun. **76**, 507 (1990)
9. C.-P. Heidmann, H. Müller, W. Biberacher, K. Meimaier, C. Probst and K. Andres: Synth. Metals **41–43**, 2029 (1991)
10. H. Urayama, H. Yamochi, G. Saito, K. Nozawa, T. Sugano, M. Kinoshita, S. Sato, K. Oshima, A. Kawamoto and J. Tanaka: Chem. Lett. 55 (1988)
11. J.M. Ziman: *Principles of the Theory of Solids*, 2nd ed. (Cambridge University Press 1972) p. 327, Fig. 174
12. J. Caulfield, J. Singleton, F.L. Pratt, M. Doporto, W. Lubczynski, W. Hayes, M. Kurmoo, P. Day, P.T.J. Hendrics and J.A.A.J. Perenboom: Synth. Metals **61**, 63 (1993)
13. N. Harrison, J. Caulfield, J. Singleton, P.H.P. Reinders, F. Herlach, W. Hayes, M. Kurmoo and P. Day: J. Phys. Condens. Matter **8**, 5415 (1996)
14. F.A. Meyer, E. Steep, W. Biberacher, P. Christ, A. Lerf, A.G.M. Jansen, W. Joss, P. Wyder and K. Andres: Europhys. Lett. **32**, 681 (1995)
15. T. Sasaki, H. Sato and N. Toyota: Physica C **185–189**, 2687 (1991)
16. T. Sasaki and N. Toyota: Synth. Metals **55–57**, 2303 (1993)
17. L.M. Falicov and H. Stachowiak: Phys. Rev. **147**, 505 (1966)
18. W.G. Chambers: Phys. Rev. **165**, 799 (1968)
19. H. Shiba and H. Fukuyama: J. Phys. Soc. Jpn. **26**, 910 (1969)
20. R.W. Stark and R. Reifenberger: Phys. Rev. Lett. **26**, 556 (1971)
21. K. Machida, K. Kishigi and Y. Hori: Phys. Rev. B **51**, 8946 (1995)
22. M. Nakano: J. Phys. Soc. Jpn. **66**, 19 (1997)
23. K. Kishigi: J. Phys. Soc. Jpn. **66**, 910 (1997)
24. A.S. Alexandrov and A.M. Bratkovsky: Phys. Rev. Lett. **76**, 1308 (1996)
25. S.S.P. Parkin, E.M. Engler, R.R. Schumaker, R. Lagier, V.Y. Lee, J.C. Scott and R.L. Greene: Phys. Rev. Lett., **50**, 270 (1983)

26. R.C. Lacoe, S.A. Wolf, P.M. Chaikin, F. Wudl and E. Aharon-Shalom: Phys. Rev. B **27**, 1947 (1983)
27. K. Murata, M. Tokumoto, H. Anzai, H. Bando, G. Saito, K. Kajimura and T. Ishiguro: J. Phys. Soc. Jpn., **54**, 1236 and 2084 (1985)
28. V.N. Laukhin, E.E. Kostyuchenko, Y.V. Sushko, I.F. Shchegolev and E.B. Yagubskii: JETP Lett. **41**, 81 (1985)
29. For a recent review about UBe13, see "Electron Correlations and Materials Properties", ed. by A. Gonic, N. Kioussis and M. Ciftan (Kluwer Academic/Plenum Publishers, ISBN 0-306-46282-6, New York, 1999)
30. R.N. Lyubovskaya, R.B. Lyubovskii, M.K. Madoba and S.I. Pesotskii: JETP Lett. **51**, 361 (1990)
31. I.J. Lee, M.J. Naughton, G.M. Danner and P.M. Chaikin: Phys. Rev. Lett. **78**, 3555 (1997)
32. G.C. Papavassiliou, D.J. Lagouvardos, J.S. Zambounis, A. Terzis, C.P. Raptopoulou, K. Murata, N. Shirakawa, L. Ducasse and P. Delhaes: Mol. Cryst. Liq. Cryst. **285**, 83 (1996)
33. K. Murata, N. Shirakawa, H. Yoshino and Y. Tsubaki: Synth. Met. **86**, 2021 (1997)
34. K. Murata, H. Yoshino, Y. Tsubaki and G.C. Papavassiliou: Synth. Met. **94/1**, 69–72 (1998)
35. G.C. Papavassiliou, D.J. Lagouvardos, I. Koutselas, K. Murata, A. Graja, I. Olejniczak, J.S. Zambounis, L. Ducasse and J.P. Ulmet: Synth. Met. **86**, 2043 (1997)
36. K. Bachgaard, K. Carneiro, M. Olsen, F.B. Rasmussen and C.S. Jacobsen: Phys. Rev. Lett. **46**, 852 (1981)
37. T. Ishiguro, K. Yamaji and G. Saito: Organic Superconductors, 2nd ed. (Springer-Varlag, Berlin, 1998) Chap. 9
38. J.P. Pouget, G. Shirane, K. Bechgaard and J.M. Fabre: Phys. Rev. B **27**, 5203 (1983)
39. H. Yoshino, K. Saito, H. Nishikawa, K. Kikuchi, K. Kobayashi and I. Ikemoto: J. Phys. Soc. Jpn. **66**, 2410 (1997)
40. H. Yoshino, K. Saito, K. Kikuchi, H. Nishikawa, K. Kobayashi and I. Ikemoto: J. Phys. Soc. Jpn. **64**, 2307 (1995)
41. T. Osada, S. Kagoshima and N. Miura: Phys. Rev. Lett. **77**, 5361 (1996)
42. M.J. Naughton, I.J. Lee, P.M. Chaikin and G.M. Danner: Synth. Met. **85**, 1481 (1997)
43. I.J. Lee and M.J. Naughton: Phys. Rev. B **57**, 7423 (1998)
44. A.G. Lebed and N.N. Bagmet: Synth. Met. **85**, 1493 (1997)
45. A.G. Lebed and N.N. Bagmet: Phys. Rev. B **55**, R8654 (1997)
46. H. Yoshino and K. Murata: J. Phys. Soc. Jpn. **68**, 3027 (1999)
47. P.M. Grant: J. Phys. (Paris) **44**, C3–847 (1983)
48. S. Uji, T. Terashima, H. Aoki, J.S. Brooks and M. Tokumoto: Phys. Rev. B **53**, 14399 (1996)
49. S. Uji, J.S. Brooks, S. Takasaki, J. Yamada and H. Anzai: Solid State Commun. **103**, 387 (1997)
50. B. Gallois, D. Chasseau, J. Gaultier, C. Hauw, A. Filhol and G. Bechgaard: J. Phys. (Paris) **44**, C3–1071 (1983)

15 High-Field Successive Phase Transitions of Spin-Density-Wave Organic Conductors α-(BEDT-TTF)$_2$MHg(XCN)$_4$ [M = K, Rb and NH$_4$ and X = S and Se]

T. Sasaki, N. Toyota and T. Fukase

15.1 Introduction

15.1.1 The α-(BEDT-TTF)$_2$MHg(XCN)$_4$ Family

An isostructural family of organic conductors α-(BEDT-TTF)$_2$MHg(XCN)$_4$, where BEDT-TTF is bis(ethylenedithio)tetrathiafulvalene, has been the subject of intensive study owing to the variety of the ground states [1]; metals (X = Se and M = K [2] or Tl [3]), superconductors [4] (X = S and M = NH$_4$) and antiferromagnetic metals [5] (X = S and M = K, Tl or Rb). The crystal structure of α-(BEDT-TTF)$_2$KHg(SCN)$_4$ is shown in Fig. 15.1. Other members have the same structure with M and X atoms substituted. The thick anion layer (MHg(XCN)$_4$) and the so-called α packing pattern of the BEDT-TTF molecules (Fig. 15.1b) are characteristic of this family.

15.1.2 α-(BEDT-TTF)$_2$KHg(SCN)$_4$: A Novel Density-Wave Metal

Within the family, much attention has been focused on the antiferromagnetic metals which have a unique antiferromagnetic state (AF phase) below T_A = 8–10 K in zero magnetic field and magnetic field-induced successive phase transitions [6]. In Fig. 15.2, the temperature dependence of the zero-field resistance, the Hall coefficient and the spin susceptibility of α-(BEDT-TTF)$_2$KHg(SCN)$_4$ are shown. The phase transition is associated with a shoulder-type anomaly in the resistance. The behavior of the Hall coefficient and the static spin susceptibility below T_A suggests that a gap opens at the Fermi level. These behaviors and the results of muon spin resonance (μSR) [10] are indicative of antiferromagnetic ordering due to Fermi surface nesting. Hence, we have proposed that the transition from the Pauli paramagnetic phase (P phase) to the AF phase at T_A is a spin-density-wave (SDW) transition [10,11].

As expected from the discussion above, the topology of the Fermi surface (FS) (Fig. 15.3) obtained by the band structure calculations [2,12], which consists of a pair of open-planar and closed-cylindrical (α-orbit) parts, is likely to possess a nesting instability for a periodic modulation such as a SDW or charge-density-wave (CDW).

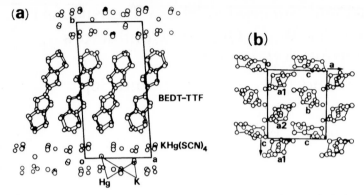

Fig. 15.1. (a) Crystal structure of α-(BEDT-TTF)$_2$KHg(SCN)$_4$ projected along the c direction. **(b)** The arrangement of BEDT-TTF molecules in the unit cell viewed along the molecular long axis

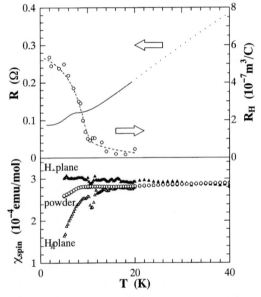

Fig. 15.2. Temperature dependence of the zero-field resistance, the Hall coefficient and the spin susceptibility of α-(BEDT-TTF)$_2$KHg(SCN)$_4$

The reconstructed FS model has been proposed as a result of detailed studies of angle-dependent magnetoresistance oscillations (ADMRO) [13–16]. The Brillouin zone is reconstructed by the new nesting periodicity below T_A [13] and then the cylindrical α-orbits are also reconstructed to the multicon-nected orbit in the extended Brillouin zone. Many experiments on Shubnikov–de Haas (SdH) and de Haas–van Alphen (dHvA) effects also support the re-

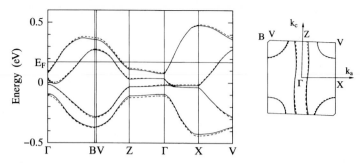

Fig. 15.3. Band structure calculations of α-(BEDT-TTF)$_2$KHg(XCN)$_4$ on the basis of the crystal structure at room temperature. The solid and broken curves correspond to X = Se and S, respectively

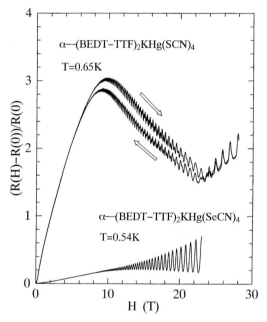

Fig. 15.4. Magnetoresistance of α-(BEDT-TTF)$_2$KHg(XCN)$_4$ with X = S and Se in a magnetic field perpendicular to the conducting plane

construction model [13,17]. The main oscillation frequency of 670 T in the AF phase, which is the same frequency as the α-orbit expected in the P phase, originates in a magnetic breakdown (MB) orbit on the multiconnected α-orbit caused by the nesting periodicity [13].

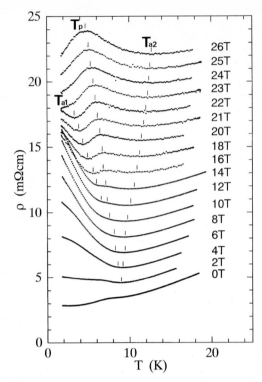

Fig. 15.5. Temperature dependence of the resistivity of α-(BEDT-TTF)$_2$KHg(SCN)$_4$ in fields perpendicular to the conducting plane. The curves are shifted by $1\,\text{m}\Omega$ cm from each other starting with the data at $H = 0\,\text{T}$

15.1.3   Magnetic Phase Diagram of α-(BEDT-TTF)$_2$KHg(SCN)$_4$

At a high magnetic fields, ca. 23 T, in the case of α-(BEDT-TTF)$_2$KHg(SCN)$_4$, a sharp kink-structure of the magnetoresistance (MR) appears [18]. Figure 15.4 shows the magnetoresistance of α-(BEDT-TTF)$_2$KHg(XCN)$_4$ with X = S and Se in a magnetic field perpendicular to the conducting plane. Several remarkable features of the MR (including a kink at 23 T and a negative slope with hysteresis) are seen in the curve of α-(BEDT-TTF)$_2$KHg(SCN)$_4$, but not in the normal metal α-(BEDT-TTF)$_2$KHg(SeCN)$_4$. The kink-structure has been considered to be a phase transition between the AF and P phases. That is, the P phase is restored from the AF phase by applying high magnetic fields [19]. Recent experiments in α-(BEDT-TTF)$_2$KHg(SCN)$_4$, however, indicate that the high-magnetic field phase above 23 T is an unknown magnetic metal phase (M* phase) which is different from either the AF or the P phase [20,21].

Figure 15.5 shows the temperature dependence of the resistivity in a magnetic field normal to the conducting plane. A shoulder-type anomaly is observed around $T_A = 8\,\text{K}$ for $H = 0\,\text{T}$. With magnetic fields applied, the magnetoresistance increases rapidly below the temperature $T_{a1}(H)$ which corresponds to the kink transition and has been so far interpreted as an onset temperature for a density-wave phase. The temperature T_{a2}, defined by the shallow minima, shifts to higher temperatures with increasing fields. Below T_{a2}, a broad peak appears at T_p. These MR structures clearly indicate that the high magnetic field phase (M* phase) above 23 T is not a simple metal phase like the P phase and α-(BEDT-TTF)$_2$KHg(SeCN)$_4$. In the M* phase, the magnetic torque shows hysteresis with sweeping magnetic field and temperature [22,23]. From the viewpoint of the FS topology, ADMRO measurements show that the FS in the M* phase is different from that in the AF phase, and similar to that in the P phase [16], while SdH and dHvA oscillations have the same main frequency of 670 T in both the AF and M* phases [18,20]. However, it is noted that the magnitude of higher harmonics of the SdH and dHvA oscillations in the M* phase are rather reduced in comparison with that in the AF phase [24,25]. The change of the effective mass m_c and/or the g-value in the AF and the M* phases has been proposed to explain the change of the higher harmonics content. In addition, there are controversies about m_c and the Dingle temperature, T_D, obtained by SdH and dHvA experiments [26]. The effective mass in the AF phase has been reported to be smaller than that in the M* phase so far. Recently, it was pointed out that the effective mass in the AF phase might be underestimated [26] because of the omission of an additional temperature-dependent factor to the standard Lifshitz–Kosevich (LK) formulation [27]. The MB gap in the AF phase, which should have a temperature dependence, is suggested to be the possible origin. In the same way, a reconsideration of the Dingle temperature is required [20].

As described above, the three different phases, the AF, M* and P phases, are recognized as the antiferromagnetic (SDW) metal at low-temperature and low-magnetic field ($T < T_A, H < 23$ T), the unknown magnetic metal ($T < T_A, H > 23$ T) and the paramagnetic metal ($T > T_A$) phases, respectively, in α-(BEDT-TTF)$_2$KHg(SCN)$_4$. The magnetic phase diagram is shown in Fig. 15.6. The microscopic electronic state and parameters such as the effective mass, the g-value and so on in each phase have not been well known yet because of the complex phase transitions.

15.1.4   Subjects of this Review

As described in the above sections, the magnetic phase diagram of a novel density-wave organic conductor α-(BEDT-TTF)$_2$KHg(SCN)$_4$ has been constructed. However, the physical properties of each phase and the origin of the phase transition are problems now under active discussion. In the following sections, our recent studies [28] on the spin-splitting-zero phenomena in the de Haas–van Alphen effect are reviewed in connection with the magnetic phase diagram proposed so far.

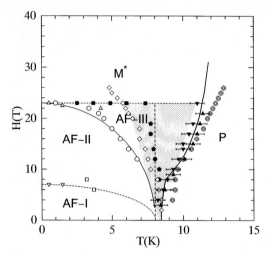

Fig. 15.6. The magnetic phase diagram of α-(BEDT-TTF)₂KHg(SCN)₄. The meaning of the marks are as follows: the open circles and triangles signify T_{a1}; the double circles, T_{a2}; the open diamonds, T_p; and the filled triangles and its reverse, T_{a2}, as determined by torque measurements

15.2   Experiment and Analysis

15.2.1   Samples and Measurements

Single crystals of α-(BEDT-TTF)₂MHg(XCN)₄ (M = K or NH₄, and X = S or Se) were grown by the electrochemical oxidation method [2,20]. The crystals, having a typical size of a few mm²× ∼ 0.3 mm, were grown on a platinum anode with a constant current of 0.5 μA. The well-developed facet is the crystallographic a–c plane.

The magnetic torque was measured with a capacitive cantilever-beam torque meter. The moving electrode and the ground plate were surrounded by a small metallic case used as a capacitive guard. This unit could be rotated smoothly in the magnetic field and at low temperatures down to about 0.5 K. A single crystal was fixed on the moving plate with a small amount of grease. The capacitance was measured by using a decade capacitance bridge (General Radio 1615A) and a lock-in amplifier (PAR 124A) or an auto-capacitance bridge (AH2500A).

Magnetic torque and magnetoresistance measurements were carried out using the 30 T hybrid magnet at the High Field Laboratory, IMR, Tohoku University. The temperature in the magnetic field was measured by using a Cernox thermometer (Lake Shore Cryotronics, Inc.), which was calibrated with a capacitance thermometer.

15.2.2   De Haas–van Alphen Oscillations of the Magnetic Torque

Magnetic torque per unit volume is given by the expression $\tau = M \times H$, where the magnetization is related to the susceptibility tensor $\hat{\chi}$ by $M = \hat{\chi} H$. The torque in the a–b plane is described by

$$\tau = \frac{1}{2} H^2 (\chi_{aa} - \chi_{bb}) \sin 2\theta, \tag{15.1}$$

where χ_{aa} and χ_{bb} are the diagonal elements of $\hat{\chi}$ and θ is the angle between the magnetic field and the b-axis. Here, the off-diagonal elements are neglected and the rectangular coordinate axes are used for simplicity. The magnetic oscillations of the torque appear through the oscillation of χ_{bb} in the present case. The first harmonic of the oscillation part of the torque based on the semiclassical Lifshitz–Kosevich (LK) theory [27] is given by

$$\tau_{osc} = -\frac{1}{F} \frac{dF}{d\theta} M_{osc} H, \tag{15.2}$$

$$M_{osc} \propto TFH^{-1/2} \frac{\exp[-\lambda(m_c/m_0)T_D/H]}{\sinh[\lambda(m_c/m_0)T/H]}$$
$$\times \cos[\pi g(m_c/m_0)/2] \sin[2\pi(F/H - 1/2) \pm \pi/4], \tag{15.3}$$

where F is the oscillation frequency, m_c and m_0 are the cyclotron effective mass and the free electron mass, $\lambda \equiv 2\pi^2 m_0 c k_B / e\hbar = 14.69\,\mathrm{T/K}$, c is the light velocity, k_B the Boltzmann constant, $\hbar$ the Planck constant, g the electron g-value, and T_D the Dingle temperature. The extremal cross-sectional area A of the FS in the plane normal to the applied magnetic field H is obtained from the relation $F = (c\hbar/2\pi e)A$. In a way similar to the magnetic oscillations with the period of $1/H$, it is noted that the oscillation with a period of $1/\cos\theta$ in constant magnetic field is superimposed on the $\sin 2\theta$ background torque curve in the case of the two-dimensional (2D) FS with $F(\theta) = F(0°)/\cos\theta$. This kind of the oscillation has amplitude nodes at the magnetic field directions where the spin-splitting-zero (SSZ) conditions, $g(m_c/m_0) = 2n + 1(n = 0, 1, 2, \ldots)$, are satisfied. In the case of the 2D-FS, the SSZs appear with a period of $1/\cos\theta$ because the angle dependence of the effective mass is expected to be $m_c(\theta) = m_c(0°)/\cos\theta$.

15.3   Spin-Splitting Phenomena in dHvA Oscillations

15.3.1 α-(BEDT-TTF)$_2$KHg(SeCN)$_4$: Normal Metal

Figure 15.7 shows the magnetic torque curves in a small single crystal (0.52 mg) of α-(BEDT-TTF)$_2$KHg(SeCN)$_4$ at $T = 0.52\,\mathrm{K}$. The background torque curves show the $\sin 2\theta$ behavior, which is indicated by the broken curves. The inset figure demonstrates the H^2-dependence of the torque amplitude of the $\sin 2\theta$ curves, which are described by (15.1). Clear dHvA oscillations are superimposed on the background torque curves, and are periodic in $(\cos\theta)^{-1}$.

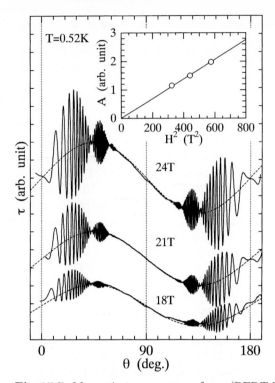

Fig. 15.7. Magnetic torque curves for α-(BEDT-TTF)$_2$KHg(SeCN)$_4$. The broken curves show the $\sin 2\theta$ dependence resulting from the anisotropy of the static susceptibility. The inset demonstrates the H^2 dependence of the torque amplitude

Table 15.1. The $g(m_c/m_0)$ value obtained by the SSZ method in α-(BEDT-TTF)$_2$A, with A = KHg(SeCN)$_4$ and NH$_4$Hg(SCN)$_4$

α-(BEDT-TTF)$_2$A	$g(m_c/m_0)$	m_c/m_0	g
A=KHg(SeCN)$_4$	3.63	1.9	1.91
A=NH$_4$Hg(SCN)$_4$	4.48	2.5	1.80

The periodicity in $(\cos\theta)^{-1}$ results from the cylindrical FS with cross sectional area corresponding to $F(0°) = 670$ T. This observation is consistent with band structure calculations [2] and is in good agreement with previous SdH [2] and dHvA [29] measurements. Nodes of the oscillation amplitude are seen at $\theta_{SSZ} = 43°, 58°, \ldots, 122°, 137°$. At these angles, the SSZ conditions mentioned above are satisfied. The SSZ angles do not change with magnetic field. This is quite reasonable because the gm_c value does not depend on the magnitude of the magnetic field in general.

The linear dependence of $(\cos\theta_{SSZ})^{-1}$ on $2n + 1$, where n is an integer, represents the two-dimensional angle dependence of the effective mass with

$m_c(\theta) = m_c(0°)/\cos\theta$. Here we assume that the g value does not change much with the angle. The values of $g(m_c/m_0)$, m_c/m_0 were obtained from the temperature dependence of the dHvA oscillation amplitude [see (15.3)], and the resultant g values are summarized in Table 15.1 with the results for the organic superconductor α-(BEDT-TTF)$_2$NH$_4$Hg(SCN)$_4$. These values are in good agreement with previous reports [29,30].

15.3.2 α-(BEDT-TTF)$_2$KHg(SCN)$_4$: AF Metal

Figure 15.8 shows the magnetic torque curves in a single crystal of α-(BEDT-TTF)$_2$KHg(SCN)$_4$. The torque curve at 24 T and in the angle range between $0°$ ($120°$) and about $60°$ ($180°$) is considered to be measured in the M* phase from previous studies on the angle dependence of the magnetic phase diagram [9,31–33]. By the same consideration, other curves below 20 T are measured in the AF phase. The curve at 22 T is expected to be in the boundary region between the AF and the M* phases. Characteristic features which are not observed in the P phase of the normal metal α-(BEDT-TTF)$_2$KHg(SeCN)$_4$ are seen on the torque curves both in the AF and the M* phases. First, slow oscillatory structures are observed in $\pm \sim 30°$ region centered at $\theta = 90°$ which are not observed in α-(BEDT-TTF)$_2$KHg(SeCN)$_4$ and α-(BEDT-TTF)$_2$NH$_4$Hg(SCN)$_4$. The angles at which the structures appear shift with magnetic field and do not follow possible trial trigonometric functions such as $(\cos\theta)^{-1}$, $\tan\theta$ and so on. Both features imply that neither ADMRO nor small 2D-FS are the possible origin of the structure. We shall return to this point later. Secondly, the SSZ angles are different in the AF and the M* phases.

Figure 15.9 shows the integer plot of the SSZ angles in several magnetic fields. The different slopes of the two straight lines fitted to the data for 24 T and lower magnetic fields demonstrate the change of the SSZ periodicity at the phase boundary between the AF and the M* phases. The $g(m_c/m_0)$ values obtained from the slopes are 3.63 in the M* phase and 4.7 in the AF phase. These values are summarized in Table 15.2. In the AF phase, SSZ corresponding to $2n+1 = 5$ could be expected to appear around $(\cos\theta)^{-1} \sim 1$, that is, $\theta \sim 0°$ or $180°$. But the present torque experiments do not distinguish the node of the oscillation amplitude around the expected angle because only a few oscillations are observed along these directions. The splitting-wave form, however, has been observed in SdH and dHvA oscillations [25] in a magnetic field perpendicular to the plane ($\theta \sim 0°$). This observation also suggests the reliability of SSZ which occurs at $\theta \sim 0°$ because the enhancement of the higher harmonics in the oscillations takes place at the SSZ angle in general.

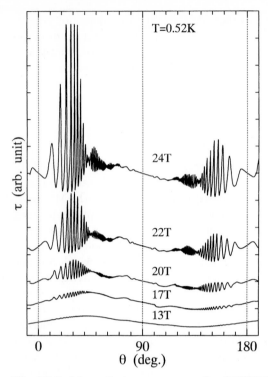

Fig. 15.8. Magnetic torque curves of α-(BEDT-TTF)$_2$KHg(SCN)$_4$. The curve at 24 T is measured in the M* phase and other curves in the AF phase

Table 15.2. The $g(m_c/m_0)$ value obtained by the SSZ method in α-(BEDT-TTF)$_2$KHg(SCN)$_4$ in the AF and the M* phases

	$g(m_c/m_0)$	m_c/m_0	g
M* phase	3.63	1.86	1.95
AF phase	4.48	see discussions in text	

15.4   Effective Mass and g-Factor in the Magnetic Phases

We shall discuss the gm_c value in the AF and the P phases of α-(BEDT-TTF)$_2$KHg(SCN)$_4$ in comparison with the value in the P phase of α-(BEDT-TTF)$_2$KHg(SeCN)$_4$. The first point to note is that the same $g(m_c/m_0)$ value of 3.63 is obtained in the M* phase of α-(BEDT-TTF)$_2$KHg(SCN)$_4$ and the P phase of α-(BEDT-TTF)$_2$KHg(SeCN)$_4$. Almost the same value of the effective mass, about 1.9m_0 in both phases, has been obtained on the basis of the LK-analysis in the temperature dependence of the SdH and dHvA oscillation amplitude [2,20,29]. The g values in both phases are estimated to be about 1.9

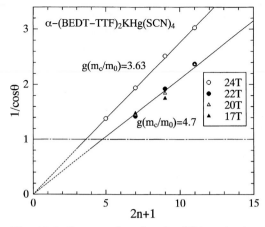

Fig. 15.9. Integer plots for the SSZ angles in α-(BEDT-TTF)$_2$KHg(SCN)$_4$. The slope of the straight lines, corresponding to the $g(m_c/m_0)$ value, are different in the AF (17, 20, 22 T) and the M* phases (24 T)

from the $g(m_c/m_0)$ value in the present results and the reported m_c values. It is worth noting that the gm_c values obtained with independent methods, the SSZ method and the fitting to the LK-formulation are consistent with each other. This suggests that no additional temperature-dependent factor in the dHvA oscillation amplitude needs to be taken into account in these phases, while such an additional factor may become important for the MB orbits in the AF phase of α-(BEDT-TTF)$_2$KHg(SCN)$_4$. Such an explanation was recently proposed [26] and will be mentioned later.

Next, we discuss the change of the gm_c value in the AF and the M* phases of α-(BEDT-TTF)$_2$KHg(SCN)$_4$. The important point is that the SSZ method and the LK-analysis show a different tendency of the enhancement of the gm_c or m_c value below and above the phase transition. In the present SSZ method, the $g(m_c/m_0)$ value (~ 4.7) obtained in the AF phase is larger than that (~ 3.63) in the M* phase. In contrast, a smaller effective mass ($m_c \sim 1.6m_0$) obtained in the AF phase than that ($\sim 1.9m_0$) in the M* phase has been reported on the basis of the LK-analysis of the temperature dependence of the oscillation amplitude [20,26]. This discrepancy may be explained by a model which requires an additional temperature factor in the LK formulation in the AF phase. In the AF phase, the observed α-orbit results from the MB effect on the multiconnected FS which is reconstructed by the SDW formation. The MB gap may have a temperature dependence because this gap is opened by the periodic SDW potential, which is temperature dependent. In addition, the magnitude of the SDW gap, which is expected to be ~ 10 K, is comparable to the temperature at which the experiments are done. The gap would become small effectively due to thermal excitations near the Fermi level. The detail of the formulation is discussed in [26]. This consideration implies that the effective mass, m_c^{obs}, and the Dingle temperature, T_D^{obs}, obtained so far by using the standard LK-analysis in the AF phase are underestimated and overestimated with respect to the correct values, with the relations $m_c^{\text{obs}} \simeq m_c(1 - laH_{\text{MB}}^0 m_0 / \lambda m_c)$ and $T_D^{\text{obs}} \simeq T_D + lH_{\text{MB}}^0 m_0 / \lambda m_c$, respectively. In short, a reasonable effective mass in the AF phase is obtained by the SSZ method. On the other hand, the underestimated value has been calculated by the inappropriate use of the LK-analysis in the AF phase.

Let us then consider the enhancement of the gm_c value in the AF phase on the basis of the results obtained by the SSZ method. Here we present two possible mechanisms. One is the enhancement of the many-body interactions of both g and m_c [27], and another is the exchange interaction between the conduction spins through SDW moments [25]. The renormalization by the electron-electron (EE) and electron-phonon (EP) interactions for the effective mass and the g value is simply $m_c = m_b(1 + A_0^{\text{EP}})(1 + A_1^{\text{EE}})$ and $g = g_s / [(1 + B_0^{\text{EE}})(1 + B_0^{\text{EP}})]$, where A_0, A_1 and B_0 are the coefficients of the Legendre expansion of the spin-symmetric and spin-asymmetric parts, respectively, of the Landau scattering function within the framework of the Landau theory of Fermi liquids. Here m_b is the band mass calculated from the band structure and g_s is the g value measured by spin-resonance experiments including spin-orbit coupling but not many-body effects. The combination value of $g(m_c/m_0)$, which is obtained experimentally, depends only on the EE interaction and not on the EP interaction because of $A_0^{\text{EP}} = B_0^{\text{EP}}$ in general. Thus $g(m_c/m_0) = g_s m_b(1 + A_1^{\text{EE}})/(1 + B_0^{\text{EE}})$, where $A_1^{\text{EE}} > 0$ and $B_0^{\text{EE}} < 0$. This means that the enhancement of the EE interaction would give a larger gm_c value, although it is difficult to distinguish the respective magnitudes of the renormalization on the effective mass and the g-value. It seems reasonable to suppose that the EE interaction is enhanced in the AF phase. The observed SdH (and also dHvA) oscillation frequencies in the AF and the M* phases are the same [18,20]. This means that the carrier number corresponding to the observed orbit is the same in the two cases of the closed orbit in the M* phase and the MB one in the AF phase. On the other hand, the total carrier number is expected to be small in the AF phase where the SDW gap is opened on the planar part of the FS. The small carrier number in the AF phase is also supported by Hall effect experiments [34]. Thus, the effective EE interaction on the carrier orbiting in the AF phase is supposed to be enhanced because of a lower screening effect on the carrier.

Another possible scenario is the magnetic exchange interaction for the g value. This effect has been applied to explain a magnetic field dependence of the splitting wave form of the SdH and dHvA oscillations. The detail of the discussion can be found in [25].

15.5   Problems for the Future

Finally, we would like to mention the problems still faced for a series of α-(BEDT-TTF)$_2$MHg(XCN)$_4$, especially for a novel conductor α-(BEDT-TTF)$_2$KHg(SCN)$_4$. First, what kind of the ground state is realized below T_A? A SDW model has been proposed from the spin susceptibility [11] and μSR [10] results. But nuclear magnetic resonance (NMF) experiments [7] do not detect an internal magnetic field which should be observed in static magnetic order such as a SDW although the density of states decreases below T_A. This suggests a CDW. And magnetic torque measurements show almost no magnetic anisotropy in the conducting plane [23,35]. Those results are difficult to understand in the simple SDW picture. The CDW model is supported by recent experimental [9,36,37] and theoretical [8,38] work in explaining mainly the shape of the magnetic phase diagram in high magnetic fields. No evidence for a superlattice of CDW, however, has been found by X-ray experiments [39]. The possibility of a SDW-CDW hybrid model has been discussed in [31] on the basis of ADMRO results. It should be noted that (TMTSF)$_2$PF$_6$, which has been thought to be a typical SDW system, is found to have an apparent charge modulation [40]. Such coexistence of a SDW and a CDW is discussed theoretically [41,42]. It is necessary to consider further the magnetic phase diagram from the viewpoint of the coexistence of a SDW and a CDW.

Secondly, what is the physical (electric and magnetic) nature characteristic of the high magnetic field phase, the M* phase? All experimental results, as far as we know, concerning the Fermi surface, ADMRO [16], the in-plane current dependence of MR [15] and the present work [28] indicate that the FS (topology, anisotropy and the effective mass) in the M* phase is the same as that in the P phase. The models [8,38] for the high magnetic field phase suggest that a spatially modulated CDW may be realized. In this hypothesis, one can naively come to the idea that the FS in the M* phase is different from that in the P phase. But the experimental results do not agree with the idea. It has been suggested that this salt exhibits a bulk quantum Hall effect (QHE) at high magnetic fields [43–45]. While it has been argued that a density wave stabilizes a QHE [46], there has not been an experiment performed that can establish QHE in high magnetic fields. In addition, one recent paper [37], which seems to be somewhat exotic in its proposal for the high magnetic field phase, suggests that both the transport and the magnetic properties at high magnetic fields closely resemble those of a type-II superconductor rather than the usual behavior expected for a density wave.

Thirdly, what is the origin of the large higher-order harmonics observed in dHvA and SdH oscillations only in the AF phase? These have been interpreted as spin-splitting [11,17,24,25,32,33] and as the frequency-doubling effect that accompanies a pinned CDW or SDW phase [47]. This problem also closely connects to the electronic states of each exotic phase.

Clearly, many challenges remain for the series of α-(BEDT-TTF)$_2$MHg (XCN)$_4$, especially for the novel ground state of α-(BEDT-TTF)$_2$KHg(SCN)$_4$ at high magnetic fields.

15.6   Conclusion

Our recent studies on the field-induced successive phase transitions of the organic conductors α-(BEDT-TTF)$_2$MHg(XCN)$_4$ [M = K, Rb and NH$_4$ and X = S and Se] are reviewed. These organic conductors have two kinds of Fermi surface: one is a closed cylindrical surface and the other a pair of open planar surfaces. The planar parts have a nesting instability resulting in a transition from a paramagnetic metal to an antiferromagnetic-like metal in zero magnetic field, i.e. a spin-density-wave or novel density-wave transition. In magnetic fields, transitions occur successively. A magnetic phase diagram has been proposed and improved by using new techniques for measurements and detecting different physical properties. In this review the spin-splitting-zero phenomena in the de Haas–van Alphen effect are considered in connection with the magnetic phase diagram proposed so far.

Acknowlegements

This work was carried out at the High Field Laboratory for Superconducting Materials, IMR, Tohoku University and partially supported by a Grant-in-Aid for Scientific Research from the Ministry of Education, Science, and Culture of Japan. The authors would like to thank Dr. W. Biberacher, Dr. M. Kartsovnik, Dr. F.L. Pratt and Dr. K. Kishigi for stimulating discussions and Professor N. Kobayashi for his encouragement.

References

1. For a recent review, see *Proc. Int. Conf. Science and Technology of Synthetic Metals*, Montpellier, France 1999 [Synth. Met. **103** (1999)]
2. T. Sasaki, H. Ozawa, H. Mori, S. Tanaka, T. Fukase and N. Toyota: J. Phys. Soc. Jpn. **65**, 213 (1996)
3. L.I. Buravov, N.D. Kushch, V.N. Laukhin, A.G. Khomenko, E.B. Yagubskii, M.V. Kartsovnik, A.E. Kovalev, L.P. Rozenberg, R.P. Shibaeva, M.A. Tanatar, V.S. Yefanov, V.V. Dyakin and V.A. Bondarenko: J. Phys. I (France) **4**, 441 (1994)
4. H.H. Wang, K.D. Carlson, U. Geiser, W.K. Kwok, M.D. Vashon, J.E. Thompson, N.F. Larsen, G.D. McCabe, R.S. Hulscher and J.M. Williams: Physica C **166**, 57 (1990)
5. On this subject, there have been many articles published so far. Thus, readers are referred to [20] containing the references for early works.

6. The nature of the low-temperature ground state has not been understood well. NMR studies have not detected any indication of antiferromagnetic order [7]. Recent works propose a possibility of the charge-density-wave state [8,9]. In this paper, however, we use the antiferromagnetic (AF) phase for the low-magnetic field low-temperature state.

7. K. Miyagawa, A. Kawamoto and K. Kanoda: Phys. Rev. B **56**, R8487 (1997)

8. R.H. McKenzie: preprint cond-mat/9706235

9. N. Biskup, J.A.A.J. Perenboom, J.S. Brooks and J.S. Qualls: Solid State Commun. **107**, 503 (1998)

10. F.L. Pratt, T. Sasaki, N. Toyota and K. Nagamine: Phys. Rev. Lett. **74**, 3892 (1995)

11. T. Sasaki, H. Sato and N. Toyota: Synth. Met. **41–43**, 2211 (1991)

12. H. Mori, S. Tanaka, M. Oshima, G. Saito, T. Mori, Y. Maruyama and H. Inokuchi: Bull. Chem. Soc. Jpn. **63**, 2183 (1990)

13. M.V. Kartsovnik, A.E. Kovalev and Kushch: J. Phys. I (France) **3**, 1187 (1993); M.V. Kartsovnik, A.E. Kovalev, V.N. Laukhin and S.I. Pesotskii: J. Phys. I (France) **2**, 223 (1992)

14. Y. Iye, R. Yagi, N. Hanasaki, S. Kagishima, H. Mori, H. Fujimoto and G. Saito: J. Phys. Soc. Jpn. **63**, 674 (1994)

15. T. Sasaki and N. Toyota: Phys. Rev. B **49**, 10120 (1994)

16. A.A. House, S.J. Blundell, M.M. Honold, J. Singleton, J.A.A.J. Perenboom, W. Hayes, M. Kurmoo and P. Day: J. Phys.: Condens. Matter **8** 8829 (1996)

17. S. Uji, T. Terashima, H. Aoki, J.S. Brooks, M. Tokumoto, N. Kinoshita, T. Kinoshita. Y. Tanaka and H. Anzai: Phys. Rev. B **54**, 9332 (1996)

18. T. Osada, R. Yagi, A. Kawasumi, S. Kagoshima, N. Miura, M. Oshima and G. Saito: Phys. Rev. **B41**, 5428 (1990)

19. T. Sasaki and N. Toyota: Solid State Commun. **82**, 447 (1992)

20. T. Sasaki, A.G. Lebed, T. Fukase and N. Toyota: Phys. Rev. B **54**, 12969 (1996)

21. M.V. Kartsovnik, W. Biberacher, E. Steep, P. Christ, K. Andres, A.G.M. Jansen and H. Müller: Synth. Met. **86**, 1933 (1997)

22. P. Christ, W. Biberacher, A.G.M. Jansen, M.V. Kartsovnik, A.E. Kovalev, N.D. Kushch, E. Steep and K. Andres: Surf. Sci. **361–362**, 909 (1996)

23. P. Christ: PhD Thesis, Technischen Universität München, ISBN 3–933083–19–2 (1998)

24. S. Uji, J.S. Brooks, M. Chaparala, L. Seger, T. Szabo, M. Tokumoto, N. Kinoshita, T. Kinoshita, Y. Tanaka and H. Anzai: Solid State Commun. **100**, 825 (1996)

25. T. Sasaki and N. Toyota: Phys. Rev. **B48**, 11457 (1993)

26. T. Sasaki, W. Biberacher and T. Fukase: Physica B **246–247**, 303 (1998)

27. D. Shoenberg: *Magnetic Oscillations in Metals* (Cambridge University Press, Cambridge, 1984)

28. T. Sasaki and T. Fukase: Phys. Rev. B **59**, 13872 (1999)

29. J. Wosnitza, G. Goll, D. Beckmann, E. Steep, N.D. Kushch, T. Sasaki and T. Fukase: J. Low Temp. Phys. **105**, 1691 (1996)

30. J. Wosnitza, G.W. Crabtree, H.H. Wang, K.D. Carlson, M.D. Vashon and J.M. Williams: Phys. Rev. Lett. **67**, 263 (1991)

31. T. Sasaki and N. Toyota: Synth. Met. **70**, 849 (1995)

32. F.L. Pratt, J. Singleton, M. Doporto, A.J. Fisher, T.J.B.M. Janssen, J.A.A.J. Perenboom, M. Kurmoo, W. Hayes and P. Day: Phys. Rev. B **45**, 13904 (1992)

33. J. Caulfield: PhD Thesis, Physics Department, University of Oxford, 1994
34. T. Sasaki, S. Endo and N. Toyota: Phys. Rev. B **48**, 1928 (1993)
35. P. Christ, W. Biberacher, W. Bensch, H. Müller and K. Andres, Synth. Metals **86**, 2057 (1997)
36. P. Christ, W. Biberacher, M.V. Kartsovnik, E. Steep, E. Balthes, H. Weiss and H. Müller: JETP Lett. **71**, 303 (2000)
37. N. Harrison, L. Balicas, J.S. Brooks and M. Tokumoto: preprint cond-mat/0004083
38. D. Zanchi, A. Bjelis and G. Montambaux: Phys. Rev. B **53**, 1240 (1996)
39. S. Kagoshima: private communications
40. J.P. Pouget and S. Ravy: Synth. Metals **85**, 1523 (1997)
41. N. Kobayashi, M. Ogata and K. Yonemitsu: J. Phys. Soc. Jpn. **67**, 1098 (1998)
42. S. Mazumdar, S. Ramasesha, R. Torsten and D.K. Cambell: Phys. Rev. Lett. **82**, 1522 (1999)
43. N. Harrison, A. House, M.V. Kartsovnik, A.V. Polisski, J. Singleton, F. Herlach, W. Hayes and N.D. Kushch: Phys. Rev. Lett. **77**, 1576 (1996)
44. S. Hill, P.S. Sandhu, J.S. Qualls, J.S. Brooks, M. Tokumoto, N. Kinoshita, T. Kinoshita and Y. Tanaka: Phys. Rev. **55**, R4891 (1997)
45. M.M. Honold, N. Harrison, J. Singleton, H. Yaguchi, C. Mieke, D. Deckers, P.H.P. Reinders, F. Herlach, M. Kurmoo and P. Day: J. Phys. Condens. Matter **9**, L533 (1997)
46. S. Hill, S. Uji, M. Takashita, C. Terakura, T. Terashima, H. Aoki, J.S. Brooks, Z. Fisk and J. Sarrao: Phys. Rev. **58**, 10778 (1998)
47. N. Harrison: Phys. Rev. Lett. **83**, 1395 (1999)

16 NMR/NQR Studies on Magnetism of Spin Ladder $Sr_{14-X}A_XCu_{24}O_{41}$ (A = Ca and La)

K. Kumagai, S. Tsuji, K. Maki, T. Goto and T. Fukase

Hole-doped cuprates with low-dimensional Heisenberg spin systems have been investigated intensively after the discovery of high-T_c superconductivity. The predictions of spin gap and superconductivity in carrier-doped ladder systems, i.e. coupled chain systems, have stimulated a renewal of studies on quantum behavior in low-dimensional systems [1,2]. The two-leg ladder cuprate, $Sr_{14}Cu_{24}O_{41}$, is a unique system in which the carriers are controlled. $Sr_{14}Cu_{24}O_{41}$ consists of CuO_2 layers with one-dimensional (1D) chains of edge-sharing clusters and Cu_2O_3 layers with a two-leg ladder configuration, as shown in Fig. 16.1 [3]. Holes are transferred from the chain to the ladder site by the substitution of Ca for Sr and the system becomes conductive with increasing x on $Sr_{14}Cu_{24}O_{41}$ [4]. In this system, the Cu-O chain is regarded as a nonconductive charge reservoir. When high pressure is applied, the compounds become more conductive and superconductivity appears finally around $x = 12$ under a high pressure of $3 \sim 4\,GPa$ [5]. It is important to investigate the magnetic nature without any holes or with dilute holes in the ladder system. For La^{3+}-substitution for Sr^{2+}, the number of holes in the system decreases. The formal valence of Cu in both the chain and the ladder is 2+ for $Sr_8La_6Cu_{24}O_{41}$, i.e. here are no holes in the system. Thus, the system is expected to be completely insulating with long-range magnetic ordering of the Cu spins.

Unexpectedly, magnetic order was observed under ambient pressure for the highly-doped $Sr_{2.5}Ca_{11.5}Cu_{24}O_{41}$ [6]. The heat capacity shows a sharp peak around $T = 2.3\,K$ and neutron diffraction measurements suggest the development of magnetic Bragg peaks. The magnetic moments seem to be small and their magnetic structure seems to be complicated. The controversy about which sites of Cu spins are responsible for the magnetic order is not yet settled, although this magnetic order is an important issue for understanding unconventional superconductivity under high pressure.

In addition to neutron scattering measurements, a local probe using NMR/NQR is a powerful tool for investigating the magnetic behavior and spin dynamics, especially to obtain local information about the valence state of the Cu sites [7–9]. In this paper, we summarize hole-doping effects on Cu-NMR/NQR of single crystals of $Sr_{14-x}Ca_xCu_{24}O_{41}$. We report field-induced staggered moments in the ladder site of $Sr_{14-x}La_xCu_{24}O_{41}$ [10]. Finally, we discuss the magnetic nature and the origin of the magnetic order of highly-doped $Sr_{2.5}Ca_{11.5}Cu_{24}O_{41}$.

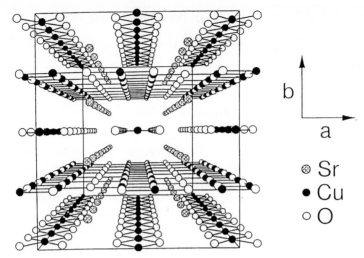

Fig. 16.1. Crystal structure of $Sr_{14}Cu_{24}O_{41}$

16.1 Experimental

Single crystals were prepared by a traveling-solvent floating-zone method under high oxygen pressure. Cu-NMR and NQR were measured by a conventional phase coherent pulse method with superconducting magnets of 9 T at Hokkaido University and also of 20 T at the MRI, Tohoku University. Heat capacities were measured in an Oxford Instruments cryostat with a superconducting magnet of 12 T and magnetic susceptibilities were measured with a SQUID magnetometer.

16.2 Hole-Doping Effects on Spin Gaps in $Sr_{14-x}Ca_xCu_{24}O_{41}$

For single crystals of $Sr_{14}Cu_{24}O_{41}$, the NMR and NQR spectra are sharp enough to resolve some peaks [10]. The spectra of Cu-NMR for each direction of external field are well fitted by a perturbation calculation with the electric quadrupole interaction. For the chain site, the spectra are classified for three kinds of signals for which the NMR shifts show a distinct temperature dependence as shown in Fig. 16.2. From the NMR shift and magnetic susceptibility at low temperature, the hyperfine coupling constants, A_{hf}, of Cu were evaluated to be -220, -25 and $-12\,\mathrm{kOe}/\mu_B$. The signal ($\nu_\rho \sim 18\,\mathrm{MHz}$) with the largest A_{hf} is attributed to Cu^{2+} ($S = 1/2$, the D-state). The other two components ($\nu_\rho \sim 33\,\mathrm{MHz}$, the A- and B-states) with small A_{hf} are attributed to Cu_{3+} (non-magnetic state or Zhang-Rice singlet) [8]. Moreover, among the signals from the non-magnetic Cu sites (A and B), T_2 of one of them shows Gaussian behavior and T_2 of the other one shows Lorentzian behavior. T_{2G}

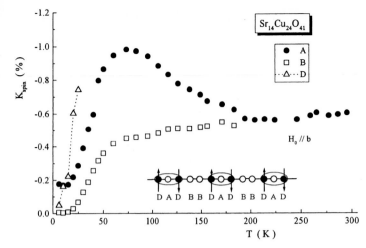

Fig. 16.2. Temperature dependence of the NMR shift of the chain Cu of single crystal $Sr_{14}Cu_{24}O_{41}$ for H‖b. The inset shows the proposed spin and hole arrangement for the chain Cu of $Sr_{14}Cu_{24}O_{41}$. ↑ and↓ represent dimer states (D) of Cu^{2+} (S=1/2), o are for Cu^{3+} (S = 0 with a hole) (A and B)

is ascribed to the dominating dephasing term via dipole interactions among nuclear spins. Thus, the signals with the Gaussian T_2 arise from the Cu^{3+} with two-adjacent Cu^{3+} configuration (the B-site), and the other arises from the separated single-Cu^{3+} configuration (the A-site).

The present NMR results show that the charge order of Cu^{2+} and Cu^{3+} is realized in the chain Cu at low temperature. The proposed charge and spin configuration of the chain in $Sr_{14}Cu_{24}O_{41}$ is shown in the inset of Fig. 16.2. The spin gap formation in the 1D-chain of $Sr_{14}Cu_{24}O_{41}$ is attributed to the charge order of Cu^{2+} and Cu^{3+} in the chain. The spin-dimers of the magnetic Cu^{2+} are realized in the particular array of Cu^{2+} and Cu^{3+} states. Contrary to the case of the chain site, we observed only Cu-NQR with a unique magnetic state for the ladder site, namely, the unique T-dependence of the Knight shift. $1/T_{2G}$ of the ladder Cu-NMR is very large and is attributed to indirect nuclear spin coupling through strongly correlated Cu^{2+} ($S = 1/2$) spins [11]. $1/T_{2G}$ increases gradually and tends to be temperature-independent with decreasing temperature below 70 K. Using the formula obtained from the Monte Carlo calculation and the parameters $A_{hf}(Q) \sim A_{hf}(0) = -120\,kOe/\mu_B$ and $J = 1300\,K$, we evaluate the magnetic correlation length, i.e. the average distance of pairing spins, to be $\xi = 3$ (in units of Cu atomic distance) at the lowest temperature. As the inter-ladder coupling is small, magnetic ordering does not occur, but the quantum-fluctuated resonating valence bond (RVB) state with spin singlet pairs is realized in the ladder site of $Sr_{14}Cu_{24}O_{41}$. Previous NMR studies [7–9] on the nuclear spin-lattice relaxation rate, $1/T_1$, of $Sr_{14}Cu_{24}O_{41}$ have also provided clear evidence for the formation of a gap for low-lying spin excitations in the ladder sites. Here, the spin gap of the Cu ladder site obtained from the T_1 measurement is large ($\Delta \sim 1000\,\mathrm{K}$) for $Sr_{14}Cu_{24}O_{41}$ and decreases rapidly with x, which is in sharp contrast to the x-independent spin gap ($\Delta \sim 400\,\mathrm{K}$) obtained from neutron scattering [12], Raman scattering [13] and thermal conductivity [14] measurements. We see two distinct energy scales for the activation energy. For the case where holes are transferred to the ladder site from the chain site for the substitution of Ca for Sr, the doped holes create single-hole rungs or two-hole rungs. In the ground state for two holes, holes enter a bound state of two holes in the same rung [15]. Therefore, there exist two types of magnetic excitations. One is the singlet-triplet magnon excitation and the other is the excitation for dissociation to single holes from paired holes. Most of Cu spins in the ladder form spin-singlet pairs at low temperature and hole pairs are dissociated to single holes at high temperature. The dissociation energy of hole pairs is considered to be equal to the binding energy of spin singlet pairs, i.e. the spin gap energy, which is monitored by T_1 at high temperature. Recent electrical resistivity and thermal conductivity measurements [14] also show two kinds of activation energy for spin excitations. The hole-depairing excitations are effective at high temperature and depend largely on x, i.e. the hole doping level. However, magnon scattering for singlet-triplet excitations is independent of the hole doping, which may be monitored by neutron and Raman scattering measurements.

16.3   Staggered Moments in $Sr_{14-x}La_xCu_{24}O_{41}$

We have confirmed a long-range antiferromagnetic ordering of Cu moments in $Sr_{14-x}La_xCu_{24}O_{41}$ for $3 \leq x \leq 6$ by heat capacity and magnetic susceptibility measurements [10]. T_N decreases with decreasing x from $x = 6$ (no holes) ($T_N = 16\,\mathrm{K}$ for $x = 6$, $T_N = 12\,\mathrm{K}$ for $x = 5$ and $T_N = 3\,\mathrm{K}$ for $x = 3$) corresponding to the increase of holes in the system. The appearance of a sizeable amount of the T-linear term of the heat capacity for the magnetically-ordered system is evidence of the collapse of the spin gap and is characteristic of a 1D Heisenberg antiferromagnetic (AF) system, where the C/T term is expected to be $2Nk_B/3J$. From this relation, the exchange coupling constant, J, is estimated to be of the order of 100 K. Thus, we may conclude that the long-range magnetic ordering of Cu spins appears at the chain sites.

As shown in Figs. 16.3 and 16.4, the Cu-NMR spectrum of the ladder splits largely into two peaks when an external field is applied [10]. It should be noted that the staggered fields appear differently according to the direction of the applied external field. The temperature dependence of the staggered field along the a-axis obeys a Brillouin function below T_N as shown in the inset of Fig. 16.3. On the other hand, the temperature dependence of the field-induced staggered field obeys the Curie–Weiss behavior for the case of

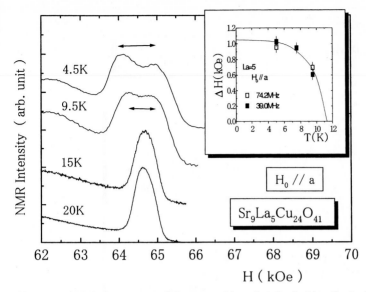

Fig. 16.3. NMR spectrum of the central line for the ladder Cu in $Sr_9La_5Cu_{24}O_{41}$ as a function of external field for $H\|a$-direction. The inset shows the temperature dependence of the field-induced staggered field at the ladder Cu site. Solid line indicates the Brillouin function of $S = 1/2$

$H//b$, as shown in Fig. 16.4. The Brillouin function type of behavior of the staggered field below T_N is not observed in the cases of $H//b$ and $H//c$.

Because of the incommensurability of ladder and chain Cu sites [16], the dipole fields from the ordered Cu moments in the chains must broaden out at the ladder site when the spins align antiferromagnetically within the chain. Thus, the observable staggered field at the ladder site indicates that the magnetic ordering of Cu spins is ferromagnetic within the chains and is antiferromagnetic between chains. Under this restriction for the dipole sum calculation, we can show that the ordered Cu moments of the chain induce hyperfine fields of $500 - 700\,$Oe at the ladder Cu site only in the case that the ordered Cu spins are aligned along the b-axis in the chain. In fact, only the magnetic susceptibility along the b-axis decreases below T_N, indicating that the spin direction in the antiferromagnetic state is along the b-axis for $Sr_{14x}La_xCu_{24}O_{41}$. Therefore, the staggered field at the ladder site observed experimentally along the a-axis is considered to originate from the dipole fields from ordered chain Cu moments.

Next, we consider the other source of field-induced staggered field far above T_N. In the case of $H//b$-axis, its magnitude at low temperature is larger than that expected from dipole field contributions (on the order of $1\,$kOe) due to the ordered Cu moments in the chains. The magnitude of the staggered field for $H//b$ increases with increasing external field, as shown in

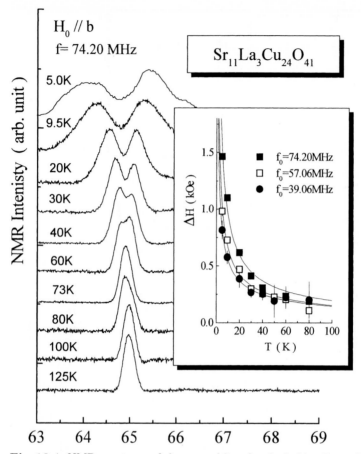

Fig. 16.4. NMR spectrum of the central line for the ladder Cu in $Sr_{11}La_3Cu_{24}O_{41}$ as a function of external field for $H\|b$-direction. The inset shows the temperature dependence of the field-induced staggered field at the ladder Cu site for several external fields (resonance frequencies). Solid line indicates the Curie-Weiss function

Fig. 16.5. Moreover, the magnitude of the induced fields for each direction is proportional to the hyperfine coupling constants, which are deduced from the analysis of the Knight shift and the magnetic susceptibility, namely, $\Delta H^a : \Delta H^b : \Delta H^c = 1 : 4 : 1 = A_{hf}^a : A_{hf}^b : A_{hf}^c$. This experimental result implies that the field-induced staggered moments originate from the magnetic nature of the ladder Cu spins themselves, but not from the transferred hyperfine field from the ordered Cu spins in the chains.

We are aware of the large distribution of T_2 for the split NMR spectrum due to staggered moments. $1/T_2$ is large at the central position and is small at the edge position. Thus, the alternative local hyperfine field at the ladder is spatially-distributed. Recently, staggered moments induced around hole

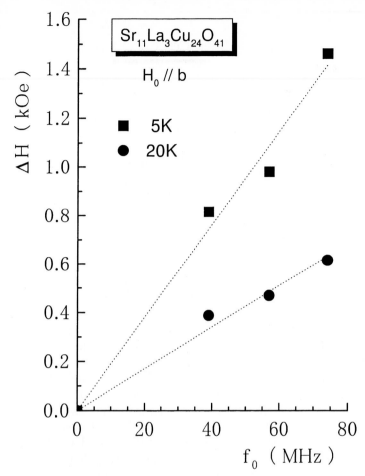

Fig. 16.5. NMR frequency (i.e. external field) dependence of the field-induced staggered field for $Sr_{11}La_3Cu_{24}O_{41}$ (for $H//b$)

or impurity centers have been proposed [17,18]. Quantum Monte Carlo simulations show the impurity contribution to the static function of the local susceptibility. The spins become quite large near the hole or impurity centers and an oscillatory or staggered susceptibility appears. The ordered moments decay exponentially along the ladder, which produces spatial inhomogeneity of the staggered field. We conclude that the field-induced staggered moments originate from the edge effect around doped-hole centers at the ladder site.

16.4   Magnetism of $Sr_{2.5}Ca_{11.5}Cu_{24}O_{41}$

As described previously, the rapid decrease of magnetic susceptibility with decreasing temperature below $80\,K$ is due to the formation of a singlet spin

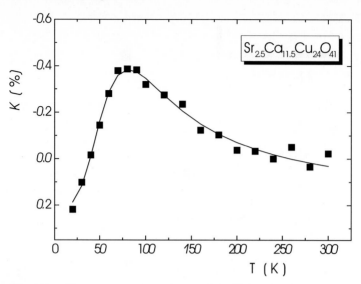

Fig. 16.6. Temperature dependence of the ^{63}Cu-NMR shift at the chain site of the single crystal $Sr_{2.5}Ca_{11.5}Cu_{24}O_{41}$ (for $H//b$)

ground state in the 1D Cu-chains. The existence of the gap behavior in the magnetic susceptibility was obtained in a highly-doped sample of $X = 11.5$. The decrease of the spin susceptibility at low temperature is observed more clearly in the Cu-NMR shift for $Sr_{2.5}Ca_{11.5}Cu_{24}O_{41}$, as shown in Fig. 16.6. The decrease of the spin part of the Cu-NMR shift, x_{spin}, for $Sr_{2.5}Ca_{11.5}Cu_{24}O_{41}$ is fitted well by the formula of the dimer model with a T-independent $x_{orb} = 0.025\%$, $A_{hf} = -19\,kOe/\mu_B$ (the exchange coupling constant along the leg $J = 130\,K$) and J' (along the rung) $= 89\,K$ [19], indicating that the spin dimers in the chain site survive up to the range near $x = 11.5$.

For very low temperature, we have found anomalies in the magnetization of $Sr_{14-x}La_xCu_{24}O_{41}$. Figure 16.7 shows the magnetic susceptibility as a function of temperature measured at $H_{ext} = 50\,G$ for $Sr_{14-x}La_xCu_{24}O_{41}$. As clearly seen, x decreases only for $H//c$-axis, which indicates the occurrence of an antiferromagnetic (AF) magnetic order at 2.3 K and that the direction of AF ordered moments is along the c-axis. However, when a field above 1000 Oe is applied, the anomalies associated with the cusp are smeared out completely and x increases smoothly with decreasing temperature. Interestingly, we have confirmed meta-magnetic behavior at low field as shown in the inset of Fig. 16.7. A change of magnetization around 600 Oe is observed when the field is applied along the c-axis. On the other hand, no anomalies of the magnetization are observed for $H//a$ and $H//b$. These results show that the ordered moments flip to the a-axis from the c-axis at the small value of 600 Oe.

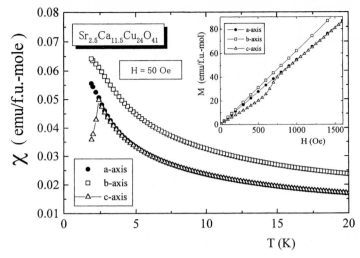

Fig. 16.7. Temperature dependence of the magnetic susceptibility of the single crystal $Sr_{2.5}Ca_{11.5}Cu_{24}O_{41}$ at $H_{ex} = 50\,G$ applied along the a-, b- and c-axes. The inset shows the magnetization as a function of external field for single crystal $Sr_{2.5}Ca_{11.5}Cu_{24}O_{41}$ at $1.8\,K$

Furthermore, we have confirmed a sharp cusp of the heat capacity of $Sr_{14-x}La_xCu_{24}O_{41}$ at $2.3\,K$, similar to the previous report [6]. This anomaly is attributed to AF magnetic ordering of Cu moments. The cusp corresponding to the magnetic transition is shifted to lower temperatures by an external field and is smeared out completely at high fields up to $6\,T$. It should be pointed out that the magnetically-ordered state disclosed by the heat capacity and magnetic susceptibility measurements coexists with the spin dimer state of the chain. The next question is which site is responsible for the magnetic order. In order to elucidate this point as well as the origin of this phase transition, Cu-NMR/NQR was performed.

For the paramagnetic state ($T > T_N$) of doped $Sr_{14-x}La_xCu_{24}O_{41}$, the $^{63/65}$Cu-NQR spectrum of the chain (the component of $\nu_Q = 33\,MHz$) was relatively broad compared to that of the starting $Sr_{14}Cu_{24}O_{41}$, which arises partly from the inhomogeneity of the electric quadrupole interaction due to a random atomic distribution of Sr and Ca. Nevertheless, comparing the Cu-NQR spectrum above and below T_N, magnetic broadening appears obvious below T_N as shown in Fig. 16.8. The spectrum below T_N is not fitted by a unique Gaussian or Lorentzian component, but is fitted well by two components. One is the signal from the Cu nuclei which do not feel any magnetic hyperfine field and the other is the signal for which a finite hyperfine field exists along the c-axis. The fitting to the observed spectrum, as shown by the solid line in Fig. 16.8, is given by the sum of dotted lines obtained from the 2nd-order perturbation calculation (the split component) and the broken

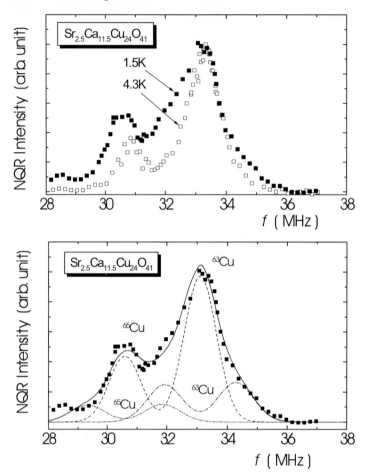

Fig. 16.8. $^{63/65}$Cu-NQR spectra around 33 MHz of the single crystal Sr$_{2.5}$Ca$_{11.5}$Cu$_{24}$O$_{41}$. (a) the spectra are normalized at the peak obtained above T_N and below T_N. (b) the spectrum with fitting lines (dotted line and broken lines) for $T \leq T_N$

line (the unchanged component). The transferred hyperfine field is uniform at the non-magnetic chain Cu^{3+} sites.

As mentioned already, holes in the compound are transferred to the ladder site from the chain site with increasing x. The number of holes is reduced and the magnetic spin states increase at the chain Cu site. The remaining holes in the chains are localized at low temperature. The number of holes in the chain for $x = 11.5$(Ca) is expected to be identical to the one for $x = 3$(La) ($T_N = 1.6$ K). The Cu-O-Cu bond with the bond angle of nearly 90° is weakly ferromagnetic. However, the Cu-O-O-Cu bond through O–2$p\pi$ coupling is antiferromagnetic and is large.

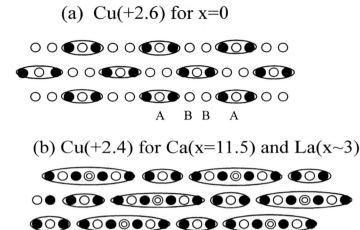

(a) Cu(+2.6) for x=0

                    A   B   B   A

(b) Cu(+2.4) for Ca(x=11.5) and La(x~3)

        D A D   B'   B''   B'

●	D : Cu^{2+}	◎	B'': Cu^{3+}
○	A : Cu^{3+}	○	B' : Cu^{3+}

Fig. 16.9. Proposed configuration for spin and charge in the chain site for **(a)** $Sr_{14}Cu_{24}O_{41}$ and **(b)** $Sr_{2.5}Ca_{11.5}Cu_{24}O_{41}$ or $Sr_{11}La_3Cu_{24}O_{41}$. A(○), B''(◎) and B'(○) represent non-magnetic (or Zhang-Rice singlet) Cu^{3+} (S=0) sites with a hole. ● represents magnetic Cu^{2+} (S=1/2) atoms

Thus, we propose the possible charge and spin configuration of the chain Cu site in $Sr_{2.5}Ca_{11.5}Cu_{24}O_{41}$ given in Fig. 16.9. The B sites (or the Zhang-Rice singlet state) of Cu^{3+} split into two distinct magnetic states [B''(◎) and B'(○)] in the chain below T_N. In that configuration we expect that the dimers survive. As the B''-site is located in a symmetric surrounding of the spin array and the B'-site is situated in a non symmetric site, the transferred hyperfine field from Cu^{2+} is canceled at the B''-site, but a finite value of H_{hf} exists at the B'-site. As the hyperfine field is determined to be 1.1 kOe, we estimate the magnitude of the moments of the chain Cu^{2+} as 0.06 μ_B using the transferred hyperfine coupling constant $-19\,kOe/\mu_B$ for the chain Cu^{3+} site.

We investigated also the NQR signal around 20 MHz as shown in Fig. 16.10, which is originated mostly from the ladder Cu in $Sr_{14}Cu_{24}O_{41}$. The spectrum below T_N is identical with that above T_N except for the low-frequency part. One can imagine that the line width seems to decrease below T_N, as opposed to the case of the NQR signals around 33 MHz. However, we confirm that the signals from the chain Cu^{2+} are superposed at around 14–18 MHz above T_N. Those signals from the chain Cu disappear at 1.5 K due to their very short T_2. Other than the contribution from the chain Cu in this range, the spectra of the ladder Cu are not affected by the magnetic order. We do not detect any difference of the Cu-NQR spectrum of the ladder through T_N. This ex-

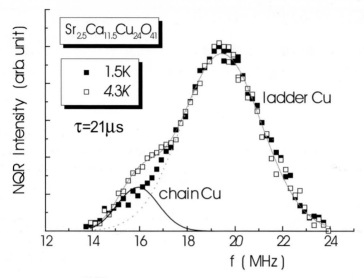

Fig. 16.10. $^{63/65}$Cu-NQR spectra around 14–24 MHz of the single crystal $Sr_{2.5}Ca_{11.5}Cu_{24}O_{41}$. The spectra obtained at $\tau = 21\,\mu s$ (pulse interval time) are normalized at the peak obtained above T_N and below T_N. The Gaussian fitting lines for the chain Cu are shown

perimental result is inconsistent with that reported previously, in which both the NQR spectra of the ladder and the chain become broader below T_N [20].

As far as the spontaneous moments are concerned, we do not detect any indication of a sizeable hyperfine field at the ladder Cu site in the Cu-NQR/NMR measurement. This leads to the conclusion that vanishing magnetically ordered moments are induced at the ladder Cu site, if they exist. Thus, we conclude that the magnetic order is attributed to the chain Cu^{2+} moments in the particular configuration among localized non-magnetic Cu^{3+} ions in highly doped $Sr_{14}Cu_{24}O_{41}$ and that the magnetic order for $Sr_{14}Cu_{24}O_{41}$ is responsible for the chain Cu spins.

16.5 Conclusion

Hole-doping effects on the magnetism of the chain and the ladder sites were investigated by Cu-NMR and NQR in single crystals of $Sr_{14-x}A_xCu_{24}O_{41}$ (A = Ca and La). The antiferromagnetic (AF) order of Cu moments was confirmed in the chain for $3 \leq x \leq 6$ is $Sr_{14-x}La_xCu_{24}O_{41}$. The magnetic order of Cu spins is ferromagnetic within the chain and is antiferromagnetic between chains. The field-induced staggered moments at the ladder Cu site appear far above T_N, and the temperature dependence obeys a Curie-Weiss behavior, suggesting that this staggered field is attributable to the magnetic nature of the ladder Cu spins themselves. A possible origin is ascribed to the disturbance of the coherence of the RVB state by doping centers in the ladder. Finally a long-range AF order in the highly-doped compound of $Sr_{2.5}Ca_{11.5}Cu_{24}O_{41}$ is discussed, based on the Cu-NMR/NQR results. This AF order is responsible for the chain Cu spins.

Acknowledgment

The author is grateful to Prof. Y. Koike for his valuable discussions and collaboration. This work is supported in part by a grant-in-aid for Scientific Research from the Ministry of Education, Science, Sport and Culture of Japan.

References

1. For reviews, E. Dagotto and T.M. Rice: Science **271**, 618 (1996) and S. Maekawa: Science **273**, 1515 (1997)
2. T.M. Rice, S. Gopalan and M. Sigrist: Europhys. Lett. **23**, 445 (1993); M. Sigrist, T.M. Rice and F.C. Zheng: Phys. Rev. **B49**, 12058 (1994)
3. E.M. McCarran, M.A. Subramanian, J.C. Calabrese and R.L. Harlow: Mater. Res. Bull. **23**, 1355 (1988)
4. M. Kato, S. Shiota and Y. Koike: Physica C **258**, 284 (1996)
5. M. Uehara, T. Nagata, J. Akimitsu, H. Takahashi, N. Mori and K. Kinoshita: J. Phys. Soc. Jpn. **65**, 2764 (1996): T. Nagata, M. Uehara, J. Goto, J. Akimitsu, N. Motoyama, H. Eisaki, S. Uchida, H. Takahashi, T. Nakanishi and N. Mori, Phys. Rev. Let. **81**, 1090 (1998)
6. J. Akimitsu, T. Nagata, H. Fujino, N. Motoyama, H. Eisaki, S. Uchida, H. Takahashi, T. Nakanishi, N. Mori, M. Nishi, K. Kakurai, S. Katano, M. Hiroi, M. Sera and N. Kobayashi: *Physics and Chemistry of Transition Metal Oxides*, Springer Series on Solid-State Sciences, Vol. 125 (Springer, Berlin 1999) p. 289
7. S. Tsuji, K. Kumagai, M. Kato, and Y. Kohike: J. Phys. Soc. Jpn. **65**, 3474 (1996); K. Kumagai, S. Tsuji, M Kato and Y. Koike: Phys. Rev. Lett. **78**, 1992 (1997)
8. M. Takigawa, N. Motoyama, H. Eisaki and S. Uchida: Phys. Rev. B **57**, 1124 (1998)
9. K. Magishi, S. Matsumoto, Y. Kitaoka, K. Ishida, K. Asayama, M. Uehara, Y. Nagata and J. Akimitsu: Phys. Rev. B **57**, 11533 (1998)
10. K. Kumagai, S. Tsuji and K. Maki: Physica B **259–261**, 1044 (1999); S. Tsuji, K. Maki and K. Kumagai: J. Low Temp. Phys. **117**, 1683 (1999)
11. C.H. Pennington and C.P. Slichter: Phys. Rev. Lett. **66**, 381 (1991)
12. S. Katano, T. Nagata, J. Akimitsu, M. Nishi and K. Kakurai: Phys. Rev. Lett. **82**, 636 (1999)
13. S. Sugai and M. Suzuki: Phys. Status Solidi (B) **215**, 653 (1999)
14. K. Kubo, S. Ishikawa, T. Noji, T. Adachi, Y. Koike, K. Maki, S. Tsuji and K. Kumagai: J. Phys. Soc. Jpn. **70**, 437 (2001)
15. M. Troyer, H. Tsunetsugu and T.M. Rice: Phys. Rev. B **53**, 251 (1996)
16. T. Ohta, F. Izumi, M. Onoda, M. Isobe, E. Takayama-Muromachi and A.W. Hewat: J. Phys. Soc. Jpn. **66**, 3107 (1997)

17. S. Eggert and I. Affleck: Phys. Rev. Lett. **75**, 934 (1995)
18. F. Fukuyama, T. Tamamoto and M. Saito: J. Phys. Soc. Jpn. **65**, 1182 (1996)
19. S. Tsuji, K. Maki and K. Kumagai: Physica B **284-288**, 1593 (2000)
20. S. Ohsugi, K. Magishi, S. Matsumoto, Y. Kitaoka, T. Nagata and J. Akimitsu: Phys. Rev. Lett. **82**, 4715 (1999)

17 NMR Study
of the Tl-Based High-T_c Cuprate
$Tl(Ba,Sr)_2(Y,Ca)Cu_2O_7$
in a Wide Hole Concentration Range
from the Antiferromagnetic
to the Overdoped Region

T. Goto, S. Nakajima, Y. Syono and T. Fukase

Many studies on high-T_c cuprates conducted in this decade have repeatedly emphasized universality in the electronic phase diagram. Undoped cuprates are, as is well known, insulators and are two-dimensional Heisenberg antiferromagnets. With doping of a small number of holes, the magnetism is drastically destroyed by the frustration effect, finite metallic conductivity appears, and the superconducting transition temperature rises with increasing hole concentration. The physical properties of metals in this lightly-doped region above T_c are quite different from those of conventional metals, such as the temperature-dependent Hall coefficient and the non-Korringa-type NMR spin-lattice relaxation rate. With further doping, T_c decreases with hole concentration finally to zero, where its electronic state is well described as a Fermi liquid.

In order to explain this universal phase diagram, there have been many theories discussed that propose that spin-charge separation exists in the two-dimensional system of high-T_c cuprates and that the spinon and the holon, the spin and the charge degrees of freedom, condense at two independent temperatures [1,2]. The superconductivity appears, according to the theories, when both form condensates. In the lightly-doped region, when the condensation temperature of the spinons is higher than that of the holons, there exists a temperature region where only the spin degrees of freedom condense. This is the so-called spin-gap phenomenon in which the low-energy spin excitation is suppressed. Those theories predict that the condensation temperature of holons is higher than that for spinons in the overdoped region and hence that the spin-gap is not observed in this region. Actually, its disappearance in the overdoped region has been confirmed [3] for some cuprates such as $YBa_2Cu_3O_7$.

However, it is not self-evident whether or not the picture is universally applicable to all the high-T_c cuprates. The authors believe in the importance of a rigorous test of this favorable-looking picture. In fact, the chemical synthesis of neither a completely overdoped phase with a zero T_c of $YBa_2Cu_3O_7$ (YBCO) nor an antiferromagnetic phase of $Tl_2Ba_2CuO_6$ (Tl2201) has been successful until now. La-based curates (La214) are of concern for the well known 1/8-issue, which is the phenomenon that T_c is unexpectedly depressed at the specific hole concentration where it might be expected to be highest. The behavior of the spin-gap in Y1248 is successfully explained by the theoretical electronic phase diagram, but is completely unexplained in other cuprates such as La214, $La_2CaCu_2O_6$ (La2126) or the Bi-based ones [4,5]. These phenomena seem to break the universality.

In order to test whether or not the universal picture is applicable for most cuprates, it is necessary to accumulate data for many kinds of cuprates. In this paper, we select three topics for review from our recent NMR work [6,7] on the Tl-based cuprate $Tl(Ba,Sr)_2(Y,Ca)Cu_2O_7$ (Tl1212) in a wide range of hole concentration from the antiferromagnetic to the over-doped region.

The first topic is the antiferromagnetism in the end member Tl1212, revealed by NMR, supporting universality. By the study of the end member, we also expect the possibility of finding a hidden key parameter determining T_c, which varies from 10 to 150 K for different materials. The objective is to investigate the magnetic character in the antiferromagnetic phase to reveal differences between the Tl-based system and other systems such as La_2CuO_4 and $YBa_2Cu_3O_6$.

The second topic is the spin-gap in the Tl-based system; we have revealed that the spin-gap also exists in systems in the lightly doped region. This supports universality in high-T_c cuprates. However, a gap-like behavior was also observed in the Knight shift and the relaxation rate for the slightly overdoped region. This seems to contradict the theory of a universal phase diagram, which predicts that the spin-gap must appear only in the lightly doped region.

The third topic is the difference in T_c between TB1212 and TS1212. Since both Ba and Sr are divalent, the nominal hole concentration is the same for the stoichiometric systems. However, the state of the as-sintered sample is quite different for TB1212 and TS1212 [8–10]. We have revealed by NMR that this difference comes from the partial atomic exchange between Tl and Ca-sites.

17.1   Experimental

Polycrystalline samples of $TlBa_2Ca_{1-x}Y_xCu_2O_{7+\delta}$ ($x = 0.95$, 1) and $TlSr_2Y_{1-x}Ca_xCu_2O_{7+\delta}$ ($x = 0.2$, 0.5) have been prepared by the solid-state reaction method from Tl_2O_3, Y_2O_3, CaO, BaO_2 and CuO with a purity of four nines [8]. The hole concentration of TB1212 is controlled by the oxygen content δ. The as-sintered sample is in the slightly overdoped region with $T_c = 80$ K. The optimally doped sample with $T_c = 100$ K is obtained by heat treatment in flowing Ar-gas. The as-sintered sample of TS1212 with $x = 0.2$ and $T_c = 34$ K belongs to the underdoped region and that with $x = 0.5$ and $T_c = 79$ K belongs to the optimally doped region. The superconducting transition temperature T_c was determined as the onset of the diamagnetic

Table 17.1. List of samples dealt in this paper

composition	gas treatment	T_C (K)	state
TlBa$_2$YCu$_2$O$_7$	as-sintered	0	Antiferromagnetic
TlBa$_2$Y$_{0.95}$Ca$_{0.05}$Cu$_2$O$_7$	as-sintered	0	–
TlSr$_2$Y$_{0.8}$Ca$_{0.2}$Cu$_2$O$_7$	as-sintered	34	lightly doped
TlSr$_2$Y$_{0.5}$Ca$_{0.5}$Cu$_2$O$_7$	as-sintered	79	optimally doped
TlBa$_2$CaCu$_2$O$_7$	Ar-annealed	100	optimally doped
TlBa$_2$CaCu$_2$O$_7$	as-sintered	80	slightly overdoped

shielding signal measured by a SQUID magnetometer in an applied field 20 Oe. The list of samples dealt with in this paper is shown in Table 17.1. Samples are confirmed to be of single phase and tetragonal crystal structure by powder X-ray diffraction. They were pulverized and mixed with epoxy resin for alignment of the c-axis in a magnetic field of 11 T.

The zero-field spectrum of $^{63/65}$Cu-NMR was obtained at 4.2 K by plotting the integrated amplitude of the spin-echo signal against the frequency between 80 and 120 MHz with a step size of 10 kHz. The frequency dependence of the observed spin-echo amplitude was compensated by the factor ω_0^{-2}. The width of the excitation and the refocusing pulses were set to be approximately 2 and 4 μs, the spectral width of which was narrow enough relative to the structure of the spectrum. The spin-lattice relaxation rate T_1^{-1} of the ^{63}Cu nuclei was measured for both the central transition line and the satellite line at zero field by the conventional saturation-recovery method with a pulse train.

Tl-NMR experiments were performed in a magnetic field around 6 T at temperatures of 4.2 − 280 K and with resonance frequencies between 100 and 140 MHz. Spectra were obtained by plotting the spin-echo amplitude against the applied field. The spin-lattice relaxation rate was measured by the saturation-recovery method with a pulse train.

17.2   Antiferromagnetic Phase of TB1212

17.2.1   Cu-NMR Spectra

Figure 17.1 shows the zero-field spectrum of $^{63/65}$Cu-NMR, where one can see six resonance lines, some of which overlap. These six lines were successfully assigned to the quadrupolar-split Zeeman signals from the single copper site with the two isotopes ^{65}Cu and ^{63}Cu. The parameters of H_0, the internal field, ν_Q, the quadrupolar frequency, and θ, the angle between H_0 and the principal axis of the electric field gradient tensor, were deduced by the numerical diagonalization of the nuclear spin Hamiltonian for $I = 3/2$. We assumed axial symmetry in the electric field gradient, which is a reasonable

Table 17.2. The static parameters of the Cu site in the antiferromagnetically ordered state. The results for other antiferromagnetic phases of the high-T_c oxides La_2CuO_4 and $YBa_2Cu_3O_6$ are also shown for comparison. The hyperfine coupling constant $|A_{ab}^{Cu}-4B^{Cu}|$ is obtained by employing the effective magnetic moment as theoretically estimated, $\mu_{3d} \simeq 0.6\mu_B$

	TlBa$_2$YCu$_2$O$_7$	La$_2$CuO$_4$ [11]	YBa$_2$Cu$_3$O$_6$ [12]
H_{Cu} (T)	8.62	7.878	7.665
$^{63}\nu_Q$ (MHz)	20.44($\pm$1.3)	31.9	22.87
θ (deg)	81($\pm$9)	79	90($\pm$10)
$|A_{ab}^{Cu}-4B^{Cu}|$ (kOe/μ_B)	144	131.3	127.8

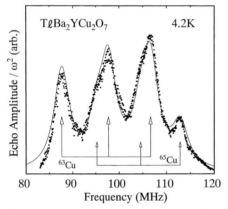

Fig. 17.1. The frequency spectrum of $^{63/65}$Cu-NMR for TlBa$_2$YCu$_2$O$_7$ at 4.2 K and zero field. The solid curve is calculated from the obtained parameters H_{Cu}, $^{63}\nu_Q$ and θ, assuming that each transition line has a Lorentzian form

assumption because of the symmetry of the crystal structure I4/mmm. The obtained parameters are shown in Table 17.2, where the results for La$_2$CuO$_4$ and YBa$_2$Cu$_3$O$_6$ are also given for comparison [11,12]. If we assume that the principal axis of the electric field gradient tensor is along the c-axis, which is also reasonable for the symmetry of the crystal structure, the direction of the $3d$ spin is found to be slightly canted out of the CuO plane.

With the parameters obtained and with the assumption of a Lorentzian form for each resonance line, we reproduced the profile of the spectrum, which is given as the solid line in Fig. 17.1, showing excellent agreement with the observed spectrum. The coefficient of the resonance line width for the Lorentzian is approximately 1.7 kOe (3.9 MHz for full width at half-maximum (FWHM) in the units of frequency) and is almost the same for both the central transition line and the satellites. This proves that the width is mainly contributed by the inhomogeneity in the magnetic field rather than the electric field gradient. That is, if it were a contribution from the latter, the width of the

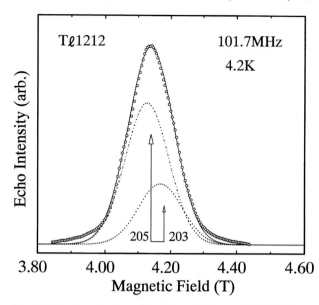

Fig. 17.2. A typical profile of the field-swept Tl-NMR spectrum for TlBa$_2$YCu$_2$O$_7$ at 101.7 MHz. The dashed curves are deconvoluted two-Gaussian forms corresponding to the two isotopes ^{207}Tl and ^{205}Tl

center peak would be much smaller than that of the satellites. Note that the Lorentzian form of the observed spectra is simply due to the distribution of the internal field rather than to the homogeneous broadening. The latter was found to be small enough to be neglected in the measurement of the spin-spin relaxation rate, $T_2^{-1} \simeq 50\,\mu s$ at 4.2 K.

The finite and not negligible line width observed here can be caused by the small number of hole carriers due to the oxygen non-stoichiometry, which has been demonstrated by NMR in La$_{1-x}$M$_x$CuO$_4$ (M = Ba, Sr) in the early stage of the high-T_c cuprate studies [13]. For the compound TlBa$_2$(Y$_{0.95}$Ca$_{0.05}$)Cu$_2$O$_7$, where a small amount of holes are explicitly introduced, no signal was observed in the frequency range 70 − 120 MHz within the signal-to-noise ratio at 4.2 K. This is due to the large inhomogeneity in the ordered field, which wiped out the entire spectrum. A similar phenomenon has been reported for La-based cuprates [13].

17.2.2   Tl-NMR Spectra

Field-swept spectra of $^{203/205}$Tl nuclei were observed at the position of nearly zero internal field; the shift at 4.2 K was $K_s \simeq 0.3\%$. A typical spectrum is shown in Fig. 17.2. One can see an extremely broad peak that was successfully deconvoluted into two Gaussians corresponding to signals from the isotopes ^{203}Tl and ^{205}Tl. The profile of the spectrum was independent of the

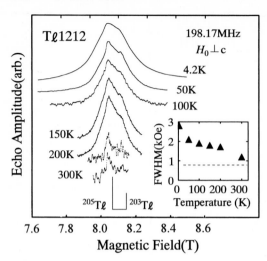

Fig. 17.3. The temperature dependence of the field-swept spectra for TlBa$_2$YCu$_2$O$_7$. At 300 K, the two lines are completely separated, indicating that the system is paramagnetic. The temperature dependence of the resonance line width of the Tl spectra is shown in the inset. The width is tentatively defined as the FWHM of the whole spectrum. The horizontal dashed line shows the separation between the resonance position of the two isotopes

angle between the applied field and the aligned c-axis of the sample. In measurements under various magnetic fields between 3 and 8 T, the separation between the two Gaussians was proportional to the resonance field, reflecting the difference in the gyromagnetic ratios of the two isotopes, $^{203}\gamma = 24.33$ and $^{205}\gamma = 24.567$. On the other hand, the width of each Gaussian, approximately 0.95 kOe, was independent of the resonance field. Profiles of the spectra at several temperature between 4.2 K and 300 K are shown in Fig. 17.3, where one can see a significant decrease in the resonance line width at higher temperatures. At 300 K, the two resonance lines were observed separately. We tentatively extracted the temperature dependence of the width determined at the FWHM of the whole spectrum, which is given in the inset of Fig. 17.3.

17.2.3 Cu-NMR Relaxation Rate T_1^{-1}

The nuclear spin relaxation rate T_1^{-1} was obtained from the recovery curve of the nuclear magnetization after saturation by a pulse train. Since the nuclear spins of ^{63}Cu and ^{65}Cu are $3/2$, the recovery of the magnetization to its thermal equilibrium follows the so-called multi-exponential function $1 - Ae^{t/T_1} - Be^{3t/T_1} - Ce^{6t/T_1}$, where A, B and C are constants that depend on the initial condition or, in other words, the occupation number of the nuclear spin energy levels right after the saturation.

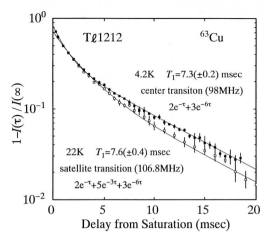

Fig. 17.4. Typical relaxation recovery curves of the Cu-NMR for TlBa$_2$YCu$_2$O$_7$ at the central transition and the satellite. The theoretical relaxation curves are also shown

If the populations of levels other than that of the saturated transition are completely unchanged, (A, B, C) is proportional to $(1,0,9)$ for the central transition between $I = \pm 1/2$, and to $(1,5,4)$ for the satellite transitions between $+1/2 \leftrightarrow +3/2$ or $-1/2 \leftrightarrow -3/2$. Next, if the populations on the levels other than that of the saturated transition are set to be in thermal equilibrium with the neighboring saturated levels, (A, B, C) is proportional to $(2,0,3)$ for the central transition and to $(3,5,2)$ for the satellites. The observed relaxation curves for both the center line and the satellites in Fig. 17.4 follow what is expected for the latter case. This indicates that there does exist a rapid relaxation process which modifies the population of the neighboring levels immediately after the saturation. Note that this relaxation process should be considered anomalously fast, because the Cu spectra spread over the frequency range as wide as few tens of megahertz. The existence of this rapid relaxation has been reported [14] also for La$_2$CuO$_4$ and YBa$_2$Cu$_3$O$_6$ and is characteristic of the antiferromagnetic phase because the relaxation curve for the superconducting phase is well described by the former case. In reports on La$_2$CuO$_4$ and YBa$_2$Cu$_3$O$_6$ by Tsuda [14], the possibility of spectral diffusion is pointed out but the detailed origin of this rapid relaxation between neighboring levels is still not clear.

The temperature dependence of the relaxation rate T_1^{-1} for the central transition line and the satellites is plotted in Fig. 17.5, where the results on La$_2$CuO$_4$ and YBa$_2$Cu$_3$O$_6$ reported by Tsuda [14] are also shown. The relaxation rate for Tl1212 is smaller than those of the other two antiferromagnets by one or two orders of magnitude. The temperature dependence of the relaxation rate is also weak compared with the other two.

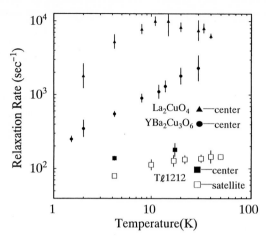

Fig. 17.5. The temperature dependence of Cu- T_1^{-1} for TlBa$_2$YCu$_2$O$_7$. Results on YBa$_2$Cu$_3$O$_6$ and La$_2$CuO$_4$ by Tsuda [14] are also shown for comparison

17.2.4   Tl-NMR Relaxation Rate T_1^{-1}

The relaxation curve of the Tl nuclei is described by the multi-exponential function 1-$I_L e^{t/T_{1L}} - I_S e^{t/T_{1S}}$, where T_{1S}^{-1} is the short component and T_{1L}^{-1} the long one. Typical relaxation curves are shown in Fig. 17.6. The two observed components in the relaxation curves indicate that there exist two Tl sites belonging to different environments, though there is only one crystallographic Tl site. This is because Tl nuclei have spins of 1/2, for which the single exponential recovery is expected in homogeneous systems.

The ratio of the two relaxation times T_{1L}^{-1}/T_{1S}^{-1} and of the two amplitudes I_L/I_S have almost no temperature dependence between 4.2 and 50 K. In order to keep the number of free parameters to a minimum, we kept T_{1L}^{-1}/T_{1S}^{-1} and I_L/I_S at fixed values of 6.3 and 0.43, respectively, in the determination of the relaxation rate by the least-squares method. We show in Fig. 17.7 the temperature dependence of the short component T_{1S}^{-1} with the result of Cu-T_1^{-1} scaled by the gyromagnetic ratios. The relaxation rate of the Tl site T_{1S}^{-1} and, hence, T_{1L}^{-1} were much smaller than the scaled relaxation rate of the Cu site $T_1^{-1}(^{63}\gamma/^{205}\gamma)^{-2}$.

17.2.5   Static Properties of the Antiferromagnetic Phase

The observed spectra of Cu-NMR at zero field, which are explained in terms of Zeeman-splitting with a quadrupole interaction, clearly demonstrate the existence of antiferromagnetic ordering in TlBa$_2$YCu$_2$O$_7$ at low temperatures. The observation of the Tl spectra at nearly zero shift confirms that the magnetic ordering is antiferromagnetic along the c-axis and is hence the three dimensional. The average value of the internal field produced by the

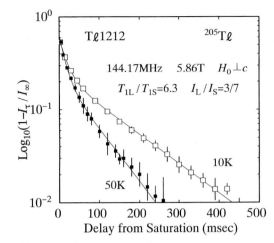

Fig. 17.6. Typical relaxation recovery curves of the Tl-NMR for TlBa$_2$YCu$_2$O$_7$. The relaxation curves were obtained by the least-squares method on the assumption that the relaxation rates have the two components, namely the short term and the long term

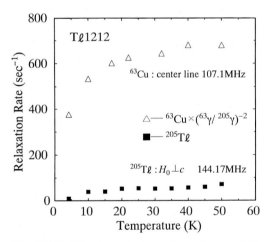

Fig. 17.7. The temperature dependence of the short component of Tl-T_{1S}^{-1} for TlBa$_2$YCu$_2$O$_7$. The relaxation rate for Cu, scaled by the gyromagnetic ratio, is also shown

ordered $3d$-spins is canceled to zero at the Tl site due to the geometrical symmetry. This static cancellation along the c-axis makes a clear contrast to the superconducting phase of Tl1212, where $3d$-spins belonging to the two adjacent CuO planes facing the TlO layer fluctuate incoherently rather than antiferromagnetically [15–17].

From the observed internal field $H_{Cu} \simeq 86.2\,kOe$, we estimated the hyperfine coupling constant of Cu site to be $|A_{ab}^{Cu}\text{-}4B^{Cu}| = H_{Cu}/\mu_{3d} \simeq 144\,kOe/\mu_B$, where A_{ab}^{Cu} and B^{Cu} denote the on-site hyperfine coupling constant [18] within the CuO plane and the transferred hyperfine coupling constant from the neighboring 3d-spin, respectively. The value of the theoretical prediction $\mu_{3d} \simeq 0.6\mu_B$ was adopted for the magnetic moment of the 3d-spins [19]. All of the obtained parameters for the Cu site except for the quadrupolar frequency are comparable to the antiferromagnetic phase of other high-T_c cuprates, as shown in Table 17.2. Here we reach the first conclusion that the existence of three-dimensional antiferromagnetism was confirmed in the Tl-based cuprates, and that the static parameters within the CuO planes are universal with respect to the antiferromagnetic phase of most high-T_c oxides. This is a reasonable consequence of the fact that the static character of the ordered state is rooted in the nature of the Cu-3d spins.

It is not possible that only the quadrupolar frequency $^{63}\nu_Q$ shows a significant difference for the three systems. Since $^{63}\nu_Q$ sensitively depends on the oxygen coordination number, a direct comparison of the observed values seems to be difficult. However, as a qualitative argument, we can see that $^{63}\nu_Q$ is smaller for the system with a smaller tolerance factor between the CuO plane and the block layer [20]. The transferred hyperfine coupling constant at the Tl site can be obtained from the observed width of the resonance lines by the following procedure. While the mean value of the ordered field produced by 3d-spins is canceled at the Tl site as stated above, its inhomogeneity δH_{Cu} is expected to reside in and contribute to the inhomogeneous width of Tl-NMR. This is confirmed by the fact that the observed profile of Tl-spectra was independent of the angle between the sample axis and the applied field, indicating that the internal field at the Tl site is random. Since there are two contributions to the width, from the 3d-spins above and below the Tl site, the inhomogeneity in the hyperfine field at the Tl site is given as

$$\delta H_{Tl} \simeq \sqrt{2}A^{Tl}\delta H_{Cu}/|A_{ab}^{Cu} - 4B^{Cu}|, \qquad (17.1)$$

where A^{Tl} is the transferred hyperfine coupling constant for the Tl site. The factor $\sqrt{2}$ in this formula is due to the fact that δH_{Tl} is the sum of two random variables. Substituting the observed values of δH_{Tl} and δH_{Cu} into (17.1), we obtain $A^{Tl} \simeq 56\,kOe/\mu_B$, which is too large to be explained in terms of the classical dipole-dipole interaction and, hence, suggests the existence of a supertransferred hyperfine interaction from the Cu to the Tl site via the apical oxygen. So far, there has been reported, much experimental evidence that suggests little spin transfer from the Cu to the adjacent non-copper layers in La-based and Y-based cuprates. For example, an extremely small hyperfine field of $1\,kOe$ at the La site in La_2CuO_4 was reported by Nishihara [21] and explained [22] in terms of the anti-bonding between the $2p_{\sigma z}$ state of apical oxygen and the $3d_{x^2-y^2}$ state, which is believed to be the ground state of the Cu hole. Kanamori et al. [22] have shown by a cluster model calculation that the supertransferred hyperfine interaction to the La site through the apical oxygen does not exist if one assumes the 6 s band in the La atom. Also, for the superconducting phase of YBa$_2$Cu$_3$O$_7$, a very small relaxation rate and Knight shift of the apical oxygen have been reported [23], suggesting the unlikeliness of a superexchange interaction between the plane site Cu and the block layer.

On the other hand, in the Tl-based systems of Tl1212 and Tl2201 (Tl$_2$Ba$_2$ CuO$_{6-\delta}$), the existence of the supertransferred hyperfine interaction has been suggested in the early stage of the study to explain the observed large relaxation rate of the Tl site [24,25]. Brom et al. [25] suggest the existence of a small atomic distortion so as to avoid the problem of the anti-bonding between $p_{\sigma z}$ and $3d_{x^2-y^2}$. According to them, it is possible to explain the large relaxation rate at the Tl site if the apical oxygen is moved only 0.17 A from the plumb line passing the Cu site. However, this idea does not seem to be likely because the structural distortion is more significant in La-based systems, where the existence of a large tilting of the CuO$_6$ octahedra is reported [26]. Still another idea, which seeks the origin of the large hyperfine coupling constant in the atomic character of Tl, is also unlikely because it contradicts the theoretical calculation by Kanamori [22], which proposes that the hyperfine interaction is small, assuming the 6 s band. So far, the experimental evidence for the large hyperfine coupling between the block layer and the Cu site has been reported only for Tl-based systems. Since this issue is closely related to the ground-state symmetry of the Cu-3d band, a theoretical reinvestigation seems to be necessary.

Finally in this section, we give a detailed account for the observed spin canting of approximately 8° in the Tl1212 system. The existence of the spin canting has been reported also for La$_2$CuO$_4$ by NMR [11] and neutron scattering [27] experiments and is interpreted in terms of the Dzyaloshinsky-Moriya interaction between 3d-spins. Generally, the Dzyaloshinsky-Moriya interaction, the form of which is $\boldsymbol{D} \cdot \boldsymbol{S}_i \times \boldsymbol{S}_j$, is induced by the spin-orbit interaction between the two spins when the middle point of the two is *not* an inversion center. Since the interaction constant $\boldsymbol{D}$ can be determined to some extent by the symmetry, we try to examine the case of Tl1212. First, the CuO planes in Tl1212 are of the bilayer type, so that there is no inversion center at the middle point of the two nearest-neighbor 3d spins. By consideration of the symmetry [28,29], one can easily find $\boldsymbol{D}$ to be proportional to $(0,0,d_x)$, where d_x is constant. Therefore, the interaction form $\boldsymbol{D} \cdot \boldsymbol{S}_i \times \boldsymbol{S}_j$ contains the spin component of S_z, which is consistent with the experimental observation of the canting out of the CuO plane.

In the case of La$_2$CuO$_4$, on the other hand, the inversion center at the middle point of the two nearest-neighbor spins is lost only when the large buckling in the CuO planes is brought by the structural phase transformation [26] around 500 K from the tetragonal phase ($I4/mmm$) to the orthorhombic phase ($Cmca$). It has been argued that this orthorhombic distortion [26] plays a crucial role for the occurrence of the Dzyaloshinsky-Moriya interaction and, hence, of the spin canting. Noting that the spin canting in Tl1212 is driven by the intrinsic crystal symmetry rather than by the structural instability, one can see that the mechanism of the spin canting is quite different for Tl1212 and La_2CuO_4, though they show similar experimental results.

17.2.6 Dynamical Properties of the Antiferromagnetic Phase

The mechanism of the nuclear spin relaxation by magnon processes in antiferromagnets had been studied intensively by Beeman and Pincus [30] in 1968. For the antiferromagnetic phase of high-T_c cuprates, Chakravarty et al. [31] investigated theoretically the two-or three-magnon process by taking the two-dimensionality into account to report the strong temperature dependence of T^2 or T^3 for the nuclear spin relaxation rate. However, according to the report by Tsuda et al. [14], the temperature dependence of T_1^{-1} for both $YBa_2Cu_3O_6$ and La_2CuO_4 is much weaker than expected from the theoretical prediction. Also, as is clear for Tl1212, the observed temperature dependence of T_1^{-1} is not explained in terms of the conventional magnon theory. Therefore, we would like to present mainly qualitative arguments here.

First, we show that the dominant relaxation mechanism for both the Cu and the Tl site is the spin fluctuation of Cu-$3d$ spins and that other mechanisms such as paramagnetic centers or the electric quadrupolar interaction are not the main contribution. The nuclear relaxation rate is generally described by Kubo's formula as

$$T_1^{-1} \propto T\gamma_n^2 \sum_q |A_q|^2 \chi''(\boldsymbol{q},\omega_0)/\omega_0 , \qquad (17.2)$$

where γ_n is the gyromagnetic ratio, A_q the hyperfine coupling constant, $\chi''(q,\omega)$ the dynamical susceptibility, and ω_0 the Larmor frequency of the nuclear spin. If the nuclear spin relaxation at both the Cu and Tl sites is driven by a single spin degree of freedom, one expects that the scaled relaxation rate $T_1^{-1}\gamma_n^{-2}|A^{-2}|$ for Cu and Tl must be equal. The hyperfine coupling constants for the two sites have been already obtained by the analysis of the spectra in the previous section. In Fig. 17.8, we present the scaled relaxation rates ^{63}Cu-T_1^{-1} and ^{205}Tl-T_{1S}^{-1} to indicate that both the magnitude and the temperature dependence of the two sites are nearly scaled, considering the ambiguity in the determination of hyperfine coupling constants. Therefore, we can conclude that the relaxation at both the Cu and Tl sites is driven by the single relaxation mechanism and that this relaxation mechanism is magnetic because the Tl nuclear spin is free from the electric quadrupole disturbance. The possibility of the paramagnetic center as a relaxation mechanism is also denied because the measurement of the Tl site is under the magnetic field of the approximately 6 T which is usually high enough to suppress the spin fluctuation of paramagnetic centers.

As a consequence of the above observation, we can draw another conclusion, that there is *little* antiferromagnetic spin correlation between the two

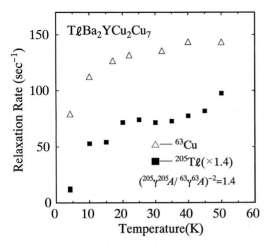

Fig. 17.8. The comparison of the relaxation rate for the Cu and Tl sites in TlBa$_2$YCu$_2$O$_7$, which are scaled by the factor of $\gamma_n^{-2}|A|^{-2}$

$3d$ spins on the adjacent CuO planes facing each other across the TlO layer. In other words, most of the nearest neighboring Cu-$3d$ spins above and below the Tl site fluctuate incoherently, so that both spins contribute to the relaxation at the Tl site. If, on the other hand, there is an antiferromagnetic spin correlation between bilayers, the magnetic field induced by the $3d$ spins will be canceled out at the Tl site due to the geometric symmetry, and the relaxation rate at the Tl site must become much smaller than observed.

In this argument we have concentrated on the short component of the relaxation rate of the Tl site, T_{1S}^{-1}. The long component in the Tl site relaxation, T_{1L}^{-1}, is explained consistently if there exists a small number of Tl sites where the geometrical cancellation partially holds. Those Tl sites are expected to contribute to the long component of the relaxation. Considering the magnetization fraction of the two relaxation components, $I_S/I_L \simeq 3.7$, the geometric cancellation holds only for a limited number ($\sim 20\%$) of Tl sites, while it is broken for the other dominant Tl sites, possibly as a result of a thermal disturbance or the inhomogeneity in the sample. This result cencerning the spin correlation is in significant contrast to that for the superconducting phase of Tl1212, where the antiferromagnetic spin correlation between bilayers is completely lost [15–17].

Next, we examine whether or not the obtained hyperfine coupling constants in the antiferromagnetic phase are the same as those in the superconducting phase. Following the procedure by Kitaoka et al. [32], we first assume that $\chi''(q, \omega_0)$ of the superconducting phase is enhanced around the antiferromagnetic vector $q_{AF} \simeq (\pi, \pi)$. Then the ratio of the relaxation rates of the Cu and Tl sites for the superconducting phase is expressed as

$$\frac{^{205}T_1^{-1} \,.205\, \gamma^{-2}}{^{63}T_1^{-1} \,.63\, \gamma^{-2}} = \frac{2(A^{\mathrm{Tl}})^2}{(A_{ab}^{\mathrm{Cu}} - 4B^{\mathrm{Cu}})^2 + (A_c^{\mathrm{Cu}} - 4B^{\mathrm{Cu}})^2}, \qquad (17.3)$$

where $^{205}\gamma$ and $^{63}\gamma$ are the gyromagnetic ratio of Tl and Cu, respectively, and A_c^{Cu} is the Cu hyperfine coupling constant parallel to the c-axis. We compare the experimentally obtained value for the left-hand side with the right-hand side calculated from the hyperfine coupling constants. Making a further assumption [32] that the on-site hyperfine coupling constants A_{ab}^{Cu} and A_c^{Cu} are nearly the same for most high-T_c cuprates with $A_{ab}^{\mathrm{Cu}} \simeq 30\,\mathrm{kOe}/\mu_{\mathrm{B}}$ and $A_c^{\mathrm{Cu}} \simeq 160\,\mathrm{kOe}/\mu_{\mathrm{B}}$, we can estimate the transferred hyperfine coupling constant B^{Cu} to be $43.4\,\mathrm{kOe}/\mu_{\mathrm{B}}$. By inserting these constants into (17.3), the right-hand side is calculated to be 0.047, which almost reproduces the observed value of $(^{205}T_1^{-1} \,.205\, \gamma^{-2})/(^{63}T_1^{-1} \,.63\, \gamma^{-2}) \simeq 0.077$ for the superconducting phase, with $T_c \simeq 78\,\mathrm{K}$, belonging to the slightly overdoped region [15–17]. This agreement indicates that the hyperfine coupling constants for the Cu site of Tl1212 do not change much from the antiferromagnetic phase to the slightly overdoped region.

Finally, we compare the Cu-T_1^{-1} of Tl1212 with that of the antiferromagnets $YBa_2Cu_3O_6$ and La_2CuO_4 to search for the hidden key parameter related to T_c. As shown in Fig. 17.5, the significant difference in the magnitude of Cu-T_1^{-1} among the three systems suggests the difference in the dynamic character of CuO planes for various systems. Since these three systems have comparable hyperfine coupling constants for the Cu site, as was revealed in 3–3, the difference in T_1^{-1} is directly related to that in $\chi''(q \simeq q_{\mathrm{AF}}, \omega_0)$, which is a measure of the spectral weight of the spin fluctuation of $3d$ spins at a low energy of $E = \hbar\omega_0$. Therefore, the smaller T_1^{-1} suggests that the center-of-mass in the spin excitation spectrum is shifted to the higher energy region. Consequently, we can conclude that the characteristic spin fluctuation energy, usually denoted [33] as Γ, of Tl1212 is higher than that of $YBa_2Cu_3O_6$ and La_2CuO_4.

Theoretical arguments on the importance the spin fluctuation with a rather high characteristic energy have been repeatedly proposed. Pines [33] and Moriya [34] reported independently that T_c for the spin fluctuation-induced superconductivity is nearly proportional to Γ. This prediction was supported first by Imai [18] and later by Kitaoka [35,36] from NMR for various high-T_c cuprates of the superconducting phase. Kitaoka extracted $\chi''(q \simeq q_{\mathrm{AF}}, \omega_0)$ from T_1 data to show that Γ in YBCO ($T_c \simeq 90\,\mathrm{K}$) is possibly higher than that in LSCO ($T_c \simeq 35\,\mathrm{K}$). He attributed this difference in Γ to the different hole carrier concentration between YBCO and LSCO. The direct measurement of the superconducting carrier concentration was performed by Uemura's μSR experiments, which agree qualitatively with Kitaoka's speculation. Combining their arguments, we note that the carrier concentration may determine the spin fluctuation, which in turn may determine T_c. Now, let us turn to our NMR results on the antiferromagnetic phase,

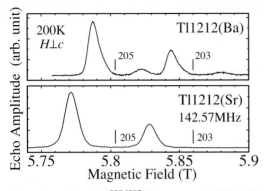

Fig. 17.9. Typical $^{203/205}$Tl-spectra for TlSr$_2$(Ca,Y)Cu$_2$O$_7$. The solid lines indicate zero-shift positions

where the intensity of the spin fluctuation increases from Tl1212, YBCO to La$_2$CuO$_4$. This order coincides with that in the superconducting phase [35], suggesting the possibility that we can predict the intensity of the spin fluctuation in the superconducting phase, and hence even T_c, by investigating the spin fluctuations in the antiferromagnetic phase.

17.3   Lightly Doped and Slightly Overdoped Phase of Tl1212

17.3.1   Spectra and Relaxation Rate of Tl-NMR

Figure 17.9 shows typical spectra for the two isotopes of ^{203}Tl and ^{205}Tl, both of which have $I = 1/2$. Though there is one crystallographic Tl site in the two systems, TB1212 shows spectra with extra peaks corresponding to another Tl site. The signal of these extra small peaks comes from the Tl atoms located at the Ca site by the partial atomic exchange between Ca and Tl. The same exchange phenomenon was reported [37] for TlBa$_2$CaCu$_2$O$_{8+\delta}$. The amplitude ratio of the Ca-site peak to the Tl-site peak is approximately 10%. The spectra of TS1212 were explained with a single Tl site, indicating that atomic exchange does not take place in this system.

Measurements of T_1 were done at the Tl-site peak for TB1212. The relaxation curve followed a single exponential function. In TS1212, the curve was fitted by a function with two relaxation rates, $1-I_S e^{t/T_{1S}}-I_L e^{t/T_{1L}}$, to obtain the four parameters I_S, T_1^S, I_L and T_1^L.

Figures 17.10 and 17.11 show the temperature dependence of the Knight shift and T_1^{-1} for TS1212. The Knight shift is temperature independent for the optimally doped sample and decreases at low temperatures for the underdoped one, the behavior of which is consistent with other cuprates like YBCO and La214. The spin-lattice relaxation curve consists of two compo-

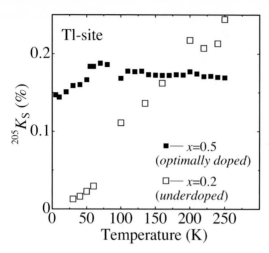

Fig. 17.10. The temperature dependence of the Knight shift for $TlSr_2(Ca,Y)Cu_2O_7$

nents. The amplitude ratios $I_L/I_S = 0.2$ for $x = 0.2$ and 0.1 for $x = 0.1$ stay constant in the measured temperature range.

The temperature dependence of these two components differ. The short component T_S^{-1} obeys the Curie–Weiss law in the higher temperature region of the normal state, reflecting the two-dimensional antiferromagnetic spin fluctuations [34,38]. It shows a reduction at the temperature T_{SG} far above T_c. Following [39], we made an Arrhenius plot of $(T_1T)^{-1}$ divided by the Curie–Weiss function as shown in Fig. 17.12. One clearly sees a gap-like behavior below T_{SG}. This result makes a clear contrast with [40], where $(T_1T)^{-1}$ showed only a gradual reduction with a broad maximum. The long component T_L^{-1} obeys the Korringa relation in the normal state, which also differs from [40], where the two components were proportional each other. Both components showed neither a kink nor an abrupt change at T_c.

For TB1212, the temperature dependence of the Knight shift $-K_S$ and T_L^{-1} are shown in Figs. 17.13 and 17.14. First, T_L^{-1} in the higher temperature region of the normal state follows the Curie-Weiss law [38]. With decreasing temperature, it starts to deviate from the law at $T_{SG} = 170\,K$ for the optimally doped sample and at $140\,K$ for the slightly overdoped one. The optimally doped sample even shows a reduction of $(T_1T)^{-1}$ in the normal state. The knight shifts shown in Fig. 17.13, where the result of Zn-doped samples is also given, are for the Tl atom located at the Ca site. The shift of the original Tl site is constant against temperature within $\pm0.1\%$. For all the samples, $-K_S$ at the Ca site is temperature independent in the highest temperature region. With decreasing temperature they start to decrease from a temperature far above T_c. This temperature coincides with the T_{SG} value determined from the relaxation data. For Zn-doped samples, the T_{SG}

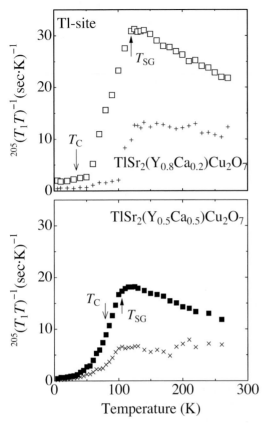

Fig. 17.11. The temperature dependence of Tl-$(T_1T)^{-1}$ for TlSr$_2$(Ca,Y)Cu$_2$O$_7$

determined from the shift shows a noticeable decrease compared to the pure samples.

The position of the Ca site is equivalent to the Y site in YBCO, so that it well probes the static spin susceptibility of the CuO plane. The negative sign of the shift comes from the fact that the transferred hyperfine coupling constant is negative [41]. The temperature where -K_S (Tl at the Cl site) starts to decrease coincides with the T_{SG} determined from T_1. The magnitude of the shift of Tl is much larger than that of Y in YBCO. This is simply because the hyperfine coupling constant of the Tl-6s orbital is very large. From the K-K plot between the Tl and Ca sites, and from the hyperfine coupling constant of the original Tl-site [6], A(Tl) = 56 kOe/μ_B, one can estimate the value of A(Tl at the Ca site) to be 280 kOe/μ_B.

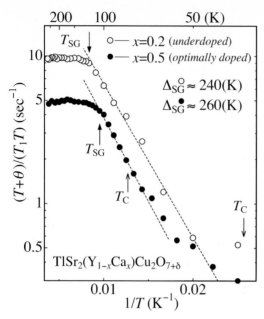

Fig. 17.12. The Arrhenius plot for $(T_1 T)^{-1}$ for $TlSr_2(Ca,Y)Cu_2O_7$ divided by the Curie-Weiss factor

17.3.2   Spin-Gap in the Tl-Based System

The steep reduction in $(T_1 T)^{-1}$ of TS1212 from T_{SG} indicates the existence of the spin-gap in underdoped solid-solution systems. For the spin-gap is easily smeared out by the small amount of impurity doping to the CuO plane [42], this result indicates that the gap is maintained when the disorder is introduced to a position other than the CuO plane. The gap-opening temperature T_{SG} decreases upon increasing the hole concentration, the behavior of which agrees qualitatively with the theoretical models [1,2] and experimental results [3] on YBCO. These two results support the case for a universal phase diagram for high-T_c cuprates.

The gap energy of $240 - 260$ K is considerably larger than that for other cuprates, e.g. 144 K for Y1248 and 113 K for 60 K-class YBCO [3,39]. Here, we must note that the analysis by the Arrhenius plot is allowed only when the symmetry of the gap is spherical. For a further discussion on the relation between the spin-gap and the superconducting gap, a more detailed theoretical analysis is necessary.

Next, in the slightly overdoped TB1212, $(T_1 T)^{-1}$ deviates from the Curie–Weiss law from the temperature TSG that is much higher than T_c. This indicates the existence of the spin-gap even in the overdoped region. For the Knight shift also starts to decrease at T_{SG}, and the possible spin-gap opens inhomogeneously in q-space. The change of the uniform spin susceptibility due to this gap must be very small, because the temperature dependence of

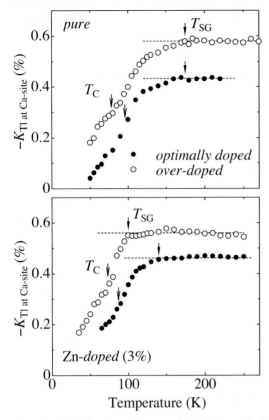

Fig. 17.13. The temperature dependence of the Knight shift for TlBa$_2$YCu$_2$O$_7$ with and without Zn-doping

the Cu-Knight shift is very small (within $\pm 0.01\%$), as we reported previously [43].

The significant reduction of the gap-opening temperature T_{SG} by Zn-doping indicates that the spin-gap in this region also tends to be smeared by impurity scattering. However, there is a difference between the spin-gap in the lightly doped region and the overdoped region. In the lightly doped region, the spin-gap is completely smeared out by introducing 3% Zn into the CuO plane. Our results indicate that, as though T_{SG} for the spin-gap is substantially reduced, the gap survives. This is consistent with the fact that the scattering by Zn is in the unitarity limit in high-T_c cuprates, where the scattering amplitude decreases with hole concentration [42,43].

The fact that the magnitude of $(T_1 T)^{-1}$ is nearly the same for the optimally doped and the overdoped samples suggests that the hyperfine coupling constant $A(\text{Tl})$ increases in the overdoped region, possibly because of an increase in the itinerancy of the doped carriers.

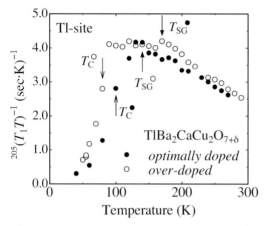

Fig. 17.14. The temperature dependence of $(T_1 T)^{-1}$ for $TlBa_2YCu_2O_7$

Finally, we comment on the difference in T_c between TB1212 and TS1212. As stated above, the state of the as-sintered sample is quite different between the two systems; $TlBa_2CaCu_2O_{7+\delta}$ is in the slightly overdoped state with $T_c = 80$ K, while TlSr2CaCu2O7+d in the completely overdoped state with $T_c = 0$. This difference is possibly due to the fact that the excess oxygen δ in the TlO layer of TB1212 is smaller [9,10] than that for TS1212. The atomic exchange between Tl^{3+} and Ca^{2+}, existing only in TB1212 and which we have revealed by NMR, may raise the potential of the TlO layer to the oxygen ion and may prevent excess oxygen.

Then, the remaining question is why this atomic exchange takes place only in TB1212 but not in TS1212. This can be considered as follows. Since the ionic radii of Sr^{2+} and Ca^{2+} are very close, the atomic exchange is expected to take place also for these two, as is observed for the Bi2212 phase. This exchange causes a distribution in the ionic radius at the Ca site, which prevents Tl from coming to the Ca site. In the Ba system, because of the considerable difference in ionic radii between Ba and Ca, the atomic exchange between the two does not take place, which then allows the Tl to come to the Ca site.

17.4   Conclusion

Our recent Cu/Tl-NMR work at the HFLSM-IMR on the Tl-based high-T_c cuprate $Tl(Ba,Sr)_2(Y,Ca)Cu_2O_7$ with a wide range of hole concentration is reviewed. In the end member $TlBa_2YCu_2O_7$, a zero-field resonance was observed, demonstrating the existence of antiferromagnetism. In the lightly-doped region of $TlSr_2(Ca,Y)Cu_2O_7$, a clear spin-gap was observed as a depression of the NMR relaxation rate. A tentative estimation of the gap energy from the data gave a value around 250 K. These two observations support the universality of the theoretical electronic phase diagram for high-T_c cuprates from the antiferromagnetic region to the overdoped region. As for the slightly overdoped samples, both the Knight shift and the relaxation rate $(T_1 T)^{-1}$ showed a significant decrease for temperatures much higher than T_c, indicating that a spin-gap exists in this region and that the gap in momentum space is inhomogeneously distributed. With regard to impurity doping of Zn, the gap-opening temperature showed a significant reduction.

Acknowledgments

The authors are grateful to Prof. M. Kataoka at Tohoku University for a kind and valuable discussion. Most of experiments for this work were performed at the High Field Laboratory for Superconducting Materials (HFLSM) at IMR, Tohoku University. This work was supported by the grant-in-aid for Scientific Research from the Ministry of Education Science and Culture.

References

1. N. Nagaosa and P.A. Lee: Phys. Rev. **B46**, 5621 (1992)
2. T. Tanamoto, H. Kohno and H. Fukuyama: J. Phys. Soc. Jpn. **63**, 2739 (1994)
3. M. Matsumura, Y. Sakamoto, T. Fushihara, Y. Itoh and H. Yamagata: J. Phys. Soc. Jpn. **B64**, 721 (1995)
4. K. Ishida, K. Yoshida, T. Mito, Y. Tokumaga, Y. Kitaoka, K. Asayama, Y. Nakayama, J. Shimoyama and K. Kishio: Phys. Rev. **B58**, R5960 (1998)
5. C. Renner, B. Revaz, J.-Y. Genoud, K. Kadowaki and O. Fischer: Phys. Rev. Lett. **80**, 149 (1998)
6. T. Goto, S. Nakajima, M. Kikuchi, Y. Syono and T. Fukase: Phys. Rev. **B53**, 3562–3570 (1996)
7. T. Goto, S. Nakajima, E. Ohshima, M. Kikuchi, Y. Syono and T. Fukase: J. Low Temp. Phys. **117**, 467–471 (1999)
8. S. Nakajima, M. Kikuchi, Y. Syono, N. Kobayashi and Y. Muto: Physica **C168**, 57 (1990); *ibid.*, **C170**, 443 (1990)
9. E. Ohshima, M. Kikuchi, M. Nagoshi and Y. Syono: Physica **C263**, 189 (1996)
10. S. Nakajima, M. Kikuchi, Y. Syono et al.: Physica **C182**, 89 (1991)
11. T. Tsuda, T. Shimizu, H. Yasuoka, K. Kishio and K. Kitazawa: J. Phys. Soc. Jpn. **57**, 2908 (1988)
12. H. Yasuoka, T. Shimizu, Y. Ueda and K. Kosuge: J. Phys. Soc. Jpn. **57**, 2659 (1988); Y. Yamada, K. Ishida, Y. Kitaoka, K. Asayama, H. Takagi, H. Iwabuchi and S. Uchida: J. Phys. Soc. Jpn. **57**, 2663 (1988)
13. Y. Kitaoka, S. Hiramatsu, K. Ishida, T. Kohara and K. Asayama: J. Phys. Soc. Jpn. **56**, 3024 (1987)
14. T. Tsuda, T. Ohono and H. Yasuoka: J. Phys. Soc. Jpn. **61**, 2109 (1992)
15. T. Goto, T. Shinohara, T. Sato, S. Nakajima, M. Kikuchi, Y. Syono, K. Miyagawa and T. Fukase: Advances in Superconductivity V (Springer, Berlin, 1993) p. 133
16. T. Goto, T. Shinohara, T. Sato, S. Nakajima, M. Kikuchi, Y. Syono and T. Fukase: Physica **C185–189**, 1077–1078 (1991)

17. T. Goto, T. Shinohara, T. Sato, S. Nakajima, M. Kikuchi, Y. Syono and T. Fukase: Jpn. J. Appl. Phys., Series 7, Mechanisms of Superconductivity (1992) p. 197

18. T. Imai: J. Phys. Soc. Jpn. **59**, 2508 (1990)

19. T.E. Manousakis: Rev. Mod. Phys. **63**, 1 (1991)

20. P. Ganguly and C.N.R. Rao: J. Solid. State. Chem. **53**, 193 (1984)

21. H. Nishihara, H. Yasuoka, T. Shimizu, T. Tsuda, T. Imai, S. Sasaki, S. Kanbe, K. Kishio, K. Kitazawa and K. Fueki: J. Phys. Soc. Jpn. **B56**, 4559 (1987)

22. M. Takahashi, T. Nishio and J. Kanamori: J. Phys. Soc. Jpn. **60**, 1365 (1991)

23. Y. Yoshinari, H. Yasuoka, Y. Ueda, K. Koga and K. Kosuge: J. Phys. Soc. Jpn. **59**, 3698 (1990)

24. H.B. Brom, D. Reefman, J.C. Jol, D.M. de Leeuw and W.A. Groen: Phys. Rev. **B41**, 7261 (1990)

25. H.B. Brom, D. Reefman and J.C. Jol: Phys. Rev. **B41**, 7261 (1990)

26. N.E. Bonesteel: Phys. Rev. **B47**, 9144 (1993)

27. M.A. Kastner, R.J. Birgeneau, T.R. Thurston, P.J. Picone, H.P. Jenssen, D.R. Gabbe, M. Sato, K. Fukuda, S. Shamoto, Y. Endoh, K. Yamada and G. Shirane: Phys. Rev. **B38**, 6636 (1988)

28. T. Moriya: Phys. Rev. **120**, 91 (1960)

29. D. Coffey, T.M. Rice and F.C. Zhang: Phys. Rev. **B44**, 10112 (1991)

30. D. Beeman and P. Pincus: Phys. Rev. **166**, 359 (1968)

31. S. Chakravarty, M.P. Gelfand, P. Kopietz, R. Orbach and M. Wollensak: Phys. **B43**, 2796 (1991)

32. Y. Kitaoka, K. Fujiwara, K. Ishida, K. Asayama, Y. Shimakawa, T. Manako and Y. Kubo: Physica **C179**, 107 (1991)

33. P. Monthoux and D. Pines: Phys. Rev. **B47**, 6069 (1993)

34. T. Moriya and K. Ueda: J. Phys. Soc. Jpn. **63**, 1871 (1994)

35. Y. Kitaoka, K. Ishida, G.-Q. Zheng, S. Ohsugi, K. Fujiwara and K. Asayama: Jpn. J. Appl. Phys., Series 7, *Mechanisms of Superconductivity* (1992) p. 185; J. Phys. Chem. Solids **53**, 1385 (1993)

36. G.-Q. Zheng, K. Magishi, Y. Kitaoka, K. Asayama, T. Kondo, Y. Shimakawa, T. Manako and Y. Kubo: Physica **B186–188**, 1012 (1993)

37. K. Fujiwara, Y. Kitaoka, K. Asayama et al.: J. Phys. Soc. Jpn. **57**, 2893 (1988)

38. T. Moriya, Y. Takahashi and K. Uchida: J. Phys. Soc. Jpn. **59**, 2905 (1990)

39. H. Yasuoka, S. Kambe, Y. Itoh and T. Machi: Physica **B199**, 278 (1994)

40. K. Magishi, Y. Kitaoka, K. Asayama et al.: Phys. Rev. **B54**, 3070 (1996)

41. H. Alloul, T. Ohno et al.: J. Less-Common Metals, **164–165**, 1022 (1990)

42. K. Ishida, Y. Kitaoka, N. Ogata, T. Kamino, K. Asayama, J.R. Cooper and N. Athanassopoulow: J. Phys. Soc. Jpn. **62**, (1993) 2803

43. T. Goto, S. Nakajima, M. Kikuchi, Y. Syono and T. Fukase: J. Phys. Soc. Jpn. **65**, 3666 (1996)

Part VI

Chemistry, Biology and Crystal Growth in High Fields

18 Magnetic Levitation

M. Motokawa

In magnetic levitation, the magnetic force acting on a material balances the gravitational force and thus results in stable levitation of the material in space without contact to a container or a magnet pole. This effect was expected when Faraday discovered in 1846 the diamagnetic effect, i.e. a repulsive force on a nonmagnetic material against the magnetic flux. It was only in 1930 that the first magnetic levitation was performed for Bi, which has a large diamagnetic susceptibility. Since then some other materials such as graphite have been tried [1]. The most striking experiment was that done by Beaugnon and Tournier [2] in 1991. They succeeded in levitating water and some diamagnetic materials with a diamagnetic susceptibility that is much smaller than those of materials tried previously. Since then magnetic levitation has become one of the hot topics in high magnetic field science. At first this effect attracted attention only due to curiosity, because the levitated objects were wine, plastics and even frogs. This was enough to surprise the general public, who are not familiar with the diamagnetic properties of materials, because usually these are negligibly small. To perform these demonstrations, we need a very high magnetic field with a strong gradient; now every high-field facility in the world is interested in magnetic levitation. When liquid He was levitated, a strange behavior of the motion was observed [3]. The Grenoble group is doing experiments on the dynamics of droplets [4], the Nijmegen group is interested in demonstrating this effect using frogs and some other materials [5–7]. At the National High Magnetic Field Laboratory of the US, experiments to observe the dynamics of granular materials and chaotic behavior are under way [8]. At Sendai, we are now interested in growing crystals, in synthesizing new materials and in material processing [9–11]. These experiments are summarized in the proceedings of the 6th International Symposium on Research in High Magnetic Fields [12]. The first experiments were for diamagnetic materials; the levitation of paramagnetic or ferromagnetic materials was considered to be difficult due to Earnshaw's theorem. As a matter of fact, one realizes that a ferromagnetic material like iron is attracted by a magnet and easily contacts the magnet pole; this means that stable levitation can never be achieved for a magnetic material. However, Kitazawa has found a trick for levitating other materials, even paramagnetic materials, in a fairly low field using a paramagnetic gas environment [13]. In addition, Geim quite recently demonstrated a method for levitating even a ferromagnetic material [14]. The advantages of crystal growth and material processing under levitation conditions are :

1. containerless crystal growth, which provides a clean environment that is free from contamination caused by a container and from stress,
2. suppression of uncontrollable heterogeneous nucleation,
3. easy supercooling and easy supersaturation and
4. high-temperature heating and melting of a material without a crucible.

These techniques are expected to open up a new technology for the synthesis of novel materials. This is in part similar to the experiments being done or planned in the space laboratory. However, magnetic levitation is much more economical than the cost of going into space. In this chapter, the details of magnetic levitation are explained and some experimental results are described. In particular, the work performed at the High Field Laboratory for Superconducting Materials of the Institute for Materials Research at Tohoku University is described with much detail. It is well known that due to the Meissner effect a superconducting material with large negative susceptibility can easily be levitated above a permanent magnet. But we shall not discuss this phenomenon because it is not related to atomic or molecular properties.

18.1   Levitation by Means of a Magnetic Field

18.1.1   General Principle

All materials are composed of nuclei and electrons; the material properties are mainly due to the behavior of the electrons. An electron has charge and spin and responds to a magnetic field via orbital motion and spin polarization. This response is phenomenologically expressed in terms of the induced magnetization M as

$$M = \frac{\chi}{\mu_0} B \,, \tag{18.1}$$

where μ_0 is the permeability of vacuum and B is the magnetic flux density. χ is the magnetic susceptibility of the material (we use hereafter susceptibility per unit mass, m^3/kg, in the SI(S) unit system, to compare directly with gravity). The origin of χ will be explained later. The magnetic energy E of a material in a magnetic field $\boldsymbol{B}$ is written as

$$E = -\frac{1}{2\mu_0} \chi B^2 \,, \tag{18.2}$$

where B is the magnitude of $\boldsymbol{B}$. The force $\boldsymbol{F}$ acting on a material in a magnetic field is the gradient of E; the z-component parallel to the field direction is expressed as

$$F_z = -\mathrm{grad}_z E = \frac{\chi}{\mu_0} B \frac{\mathrm{d}B}{\mathrm{d}z} \,. \tag{18.3}$$

Here we consider only the case of solenoid-type magnets with vertical fields like a common superconducting magnet or hybrid magnet. According to (18.3), when a material is placed at the center of a magnet, the magnetic field there is supposed to be uniform and then a force is not expected. In this case, the induced magnetic moment undergoes a torque that contributes to the orientation of the material. However, if a material is placed at an off-center position where the field is usually inhomogeneous and thus has a gradient, the material experiences a force as expressed by (18.3). The direction of the force depends on the sign of χ. If a diamagnetic materials with negative χ is placed above the center of the magnet, where the magnetic field is decreasing with increasing z, i.e. $dB/dz < 0$, the magnetic force acting on the material is positive, i.e. $F > 0$. This means that the force is in the upwards direction in a vertical field. If the diamagnetic material is placed below the center of the magnet, the force is directed downwards. In any case, the force is repulsive from the center of the magnet. When the upward force is strong enough to exceed the force due to the gravity, i.e.

$$\frac{\chi}{\mu_0} B \frac{dB}{dz} > g \,, \tag{18.4}$$

where g is the gravitational constant, the material can be levitated. Equation (18.4) can be rewritten using absolute values of field gradient and suceptibility, defining K, as

$$B \left| \frac{dB}{dz} \right| > K \equiv \frac{g\mu_0}{|\chi|} \,. \tag{18.5}$$

In the case of water, K is $1.36 \times 10^3 \, \text{T}^2/\text{m}$. This means that we need a magnet that can produce a gradient field of this magnitude to levitate water. $B|dB/dz|$ changes as a function of z and depends on the field intensity at the center of the magnet. It was calculated using the the specific data of the magnet used for the experiment. Thus $B|dB/dz|$ is a characteristic property of any given magnet. When the field intensity at the center of the magnet is high enough to produce a maximum value of $B|dB/dz|$ above the center of the magnet that is larger than K, there are two crossing points as indicated by a and b in Fig. 18.1; this was calculated using a hybrid magnet installed at the High Field Laboratory for Superconducting Materials of the Institute for Materials Research at Tohoku University. The positions of the crossing points are the places where the magnetic force balances the gravity, and the magnetic force is stronger than the gravity between these points. If the maximum value of $B|dB/dz|$ is smaller than K, there is no crossing point and magnetic levitation never occurs. At the present time, we have a magnet in our facility that produces a gradient field up to $B|dB/dz|=5.3 \times 10^3 \, \text{T}^2/\text{m}$ and therefore we can levitate a material with χ larger than $2.3 \times 10^{-9} \, \text{m}^3/\text{kg}$. In Table 18.1, χ and $K = g\mu_0/|\chi|$ of some materials are shown as examples. We can see that it will be possible to levitate materials for which the absolute value of χ is larger than that of Ag.

266   M. Motokawa

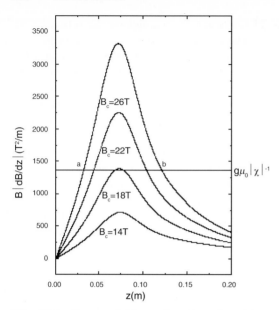

Fig. 18.1. Curves of K as a function of position z with a parameter B_c, the field at the center of the magnet. The horizontal straight line refer to the volume of $g\mu_0|\chi|^{-1}$ of water. The crossing points a and b are the positions where the force due to gravity balances. The abscissa is the distance z from the center of the magnet

Table 18.1. Susceptibilities K and $K = g\mu_0/|\chi|$ of some diamagnetic materials at room temperature. $C_{g//c}$, $C_{g\perp c}$ and C_d mean graphite for $B//c$-axis, graphite for $B\perp c$-axis and diamond, respectively. In order to levitate these materials, a field-gradient product larger than K is needed

	$C_{g//c}$	H_2O	NaCl	C_d	$C_{g\perp c}$	Al_2O_3	Ag
$-\chi(\times 10^{-9}$ m^3/kg)	286	9.05	6.50	6.16	5.03	4.56	3.58
$B(dB/dz)$ ($\times 10^3$ T^2/m)	0.043	1.36	1.89	2.00	2.44	2.70	5.11

	Au	Si	Ge	Cu	InSb	InAs	GaAs
$-\chi(\times 10^{-9}$ m^3/kg)	2.1	1.78	1.39	1.33	1.08	3.41	2.89
$B(dB/dz)$ ($\times 10^3$ T^2/m)	6.91	8.86	9.26	11.4	3.44	3.61	4.26

In order to obtain stable levitation of a material, there must be a potential minimum in the magnetic field along both the z-direction and the radial direction r. The potential energy U of a material in a magnetic field under the influence of gravity is written as

$$U = -\frac{1}{2\mu_0}\mu B^2 + gz + C, \tag{18.6}$$

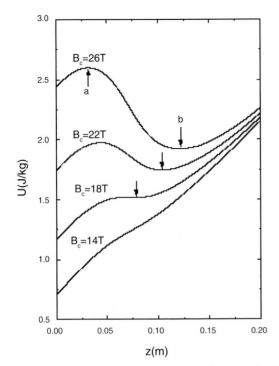

Fig. 18.2. Potential curve for water along the vertical field direction from the center of the magnet

where C is an arbitrary constant. Figure 18.2 shows the potential energy for water calculated using the real data of our magnet along the z-axis at the center of the bore, i.e. for $r = 0$. The crossing points a and b indicated in Fig. 18.1 correspond to the maximum and minimum points of the potential, respectively, as shown in Fig. 18.2. It is evident that point a is the unstable equilibrium position. Point b is the stable equilibrium position, but if $B|dB/dz|$ is equal to or smaller than K, there is no minimum in U. In this case, it is impossible to obtain stable levitation. As the center field is increased, the potential minimum becomes deeper. So it looks as if the levitation gets more stable as the central field is increased. However, the potential energy along the radial direction r calculated from the data of our magnet changes from concave to convex as shown in Fig. 18.3 and the position of the potential minimum along the z-direction is no longer the stable position with respect to the r-direction. This means that the material will move towards the inner wall of the magnet bore. So we have to determine the best position carefully according to the specific data of the magnet used and the susceptibility of the material. This condition can also be discussed using equipotential curves as shown in Fig. 18.4. The small space enclosed by the

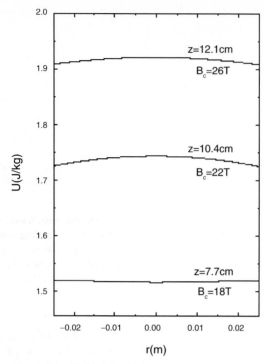

Fig. 18.3. Potential curve for water along the radial direction. When $B_c = 18\,\text{T}$, the potential is concave, but when $B_c = 22\,\text{T}$, it is convex and the material is no longer stable at the bore center

equipotential curves shown in Fig. 18.4a is the stable area, but in the case of an open equipotential curve like Figs. 18.4b, c, there is no stable point.

Magnetic levitation is considered to be almost equivalent to microgravity or zero gravity because the magnetic force acts on each atom or molecule composing the material. This point is completely different from other levitation techniques like levitation by means of blowing air, a static electric field or a high-frequency induction field for metals. In the space where microgravity is supposed to be, the potential is uniform and there is no potential minimum, which means that it is difficult to fix a material stably at a certain position without any other additional technique. Magnetic levitation is different in this respect. However, the existence of a potential minimum inevitably contradicts complete zero gravity over the finite size of a material, i.e. the gravity varies, if only slightly, as a function of position in the material. The perfectly balanced position is only at the potential minimum; at other positions in the potential well, gravity acts on the material more or less. The distribution of gravity in the potential well depends on, depth and width of the well. In the case of our hybrid magnet, the gravity is estimated to vary by up to $10^{-2}g$ in a 5 mm diameter globe. Brooks reported that it varies from 10^{-1} to $10^{-6}g$

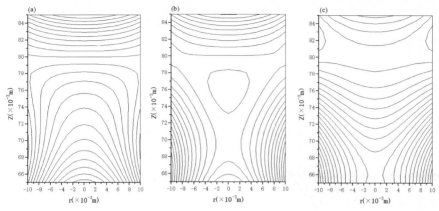

Fig. 18.4. Equipotential curves in a magnet bore. When there is an area closed by an equipotential curve like (**a**), a potential minimum along both the vertical z and radial r direction exists. In the case of no closed area, there is no potential minimum along the r direction in spite of deep potential minimum along the z direction due to too high magnetic field like (**b**) or no potential minimum along both direction due to too low field like (**c**)

for a sample size of 10 to 10^{-1} mm for the resistive magnet that is used at NHMFL [8]. It is technically possible to make the bottom of the potential well flat and achieve a space with microgravity over a few mm. By integrating (18.4), we obtain,

$$B(z) = \sqrt{B_0^2 - 2Kz}\,, \qquad (18.7)$$

where B_0 is the field strength at $z = 0$. If the magnetic field changes according to this function as the ideal case shown in Fig. 18.5, the potential is flat over the entire field range. In the case of $B_0 = 10\,\text{T}$, for example, the potential is flat from $z = 0$ to $z = 0.036\,\text{m}$. However, it seems to be quite difficult at the present time to build such a magnet due to the limited strength of commercially available superconducting wires.

18.1.2   Necessity of Negative Susceptibility

In the case of paramagnetic or ferromagnetic materials with positive χ, there is also a potential minimum along the z-direction but it is below the center of the magnet, as shown in Fig. 18.6 for the case of Al. In this case, however, the potential along the radial direction r has inevitably a maximum at the center of the magnet bore $r = 0$ and stable levitation without contact with the wall of the magnet bore never occurs except under a special condition [13,14]. The fact that magnetic levitation is possible for diamagnetic materials but not for paramagnetic materials is mathematically shown as follows:

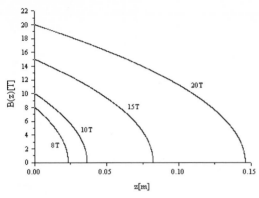

Fig. 18.5. Field distribution of (18.7) for water. If we can make a magnet with this field distribution, we obtain a potential flat space in the magnet bore tube

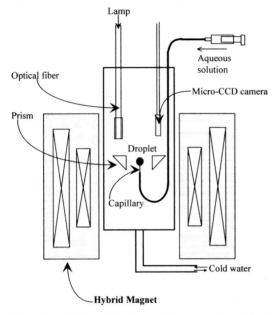

Fig. 18.6. Schematic view of the experimental setup for crystal growth

In general, the condition which must be satisfied to obtain stable levitation is as follows: (i) there must be a position where the magnetic force is zero in order to obtain an optimum of the potential, (ii) the derivative of the force must be negative for the potential to have a minimum. In our case, this is expressed as

$$\mathrm{div}(-\mathrm{grad}\ U(R)) < 0,\tag{18.8}$$

where U is the potential energy defined by (18.6) but is a function of the general coordinate R in this case. This leads to the condition

$$\chi \text{div}(\text{grad } B(R)^2) < 0. \tag{18.9}$$

On the other hand, it is easily shown that $\text{div}(\text{grad } B(R)^2) > 0$. Therefore, (18.9) is satisfied only when χ is negative. It should be emphasized that this is valid only in the case of a static magnetic field; if the magnetic field is controlled by a feedback system or if some special condition is added [13,14], this no longer holds. This is consistent with Earnshaw's theorem that was originally formulated for electric forces. In the case of the electric force, no equilibrium position is expected in a static electric field because there is no matter with $\chi < 0$.

18.1.3   Levitation Condition for Two Coexisting Materials

When a solid material is immersed in a liquid or gas in a magnetic field, the material suffers buoyancy due to Archimedes principle in addition to the magnetic force. This was first discussed by Kitazawa's group [13]. In this case, the total force per unit volume for the system composed of the solid material and the surrounding material is expressed as

$$F = \frac{1}{\mu_0}(\rho_1\chi_1 - \rho_2\chi_2)B\frac{dB}{dz} - (\rho_1 - \rho_2)g, \tag{18.10}$$

where ρ_1 and ρ_2 are the densities of the solid material and the surrounding material like liquid or gas, respectively. When $F \geq 0$ is satisfied, the solid material levitates in the surrounding material. We can consider four cases:

(i) $\rho_1\chi_1 \geq \rho_2\chi_2$ and $\rho_1 \geq \rho_2$,
(ii) $\rho_1\chi_1 \geq \rho_2\chi_2$ and $\rho_1 < \rho_2$,
(iii) $\rho_1\chi_1 < \rho_2\chi_2$ and $\rho_1 \geq \rho_2$,
(iv) $\rho_1\chi_1 < \rho_2\chi_2$ and $\rho_1 < \rho_2$.

Note from the above argument that only when $\rho_1\chi_1 - \rho_2\chi_2 < 0$, is stable magnetic levitation possible, namely, the magnetic force is attractive for the case of $\rho_1\chi_1 - \rho_2\chi_2 > 0$ and no equilibrium point is in the magnet. Even when $\rho_1\chi_1 - \rho_2\chi_2 \geq 0$, the solid material may float if $\rho_1 - \rho_2 < 0$, but this is due to the buoyancy force and no longer to magnetic levitation. Therefore cases (i) and (ii) are omitted. In cases (iii) and (iv) there is coexistence of both the magnetic levitation effect and the buoyancy force. Which is dominant depends on the combination of materials. Geim et al. found a way to levitate a permanent magnet by placing a Bi cylinder inside a superconducting magnet. They also found an equilibrium position for a permanent magnet between fingertips 2.5 m below a powerful superconducting magnet [14]. These phenomena are of great interest because this is a completely new idea which breaches the theory that any paramagnetic or ferromagnetic material cannot be levitated as has been commonly believed for a long time.

18.1.4   Origin of the Magnetic Susceptibility

As mentioned at the beginning of this chapter, any material responds to a chapter magnetic field via the susceptibility χ that is caused mainly by the motion of the electrons. This is phenomenologically divided into two categories: diamagnetic with negative χ and paramagnetic with positive χ. In ferromagnetic materials where the net magnetic moment is almost saturated, χ is not a constant because the strong response to a magnetic field is due to the motion of the domain wall and is not a microscopic phenomenon. We consider briefly how χ arises in a material.

18.1.5   Electrons in a Closed Shell of Atoms or Ions

The effect induced by these electron is similar to Lenz's law in a classical way. An externally applied magnetic field changes the atomic orbit of electrons in such a way that an additional electric current is generated; the magnetic field induced by this current is opposed to the external field. In general, this additional orbital current of electrons induces a magnetic moment expressed as

$$M_\mathrm{L} = -\frac{\mu_0 e^2}{6m} a^2 B \,, \tag{18.11}$$

where a is the radius of the orbit while e and m are the electric charge and mass, respectively. This leads to the negative susceptibility

$$\chi = -\frac{N\mu_0^2 e^2 Z}{6m} \overline{a^2} \,, \tag{18.12}$$

for N atoms or ions composed of Z electrons in a shell with an average orbital radius $\overline{a^2}$. Quantum mechanical calculation gives the same result. This value changes depending on the type of atom or ion and is inevitably negative. This phenomenon is called Larmor diamagnetism. For a molecule composed of some atoms or ions without any molecular orbital, the net susceptibility can be estimated by the sum of the susceptibilities of each atom or ion. This is called Pascal's additive law.

18.1.6   Electrons in Molecular Orbits

When an electron has a large orbit in a molecule, (18.12) suggests large negative susceptibility and Pascal's additive law is not valid. An interesting case is Benzene. When the applied field is perpendicular to the plane of the ring, the large orbital around the ring results in $\chi = -15.2 \times 10^{-9}\,\mathrm{m^3/kg}$, while $\chi = -5.6 \times 10^{-9}\,\mathrm{m^3/kg}$ when the field is applied to the ring. In the case of graphite, there also is a large anisotropy due to the carbon network, as indicated in Table 18.1.

18.1.7   Unpaired Electrons in Atoms or Molecules

Atoms containing d-electrons or f-electrons as well as organic molecules with free radicals have unpaired electrons that contribute to the magnetic moment. Oxygen also has a magnetic moment. In these systems, an external field causes orientation of magnetic moments in the direction of the field and induces a net magnetization. The induced magnetization is given by the Brillouin function, and when the thermal energy is sufficiently higher than the Zeeman energy, the susceptibility can be written as

$$\chi = \frac{C}{T}, \quad C = \frac{g^2 \mu_B^2 J(J+1)}{3K_B}. \tag{18.13}$$

This is called Curie's law and C is the Curie constant, where g, J and K_B are the so-called g-factor, the angular momentum from both spin and orbit of the unpaired electrons and the Boltzmann constant, respectively; T indicates temperature and μ_B is the Bohr magneton given by

$$\mu_B = \frac{\mu_0 e \hbar}{2m} = 1.165 \times 10^{-29}\,[\text{Wb} \cdot \text{m}], \tag{18.14}$$

where $\hbar$ is the Planck constant. The susceptibility is positive in this case and much larger than the absolute values of the diamagnetic susceptibilities even at room temperature. The measured susceptibility is the sum of this paramagnetic susceptibility and the negative diamagnetic susceptibility due to the inner shell of the magnetic atom or ion. But the diamagnetic contribution is negligibly small as compared to the paramagnetic susceptibility.

18.1.8   Conduction Electrons in Metals

Conduction electrons contribute to both paramagnetism and diamagnetism. The numbers of electrons with up-spin and down-spin are the same up to the Fermi level in zero field. When an external field is applied in the upward direction, the number of electrons with up-spin exceeds those with down-spin at the Fermi energy. This effect induces a net magnetic moment at a temperature that is sufficiently low by comparison to the Fermi temperature; the related susceptibility is

$$\chi_P = \frac{3N\mu_B^2}{2K_B T_F}, \tag{18.15}$$

where T_F is the Fermi temperature, which is approximately 10^4 K. This is called Pauli paramagnetism. On the other hand, in a magnetic field the conduction electrons occupy Landau levels, and this results in diamagnetism with a susceptibility given by

$$\chi_L = \frac{N\mu_B^2}{2K_B T_F}, \tag{18.16}$$

This is called Landau diamagnetism and this effect cannot be obtained from classical calculations. The interesting point is that the Pauli susceptibility is three times larger than the Landau diamagnetism. Thus the former effect appears to be inevitably larger than the latter and all metals ought to be paramagnetic. As a matter of fact, the susceptibilities of all the alkali metals, in which electrons are supposed to be nearly free, are positive. The electron mass used for the Bohr magneton in (18.16), however, must be replaced by the effective mass m^* and then the following relation is obtained:

$$\chi_L = -\frac{1}{3}(\frac{m}{m^*})^2 \chi_P \,, \tag{18.17}$$

If m^* is sufficiently small, the absolute value of χ_L is possibly larger than χ_P. Some metals like noble metals show negative susceptibilities as listed in Table 18.1.

18.1.9   Electrons in Superconductors

As is well known, superconductors show the Meissner effect. This leads to perfect diamagnetism below the critical field. In this case, the susceptivility is

$$\chi = -1 \,. \tag{18.18}$$

This is quite large compared to the diamagnetic susceptibilities listed in Table 18.1. In the case of type-II superconductors, this is complicated in the mixed state. Especially high-T_c superconductors show more complicated behavior like the fishing effect in a magnetic field. The details are not discussed here.

In real materials, the susceptibility is the sum of these effects. But it is not easy to calculate the susceptibilities of real materials in general, especially in the cases of metals or semiconductors, because m^* depends on the band structure and electron correlation, and it may change due to phase transitions. An interesting example is the case of Bi [15]. The large diamagnetic susceptibility drastically drops at the melting point due to the change of the band structure.

18.2   Magnetic Levitation Experiments

18.2.1   High Field Laboratory
for Superconducting Materials (IMR) at Tohoku University

Crystal Growths in Levitated Aqueous Solutions. Some ionic crystals have been grown from levitated aqueous solutions [9,10]. Aqueous solutions are placed at an appropriate position in the hybrid magnet from a thin glass capillary. The aqueous solutions become completely spherical due to surface tension when they levitate and precipitation occurs as the temperature decreases. In order to see the side view of the droplets of aqueous solutions and to observe the crystal growth in situ, a prism and a micro-CCD camera are installed in the magnet bore as shown in Fig. 18.6. The crystal growth in the levitated droplet apparently reduces heterogeneous nucleation as compared to the zero-field case and then supercooling or supersaturation easily occurs in this condition.

First we tried to grow NH_4Cl [9]. The saturated solution at $12\,^\circ C$ $(25.4\,wt\%)$ was placed at the potential minimum position of the hybrid magnet where the distance z from the center of the magnet was 7.8 cm with a central magnetic field of 18.1 T. The field at the sample position was slightly lower than this field. The droplet was a complete sphere with a diameter of 7 mm. In a supersaturated condition, first a small cross-shaped nucleus appeared at a certain place in the upper hemisphere. It moved downward to the bottom of the droplet and grew quickly in a few minutes due to supersaturation. This is shown in Fig. 18.7. It should be emphasized that only one single dendrite crystal was obtained. It is well known that many tiny crystals are simultaneously obtained in a container when crystals are grown quickly under normal conditions. In this case, the crystal grew on the bottom of the levitating droplet. In this experiment, two materials were treated and the consideration discussed in Sect. 18.1.3 must be taken into account. The susceptibilities per unit volume $\rho_1\chi_1$ and $\rho_2\chi_2$ and densities ρ_1 and ρ_2 for the NH_4Cl crystal and the saturated solution, respectively, are $\rho_1\chi_1 = -13.2 \times 10^{-6}$, $\rho_2\chi_2 = -9.6 \times 10^{-6}$ and $\rho_1 = 1.53 \times 10^3\,kg/m^3$, $\rho_2 = 1.07 \times 10^3\,kg/m^3$. So this situation corresponds to case (iii) mentioned above. When the solution is levitating,

$$F_{sol} = -\frac{1}{\mu_0}\rho_2\chi_2 B\frac{dB}{dz} + \rho_2 g = 0\,, \tag{18.19}$$

where F_{sol} is the force acting on the solution. Therefore, the force acting on the crystal is

$$F = \frac{1}{\mu_0}\rho_1\chi_1 B\frac{dB}{dz} - \rho_1 g\,. \tag{18.20}$$

Depending on the sign of this force (positive or negative) at the field where the aqueous solution is balanced, the crystal floats or sinks in the droplet. In this case, F is negative and the grown crystal sinks in the solution as shown in Fig. 18.7. When the field intensity is adjusted such that F in (18.20) is zero, the crystal floats in the middle of the aqueous solution as shown in Fig. 18.8. In this case the container must be fixed in order to prevent it from moving up due to the stronger field. This experiment suggests the possibility of growing stress-free crystals and it is interesting from the crystallographic and morphologic points of view. As shown in Fig. 18.9, when the field is increased further, F is positive and a dendrite crystal grew under the top

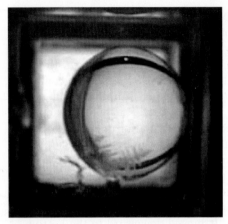

Fig. 18.7. Dendrite NH$_4$Cl single crystal grown on the bottom of the levitating solution at the central field of 18.1 T

Fig. 18.8. Levitating dendrite NH$_4$Cl single crystal floating in levitating solution at the central field of 19.1 T

of the solution that was kept on the fixed glass plate by the adhesive force between water and glass.

From these experiments, we found that crystal growth in the levitating droplet apparently reduces the number of growing crystals, implying a reduction of heterogeneous nucleation, while uncontrollable heterogeneous nucleation would occur at the container walls in the usual case. Another effect we found is that once the nucleus sinks to the bottom of the droplet, the dendrite crystal grows along the surface. This is partly due to the evaporation of the water and partly due to peculiarities of the liquid-gas interface, suggesting that mass transport at the interface would be much larger than that in the bulk solution [9].

Next KCl was tried. The saturated aqueous solution was levitated at the position 14.6 cm above the center of the magnet at a central field of 20.7 T.

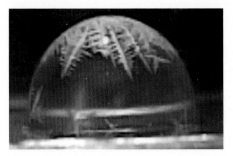

Fig. 18.9. Dendrite NH_4Cl single crystal grown under the top of the fixed solution

Decreasing the temperature from 15 °C to 2.3 °C, a tiny crystal appeared in the supercooled and supersaturated aqueous solution and precipitated on the bottom. It grew along the surface of the globe and then the shape was not a regular cube.

The third experiment was for KNO_3. The saturated aqueous solution levitated at the central field of 18.7 T. A single crystal of this material has a large anisotropy in susceptibility; therefore it is expected to grow with the a-axis perpendicular to the field direction due to the orientation effect. However, it grew with the a-axis along the surface and the a-axis of the distorted crystal is not necessarily perpendicular to the field direction. It is inferred that the characteristic of the interface between liquid and air plays a more important role for the crystal growth than the orientation effect.

Melting of Some Materials under Levitation. By using a CO_2 infrared laser, melting of some levitated materials has been achieved. A schematic view of the equipment is shown in Fig. 18.10. A solid material is levitated in the magnetic field and an infrared beam is focused on the material. The temperature control is difficult in this particular method but it is easy to heat up the material without a furnace. The first trial was for BK7 glass (mixture of SiO_2 and B_2O_5) [11]. A glass cube with 5 or 7 mm edges was placed in a platinum cage and set at the appropriate position above the center of the magnet. The magnetic field was increased up to 22.9 T at the center of the magnet; this is slightly below the field where the glass cube levitates at room temperature. As the temperature of the glass increased due to heating from the infrared light beam of a CO_2 laser, the magnitude of χ slightly increased and the glass cube levitated as shown in Fig. 18.11a. It melted in a few minutes and the shape changed from the cube to a perfect sphere as shown in Fig. 18.11b.

Next Na_2O-$2TeO_2$ glass was tried. A small glass bulk was set on the top of a rod in the magnet as shown in Fig. 18.12a. When a strong pulsed infrared laser beam was focused on it without applied field, the exploding flame was drawn upward due to convection as shown in Fig. 18.12b. When an external

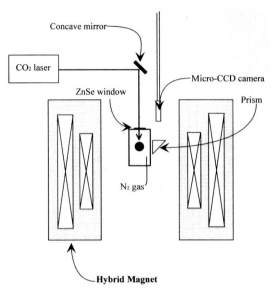

Fig. 18.10. Schematic view of the experimental setup for melting by a CO_2 laser

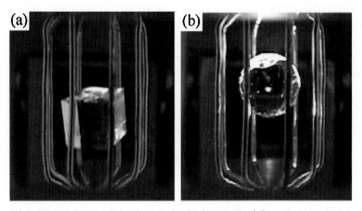

Fig. 18.11. Levitating glass cube before melt (**a**) and melted globe (**b**)

magnetic field was applied up to the level needed for levitation while the glass was still kept on the rod, it exploded almost uniformly as shown in Fig. 18.12c due to the instantaneous heating by the strong pulsed laser beam. The aim of this experiment is identical to that of experiments performed in the microgravity condition obtained by free-fall in a deep hole, by diving in an aircraft or in the space shuttle. In these experiments, the glass evaporates from the surface uniformly and emits fine particles with complete spherical shape and with diameter of the order of $1\,\mu m$, as shown in Fig. 18.13. These glass as shown particles are made radioactive in an atomic reactor and are used for medical purposes by injecting them to the site of a cancer.

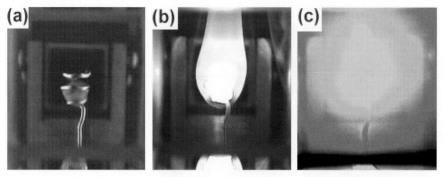

Fig. 18.12. (a) A small piece of Na_2O-$2TeO_2$ glass set on the top of a rod in a magnet. (b). Upward exploded flame due to the instantaneous heating by a strong pulsed laser beam and convection without an applied field. (c)Uniform exploded flame due to the instantaneous heating by a strong pulsed laser beam in the levitation condition

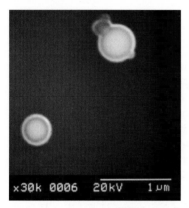

Fig. 18.13. Fine glass particle obtained from the vapor in the levitating condition

The melting of sugar is also of interest. Small pieces of crystalline sugar were placed in the magnet and a magnetic field was applied to induce levitation. Each piece levitated simultaneously, forming a bunch. When the laser beam was turned on, each piece violently shook around the original position. This may have been due to the change of the susceptibility on heating. Next a small portion from a sugar cube as used for tea or coffee was placed in the magnet. It levitated as shown in Fig. 18.14a and melted forming a globe as shown in Fig. 18.14b. In the case of levitated paraffin, it started to melt by heating, became a sphere and then went into a strange rotating motion, probably due to Marangoni convection.

These experiments have demonstrated that magnetic levitation is a useful technique for melting a material because it is isolated from a crucible and no

Fig. 18.14. Levitating sugar bunch (**a**) and melted one (**b**)

contamination is expected. Some experiments planned in the space laboratory may be replaced by this method.

18.2.2   Other Facilities in the World

At Service National des Champs Intenses (Grenoble) [2,4], the first trial of levitation of some low-x materials was performed using the hybrid magnet. In addition to Bi and Sb, levitation of wood, plastic, water, ethanol and acetone were reported. Later, a superconducting magnet was used for the levitation of liquid droplets and it was found that the levitated droplets of ethanol bounced several times from each other before merging. Some reasons were suggested as well as for the case of liquid He [3] but the real reason still remains unclear. The present author considers that the reversed magnetic dipole moment induced in each droplet may cause a repulsive force.

At the High Field Laboratory, University of Nijmegen [5–7,14], a Bitter-type magnet was used for levitation experiments and many subjects were tried including biological materials like a frog ([6] is a tutorial and is helpful for getting a picture of the history of magnetic levitation). The researchers at Nijmegen found a way to levitate even a permanent magnet by using the strongly diamagnetic Bi in a magnet.

At NHMFL, Florida State University [8], the 20 T resistive magnet with 195 mm bore was used. The maximum field-gradient product was $760\,\mathrm{T^2/m}$ and not sufficient to levitate water, plastics etc. The susceptibility of graphite-epoxy composites is large enough for obtaining levitation by this magnet. Four types of different shape, i.e. rods, beads, a sphere and a rigid body, were used and their kinetics in the potential well were studied.

At Brown University [3], by using a superconducting magnet, the dynamics of liquid helium droplets were studied. The slight deformation from a sphere of liquid in the potential well was calculated. Non-coalescence of droplets in the levitation condition was observed and ascribed to the existence of the layer of vapor between droplets. The present author considers that the induced magnetic moments aligning side by side may contribute to the repulsive force as mentioned above.

At the University of Tokyo [13], a new method was developed to levitate even a paramagnetic material by analyzing the case of two coexisting materials as mentioned in Sect. 18.1.3. The effect of buoyancy was taken into account and paramagnetic copper sulfate was successfully levitated in paramagnetic oxygen gas at high pressure. This is a combination of the magnetic levitation effect and the Archimedes principle and is called the Magneto-Archimedes Effect.

18.3   Conclusion

Magnetic levitation had been dormant since Beaugnon and Tournier's experiment in 1991; the recent revival is due to the availability of magnetic fields that are strong enough to levitate water, plastics and many other materials, even a frog. This phenomenon is not only interesting for watching; it will be quite useful for developing new materials. At Tohoku University, crystal growth in free space and the melting of some materials by an infrared laser without a crucible have been successfully performed.

References

1. W.Z. Braunbeck: Physics **112**, 753 (1939)
2. E. Beaugnon and R. Tournier: Nature **349**, 470 (1991)
3. M.A. Weilert, D.L. Whitaker, H.J. Maris and G.M. Seidel: Phys. Rev. Lett. **77**, 4840 (1996)
4. E. Beaugnon and R. Tournier: Mater. Res. Soc. Symp. Proc. **551**, 211 (1999)
5. M.V. Berry and A.K. Geim: Eur. J. Phys. **18**, 307 (1997)
6. A.K. Geim: Physics Today **51**, 36 (1998)
7. M.D. Simon and A.K. Geim: J. Appl. Phys. **87**, 6200 (2000)
8. J.S. Brooks, J.A. Reavis, R.A. Medwood, T.F. Stalcup, M.W. Meisel, E. Steinberg, L. Arnowitz, C.C. Sover and J.A.A.H. Perenboom: J. Appl. Phys **87**, 6194 (2000)
9. M. Tagami, M. Hamai, I. Mogi, K.Watanabe and M. Motokawa: J. Cryst. Growth **203**, 594 (1999)
10. M. Hamai, I. Mogi, M. Tagami, S. Awaji, K. Watanabe, and M. Motokawa: J. Cryst. Growth **209**, 1013 (2000)
11. N. Kitamura, M. Makihara, M. Hamai, T. Sato, I. Mogi, S. Awaji, K. Watanabe and M. Motokawa, Jpn. J. Appl. Phys. **39**, L324 (2000)
12. Proc. 6th Int. Symp. on Research in High Magnetic Fields, eds. F. Herlach et al., Porto, 31 July to 4 August, 2000, Physica B **294&295** (2001)
13. Y. Ikezoe, N. Hirose, J. Nakagawa and K. Kitazawa: Nature **393**, 749 (1998)
14. A.K. Geim, MD., M.I. Boamfa and L.O. Heflinger: Nature **400**, 323 (1999)
15. S. Otake, M. Momiuchi and N. Matsuno: J. Phys. Soc. Jpn. **50**, 2851 (1981)

19 Effects of a Magnetic Field on the Crystallization of Protein

G. Sazaki, S. Yanagiya, S.D. Durbin, S. Miyashita, T. Nakada, H. Komatsu,
T. Ujihara, K. Nakajima, K. Watanabe and M. Motokawa

Many studies in applying a magnetic field to control crystal growth processes
have been carried out since the 1960s; in particular, magnetic fields have
been widely applied for damping convection in semiconductor melts [1–3].
However, except for this magnetic damping effect on conducting liquids, no
significant phenomenon that could be attributed to a magnetic field effect
was found in the field of crystal growth until the late 1990s. In recent years,
rapid developments in superconducting magnets of liquid helium-free type [4]
have facilitated studies of the influence of a magnetic field on crystallization.
In particular, considerable attention has been paid to the application of a
magnetic field to the crystallization of proteins, complex biological macromolecules for which a general method of preparing large, high quality single
crystals is strongly desired. In the crystallization of protein, magnetic field effects on the orientation [5–7,9–12], crystal habit [6], number [6,8] and growth
rate [13] of crystals, convection in aqueous solution [14,15], and crystal perfection [16–18] have been reported previously. In the case of organic compounds
with low molecular weight, similar studies on the magnetic orientation of
crystals were reported [19–21].

In this review, first we outline phenomena that are observed in the crystallization of protein molecules in a homogeneous and static magnetic field
of 0–11 T; then we attempt to explain these observations quantitatively.

19.1 Crystallization of Protein in a Magnetic Field

Proteins are key molecules which govern almost all biochemical reactions in
living things, and understanding the relation between their structure and
function is necessary for the development of life science. At present, X-ray
diffraction is the most general and powerful method for analyzing the complete three-dimensional structure of a protein molecule. However, for X-ray
diffraction, a protein single crystal that is large (> 0.1 mm) and of high quality is indispensable, and the crystallization has been a bottleneck. To seek
ways of overcoming this difficulty, we initiated studies on magnetic field effects on protein crystallization [6].

We chose hen egg-white lysozyme as a model protein. Crystallization was
carried out by the batch method using a salting-out technique [22]. A given
amount of lysozyme was dissolved in 50 mM sodium acetate buffer (pH 4.5).

Sodium chloride solution of 50 mg/ml was prepared in the same 50 mM acetate buffer. A supersaturated solution was prepared by mixing equal volumes of the lysozyme solution and the sodium chloride solution. From the supersaturated solutions, 100 µl quantities were pipetted into the wells of a multi-well culture plate. To prevent evaporation, the crystallizing solution was covered by 50 µl of liquid paraffin. The initial protein concentrations ranged from 30 to 100 mg/ml. For each crystallization condition, eight wells of the same concentration were prepared, and half of these were used for a control experiment in zero magnetic field. Crystallization was carried out in a cryocooler-cooled superconducting magnet system [4]. A sample container was placed in the 52 mm room-temperature bore, and the temperature of the sample was controlled to 20.0 ± 0.1 °C. Within the volume occupied by the crystallization cells, the variation in magnitude of the field was less than 1%.

The photographs in Fig. 19.1 show top views of two wells containing tetragonal lysozyme crystals crystallized from a solution of 100 mg/ml initial protein concentration. The crystals in Fig. 19.1a were grown in zero magnetic field, whereas the crystals in Fig. 19.1b were grown in a homogeneous and static magnetic field of 10 T. The direction of the magnetic field was normal to the figure. The crystals grown at 10 T are evidently smaller in number than those grown at 0 T. Furthermore, the magnetic field forced the c-axes of the tetragonal crystal to be parallel to the direction of the magnetic field. When we look at the tetragonal lysozyme crystal from its c-axis, the (101) crystal faces appear as four squares, as shown in Fig. 19.1b. More than half of the crystals were oriented in this way. This tendency was enhanced in lysozyme solutions of higher concentration. It is worth noting that a magnetic field also affected the crystal habit. In a magnetic field of 10 T, the relative length of the crystal along the c-axis diminished, as shown in Fig. 19.1. This shows that a magnetic field affects not only the nucleation process, but also the ratio of the growth rates of the crystal faces.

Horse spleen ferritin[1] was also crystallized at 0 and 10 T (data not shown). The number of ferritin crystals that grew at 10 T was diminished to one tenth of that at 0 T [6]. The decrease in the number of ferritin crystals was more drastic than that of lysozyme crystals. The ferrihydrite core of ferritin may be responsible for this difference. The magnetic field effect on the orientation and habit could not be clearly observed in the case of the ferritin crystals because the crystals were too small in size and they appeared among a large amount of precipitate deposited prior to the crystallization.

These results clearly demonstrated that a magnetic field significantly affects the nucleation, orientation and habit (growth rate) of a protein crystal, as mentioned above. In the following sections we seek to explain the effects on the orientation and growth rate of the crystals more quantitatively.

---

[1] The iron storage protein ferritin is composed of a spherical polypeptide shell (13 nm in diameter) surrounding the inorganic core of the hydrated iron oxide ferrihydrite ($5Fe_2O_3 \cdot 9H_2O$) [23].

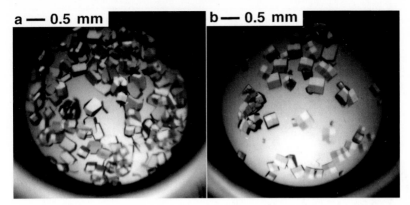

Fig. 19.1. Effect of a magnetic field on the crystallization of hen egg-white lysozyme [6]. (**a**) at 0 T, and (**b**) at 10 T. The direction of the magnetic field was normal to the figure (optical photomicrograph of transmission type). The initial lysozyme concentration was 100 mg/ml. The crystallization period was 3 days

19.2   Orientation of the Crystals

In this section, we explain the factors that determine the degree of magnetic orientation of protein crystals [11]. When a homogeneous magnetic field was applied in the vertical direction, the tetragonal lysozyme crystals were oriented such that their crystallographic c-axes were parallel to the field, as described previously. The orientation ratio R was defined as the number of crystals aligned with the c-axis within $3°$ ($\pi/60$ radians) of the magnetic field direction, divided by the total number of crystals in a cell. This criterion was arrived at by first gently detaching from the bottom of the cell those crystals that were judged by eye to be oriented. Measurement of the angles between the attachment face and the crystal edges allowed determination of the angle between the c-axes and the magnetic field direction. The largest deviation was about $3°$, and this value was considered to be the criterion for magnetic orientation.

Figure 19.2 shows the dependence of the degree of orientation R on the magnetic field and the initial protein concentration c_{init}. There is sizable scatter in the data, and the curves drawn to aid the eye should not be taken as precise fits. Nonetheless, it is clear that R increases with magnetic field, and also increases with c_{init} at any nonzero value of the field. We note that even in zero field there may be a slight tendency for the crystals to be oriented with the c-axis vertical: the zero-field R is found to be roughly 0.06, which is larger than the value of 0.02 expected for randomly oriented crystals, though probably within the experimental error.

To understand this phenomenon quantitatively, it is first necessary to know the magnetic properties of lysozyme molecules and crystals. Unfortunately, there are very few studies on the magnetic properties of proteins in

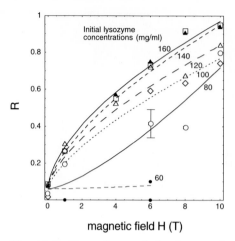

Fig. 19.2. Dependence of the degree of orientation R on the magnetic field for initial lysozyme concentrations of 60, 80, 100, 120, 140, and 160 mg/ml [11]. The curves are a guide for the eye. The data points are averages of several experiments; a typical error bar is indicated. R increases with both field strength and protein concentration

general. For the exceptional case of lysozyme, a value of the average magnetic susceptibility has been measured [24], but the anisotropy of the susceptibility had not yet been studied when the present research was started. In order to estimate the anisotropy of the susceptibility for tetragonal lysozyme crystals, we followed the method of Worcester [25], using literature values for the susceptibilities of the α-helix structure, the β-sheet structure, and aromatic residues [26], together with the known structure of the lysozyme molecule in a tetragonal crystal [27]. We obtained a value for the anisotropy $\Delta\chi = \chi_c - \chi_a = 76 \times 10^{-6}$ emu/mole, where $\chi_a\ (=\chi_b)$ and χ_c are the molar susceptibilities along the a- and c-axes of the crystal, respectively. The main contribution to $\Delta\chi$ comes from residues containing aromatic rings. As $\Delta\chi$ is positive, a crystal will tend to have its c-axis aligned with the magnetic field. This value of $\Delta\chi$ is about twice as large as that used in [7], and is about 4% of the measured average χ reported in [24].

With the $\Delta\chi$ estimated as described above, the orientation energy of a crystal comprising N molecules can be calculated from

$$E(N, \theta, H) = -\frac{1}{2}\left(\frac{N}{N_A}\right)\Delta\chi \cos^2\theta \cdot H^2, \qquad (19.1)$$

where N_A is the Avogadro constant and θ is the angle between the c-axis and the magnetic field. The probability $R(N, H)$ for a crystal of N molecules

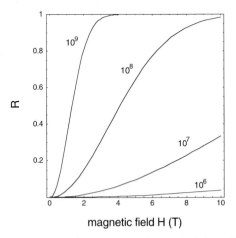

magnetic field H (T)

Fig. 19.3. Changes in the probability $R(N,H)$ for a crystal of N lysozyme molecules to have its c-axis within $3°$ of the field direction as a function of magnetic field strength H [11]. $R(N,H)$ was calculated using (19.2) for $N = 10^6$, 10^7, 10^8, and 10^9

to have its c-axis within $\pi/60$ radians of the field direction is then given by [9,28]

$$R(N,H) = \frac{\int_0^{\pi/60} \exp\left[-\frac{E(N,\theta,H)}{kT}\right] \sin\theta d\theta}{\int_0^{\pi/2} \exp\left[-\frac{E(N,\theta,H)}{kT}\right] \sin\theta d\theta} . \tag{19.2}$$

For the orientation energy in even a $10\,\mathrm{T}$ field to exceed the thermal energy kT at room temperature requires a crystal (taken as spherical) of radius at least 90 nm made up of 7×10^4 molecules. Clearly the probability of orientation of a single molecule, or even of a critical-size nucleus, which for a protein may contain 10–100 molecules, depending on the supersaturation [29], should be very small. We therefore believe that the observed high degree of orientation is due to the effect of the field on larger crystals. We calculated $R(N,H)$ using (19.2) for a series of values of N (Fig. 19.3). Comparison of Figs. 19.2 and 19.3 shows that most of the experimental data is located between the two $R(N,H)$ curves for $N = 10^7$ and 10^8. This suggests that the observed orientation ratio of Fig. 19.2 was established when N was in the range of 10^7–10^8 (crystals 0.5–1 μm in diameter).

We propose the following model to explain how the orientation of the crystal was determined. Since crystals that have attached to the container tend to adhere to it rather strongly, the orientation must be attained before crystals contact the container, i.e. while they are sedimenting from the solution. This picture is essentially in agreement with the work of Ataka et al. [7] and Wakayama [9], and the size at which the crystal orientation was fixed is of the same order as that reported by Wakayama [9]. Such a model qualita-

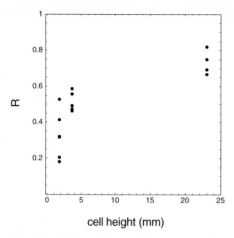

cell height (mm)

Fig. 19.4. Dependence of the degree of orientation R on the cell height (depth of the crystallizing solution) [11]. In these experiments, the initial lysozyme concentration was 80 mg/ml and the magnetic field strength was 4 T

tively explains the concentration dependence in Fig. 19.2: at higher protein concentration the crystals grow faster, and hence reach a larger size and are correspondingly more highly oriented by the time they sediment to the cell bottom.

This model also predicts that crystals grown from taller cells should be more highly oriented. We tested this by repeating the experiments with cells filled to 2 and about 13 times the original depth. The results are illustrated in Fig. 19.4. The taller cells indeed produced a higher proportion of oriented crystals.

The number of molecules N_c at which the crystal reached the bottom of the cell was estimated to be $N_c \approx 9 \times 10^8$ (4 µm in diameter). The model based on the sedimentation of the crystal is in qualitative agreement with the observations, but this value of N_c is one order of magnitude larger than the N estimated using (19.2). This inconsistency suggests that either the estimated $\Delta\chi$ is too large, or a shape-related torque on the crystal reduces R. The latter possibility is that the crystals are virtually certain to experience a gravitational torque when the crystals have contacted the bottom, since the point of initial contact is unlikely to be directly below their center of mass. We can estimate the importance of the shape effect on the crystal orientation by comparing the magnetic torque and the gravitational torque. For $H = 10\,\mathrm{T}$ and crystal radius $r = 1\,\mu\mathrm{m}$, the ratio of the gravitational torque to the magnetic torque is of the order 0.005 [11], i.e. under these conditions the gravitational torque is not large enough for this shape effect to play a major role in crystal alignment. Therefore, a crystal settled on the bottom can grow large enough to strongly adhere to the bottom while maintaining its alignment, before the effect of shape-related gravitational torque becomes important (the ratio of the torques becomes unity when $H = 10\,T$ and $r = 200\,\mu m$). However, at low magnetic field (e.g. $H = 1\,T$) the ratio is about unity for $r = 2\,\mu m$ [11] (a typical size attained during sedimentation). Considering the actual crystal shape, we would expect this effect to possibly reduce the degree of alignment somewhat at low field.

We have presented a model to account for the observed orientation of protein crystals grown in a magnetic field. Because of the small susceptibility anisotropy of most protein molecules, the orienting effect is unimportant for smaller aggregates, even those far larger than a critical nucleus. However, during the time of sedimentation of a growing crystal, it will typically reach such a size as to exhibit a relatively high degree of alignment, which will then be reflected in the final distribution of the crystal orientation. The degree of orientation thus depends on the crystal growth rate and the container geometry, as well as on the magnetic field strength, as has been confirmed experimentally. The model is in qualitative agreement with experiment, but quantitatively a higher degree of alignment is predicted than has been observed.

In our study we used lysozyme, a globular protein with small magnetic anisotropy. However, in the case of protein molecules which have much larger magnetic anisotropy, the magnetic orientation effect becomes much larger. From this viewpoint, proteins for which the magnetic field works most effectively are membrane proteins, which are also attracting much attention for their physiological importance.

19.3   Growth Rate of Protein Crystals in a Homogeneous Magnetic Field

All the results mentioned in the previous sections were obtained using ex situ observation techniques. In situ optical microscopy in a high magnetic field is necessary to measure the growth rate of protein crystals and to study the effect of the magnetic field on the growth kinetics and the habit of the crystals. Therefore, we designed and prepared an optical microscope which could be used in a high magnetic field, and studied the magnetic field effect on the growth rate of tetragonal lysozyme crystals using in situ optical microscopy [13].

Figure 19.5 shows a schematic illustration of the microscope. The lysozyme crystal in an observation cell was illuminated by white light (halogen 150 W) through an optical fiber. The field of view and location of the focal plane were adjusted using sliding stages in the X, Y and Z directions. Objective lenses of super-long working distance type ($\times 50$ and $\times 100$, NIKON Co.) were used to obtain high-resolution images of the crystals through a temperature controlled sample container. The images were recorded using a CCD camera (ELMO, UN411) and a video tape recorder. The resolution of the microscope was determined by that of the CCD camera. When objective lenses

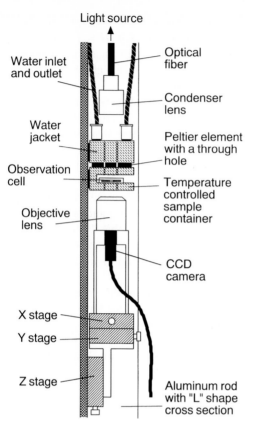

Fig. 19.5. Schematic drawing of the optical microscope for in situ observation in high magnetic fields [13]. All parts were made of paramagnetic and diamagnetic materials

of ×50 and ×100 magnification were used, the resolution limits were 1 μm and 0.5 μm, respectively. The temperature of the sample was controlled to an accuracy of ±0.1°C using a Peltier element. All parts of the microscope were made of non-magnetic materials. This microscope system was inserted into the room-temperature bore of a superconducting magnet [4]. The bore was 52 mm in diameter and its orientation was vertical. The Peltier element and the CCD camera functioned properly even at 11 T.

In situ measurement of the growth rate was carried out as follows. One seed crystal of the tetragonal form was placed in the observation cell (Fig. 19.6a) such that its crystallographic c-axis would be perpendicular to the directions of the magnetic field and gravity (Fig. 19.6b). The crystal was grown for several hours to fix it on the bottom of the cell. Then, the solution in the cell was replaced with newly prepared lysozyme solution (50–100 mg/ml). After the observation cell was attached to the temperature-controlled sample

a)

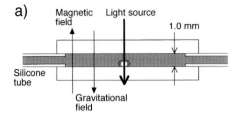

b)

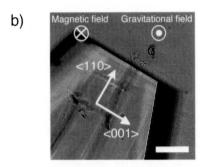

Fig. 19.6. The observation cell and orientation of the tetragonal lysozyme crystal in the cell [13]. (**a**) Longitudinal section of the observation cell. (**b**) In situ micrograph of tetragonal lysozyme crystal growing in the cell at 11 T. The crystallographic c-axis of the crystal was set perpendicular to the magnetic and gravitational fields and a {110} face was in contact with the bottom of the cell. Conditions: lysozyme, 80 mg/ml; NaCl, 25 mg/ml, in 50 mM sodium acetate buffer (pH 4.5) at 12.5 °C. Scale bar represents 50 μm

container (12.5 ± 0.1 °C) with silicone grease, the container and the sample were inserted in the bore of the magnet. The crystals and the solution were at the position of the maximum field. The inhomogeneity of the magnetic field within the observation cell was less than 0.1%. Growth of the {110} and {101} faces was observed in situ at both 0 and 11 T using the microscope described above. As far as we know, this represents the first direct measurement of crystal growth in high magnetic fields at the high resolution of 0.5–1 μm (Fig. 19.6b). Optical alignment of the microscope was not changed by the magnetic force, and the mechanical vibration from the helium compressor did not affect the image. From micrographs taken at 10–30 min intervals, we measured the distance from a selected reference point in the crystal provided by a defect or scratch on the surface to the edges of the crystal. The normal growth rates of the {110} and {101} faces were calculated from their displacements.

Growth of the {110} and {101} faces was measured for lysozyme concentrations of 50–100 mg/ml at both 0 and 11 T. As examples, the displacement of the {110} and {101} faces at 0 and 11 T in the course of time are shown in Fig. 19.7 (lysozyme: 90 mg/ml). The time at which the solution in the cell was

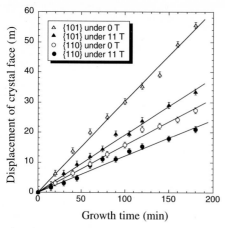

Fig. 19.7. The displacement of the {110} and {101} faces of the tetragonal lysozyme crystal at 0 and 11 T [13] as a function of time. Open and solid triangles are for {101} faces at 0 and 11 T, and open and solid circles for {110} faces at 0 and 11 T, respectively. Conditions: lysozyme, 90 mg/ml; NaCl, 25 mg/ml, in 50 mM sodium acetate buffer (pH 4.5) at 12.5 °C. The error bars shown in the figure correspond to the resolution limits of the CCD camera

replaced with newly prepared solution was taken as the zero for the growth time. As shown in the figure, the displacement at 11 T was smaller than that at 0 T, i.e. the magnetic field decreased the growth rate of the lysozyme crystal. Because of the limited observation time and spatial resolution limit of our system, we were unable to perform high-resolution growth-rate measurements on the same crystal at both 0 and 11 T. Therefore, for each experiment, different seed crystals were prepared and their growth rates were measured. Thus, the relationship between the growth rates of the {110} and {101} faces and the initial lysozyme concentration at 0 and 11 T, summarized in Fig. 19.8, shows some scatter because of the different defect distributions in the crystals. However, except in the one case of the {101} face in 80 mg/ml lysozyme solution, the growth rates at 11 T were always smaller than those at 0 T. The reduction factor varied from 0.4 at the lower concentrations to 0.9 at the higher concentrations.

Solubility is one of the most important parameters affecting the growth rate. The effect of the magnetic field on the solubility of lysozyme could be one reason the growth rate decreased at 11 T, as shown in Fig. 19.8. Therefore, we investigated the dissolution of tetragonal lysozyme crystals to estimate the solubility in a high magnetic field. The experimental procedure was as follows. Many lysozyme crystals of the tetragonal form were prepared in a 5 ml glass bottle. Then the supernatant of the lysozyme solution was decanted, and newly prepared salt solution (25 mg/ml NaCl dissolved in 50 mM sodium acetate buffer) was poured into the bottle. After the replacement of the solution

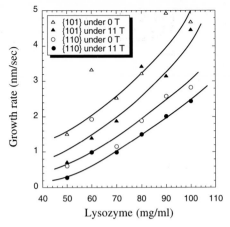

Fig. 19.8. Effects of a magnetic field of 11 T on the growth rates of the {110} and {101} faces of the tetragonal lysozyme crystals [13]. Open and solid triangles are for {101} faces at 0 and 11 T, and open and solid circles for {110} faces at 0 and 11 T, respectively. Conditions: NaCl, 25 mg/ml, in 50 mM sodium acetate buffer (pH 4.5) at 12.5 °C. The curves in the figure are guides for the eye. For each measurement, new seed crystals were prepared and their growth rates were measured in situ

in the bottle, the bottle was quickly inserted in the temperature-controlled container (20 ± 0.1 °C), which was then placed in the superconducting magnet. During the dissolution of the lysozyme crystals in a high magnetic field, 40 μl of the supernatant was withdrawn at 3–6 hour intervals. The lysozyme concentration in the supernatant was determined from the absorbance at 280 nm.

Figure 19.9 shows the lysozyme concentration during the dissolution process at 0, 5 and 10 T as a function of time. As the crystals were dissolved, the lysozyme concentration increased. The concentration measurements were stopped at 70 hours because of our limited machine time for the superconducting magnet. After the measurements, the crystals were not completely dissolved and still remained in the bottle. Although after a dissolution time of 40 hours it appeared that the concentration became nearly constant, in fact, the system may not have reached equilibrium, because of the very slow dissolution rate at concentrations near the solubility limit. In Fig. 19.9, the lysozyme concentration after 50 hours at 0 T was about 15% smaller than the previously measured solubility (14.1 mg/ml [30]). Therefore, the concentrations after 40 hours at 5 and 10 T are only lower limits to the solubilities. These values were smaller than that in zero field, as shown in Fig. 19.9. This indicates the following two possibilities:

1. the magnetic field of 5 and 10 T decreased the solubility of the tetragonal lysozyme crystals, and/or
2. the magnetic field decreased the dissolution rate of the crystal.

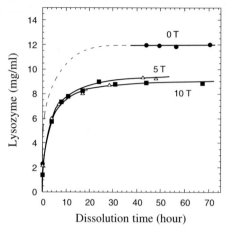

Fig. 19.9. Changes in the lysozyme concentration at 0, 5, and 10 T as a function of the dissolution time [13]. During dissolution of the tetragonal lysozyme crystals in a high magnetic field, the supernatant solution was withdrawn and the lysozyme concentration in it was determined from the absorbance at 280 nm. Conditions: NaCl, 25 mg/ml, in 50 mM sodium acetate buffer (pH 4.5) at 20.0 °C. The solid curves in the figure are guides for the eye. The concentration on the dotted curve was not measured. The error of the concentration determination was estimated as 2% of its UV absorption

    If we assume that a magnetic field decreased the solubility of the lysozyme crystals (case 1), then at a given lysozyme concentration the driving force for the crystallization must increase with increasing magnetic field. However, this tendency would be opposite to that of the observed growth rates in the magnetic field, as shown in Figs. 19.7 and 19.8. We conclude, therefore, that the decrease in the growth rate in the magnetic field could not be explained by a change in the solubility.

    On the other hand, a decrease in dissolution rate (case 2) appears reasonable. In liquid metals such as semiconductor melts, many studies reported magnetic damping of convection. Even in the case of aqueous solution, Ramachandran and Mazuruk showed that a magnetic field as small as 10 mT could affect the flow of a 10 wt% NaCl solution [31]. If we suppose that the magnetic field of 11 T damped the convection in the 25 mg/ml NaCl aqueous solution, the lysozyme concentration at the surface of a growing crystal would decrease, and thus the growth rate would also decrease. This assumption can explain not only the magnetic field effects on the growth rate but also the dissolution results of Fig. 19.9: in the magnetically damped solution, the dissolution rate of the crystals should also decrease.

19.4   Damping of Convection in NaCl Aqueous Solution by a Magnetic Field

The decrease in the growth rate in the high magnetic field suggested that a magnetic field might damp the convection in the aqueous solution. To investigate the role played by such magnetic damping effects in our protein crystallization experiments, we studied the influence of a static, homogeneous magnetic field of 10 T on the temperature-driven convection in a cell containing an aqueous NaCl solution [14].

A temperature-control unit to establish the temperature-driven convection in the observation cell (Fig. 19.10) was mounted in the microscope shown in Fig. 19.5. An objective lens of super-long working distance type ($\times 10$, NIKON Co.) was used to obtain images of the solution through the temperature-control unit and water jacket of 17 mm total length. The microscope system was inserted in the room temperature bore of a superconducting magnet (JMTD-10T100M, Japan Magnet Technology Inc.). The bore was 100 mm in diameter and its orientation was vertical. The observation cell was placed at the position of the maximum field (center of the magnet). The inhomogeneity of the magnetic field within the cell was less than 0.1%.

Figure 19.10 shows a side view of the observation cell and temperature-control unit. The cell was made of glass. It was cylindrical, 10 mm in height and 18 mm in diameter. The top and bottom of the cell were in contact with a cold block and a hot block, respectively, with temperatures T_C and T_H controlled to an accuracy of $\pm 0.1\,^\circ$C by Peltier elements. The cell was filled with 25 wt% NaCl aqueous solution (electrical conductivity: $2.1 \times 10^1\ \Omega^{-1}\mathrm{m}^{-1}$ at 25 $^\circ$C). Steady natural convection was generated by the temperature difference $\Delta T = T_H - T_C$, while the average temperature was maintained at

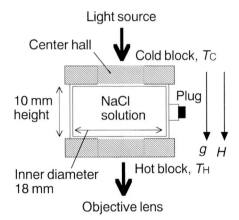

Fig. 19.10. Side view of the observation cell and the temperature-control blocks for the convection experiments [14]. The cell was cylindrical: 10 mm high and 18 mm in diameter

$(T_H + T_C)/2 = 25$ °C. The strength of the convection could be estimated from the Grashoff number $Gr = (gb\Delta Tl^3)/\nu^2$. Here g, b, l and ν are the gravitational acceleration, expansion coefficient, height of the cell and kinematic viscosity, respectively. When the temperature difference ΔT is 20 °C, the cell used in this study gives $Gr = 4 \times 10^4$. Under these conditions, the temperature difference was large enough to generate a strong convective flow in the observation cell.

Polystyrene latex particles of 5 μm diameter (density: $1.05 \, \text{g/cm}^3$) were added to the solution for flow visualization, as illustrated in Fig. 19.11. Here the particles move in the solution under the influence of an imposed temperature difference ΔT=20 °C (T_C=10 °C, $T_H = 30$ °C) with the magnetic field and the gravitational field perpendicular to the plane of the photograph. Three snapshots (Nos. 1–3) taken at 0.5 s intervals were superimposed, and the particles were tracked by their uniform displacement over each interval. Flow velocities could then be calculated after scale calibration. Since the density of the 25 wt% NaCl aqueous solution ($1.184 \, \text{g/cm}^3$ at 25 °C) was greater than that of the particles, the latex particles tended to rise to the top of the cell. However, this did not affect observation of the convective flow within the experimental time period (2–3 h), because of the small size of the particles.

Visual comparison of Fig. 19.11a, b shows that the distances of particle movement within 0.5 s intervals at 10 T (C1→C2→C3 and D1→D2→D3) were much smaller than those at 0 T (A1→A2→A3 and B1→B2→B3). Thus, a static, homogeneous magnetic field of 10 T decreased the flow rate of the temperature-driven convection. The flow rates of the convection were measured under various temperature differences with the results summarized in Fig. 19.12. The flow rate increased as expected with an increase in the temperature difference ΔT. As Fig. 19.12 shows, the homogeneous magnetic field of 10 T decreased the flow rate under all conditions to about 50% of that in the absence of a magnetic field.

In the case of liquid metals, convective flow induces a current in the liquid, and this induced current causes the magnetic damping. The electrical conductivity of 25 wt% NaCl aqueous solution was $2.1 \times 10^1 \, \Omega^{-1}\text{m}^{-1}$ at 25 °C. This value is 4-5 orders of magnitude smaller than the conductivity of the semiconductor melt. However, the magnetic field is 2–3 orders of magnitude higher than that used for the semiconductor melt, which appears to compensate for the low conductivity and results in the remarkable damping effect on the convection. Many studies have been reported previously on the magneto-hydrodynamic flow caused by a current applied in an electrolyte solution in a magnetic field [32,33]. However, as far as we know, this report is the first to show that a static and homogeneous magnetic field can damp the natural convection in an aqueous solution.

Qi et al. carried out numerical simulations of non-uniform magnetic field effects on the velocity and temperature distribution of pure water [15]. They reported that a small gradient of the magnetic field in the radial direction of a

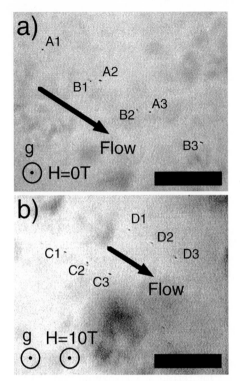

Fig. 19.11. Micrographs of polystyrene latex particles in 25 wt% NaCl aqueous solution at **(a)** 0 T and **(b)** 10 T [14]. Three snapshots (Nos. 1–3) taken at 0.5 s intervals were superimposed in each figure. The focal plane was positioned 1 mm above the bottom of the cell, where the flow is nearly parallel to the cell wall. Four particles, A and B at 0 T and C and D at 10 T, moved from upper left to lower right of the field of view. The magnetic field and gravity were perpendicular to the plane of the photograph. Scale bars represent 500 µm

solenoid-type magnet can play an important role in damping natural convection. This damping effect is due to the magnetic force caused by the magnetic field gradient. According to their rough estimation, about 10 % of the magnetic damping effect studied in this paper could arise from inhomogeneity of the field [34].

Sato et al. reported that a homogeneous magnetic field of 10 T significantly the improved the crystal perfection of orthorhombic lysozyme crystals, resulting in an increase of the maximum resolution in X-ray diffraction and higher resolution of the refined protein structure [16,17]. Other than hen egg-white lysozyme, Lin et al. also reported the enhancement in the maximum resolution of two enzyme protein crystals from snakes and humans [18]. Magnetic damping of the convection should reduce perturbation of the solute concentration distribution around the growing crystals. This would prevent

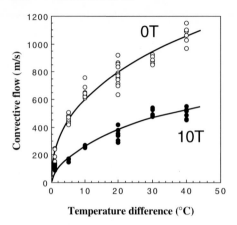

Fig. 19.12. Effect of a static, homogeneous magnetic field on the flow rate of the convection driven by the temperature difference [14]. Open points correspond to flow rates at $0\,T$, and solid lines to those at $10\,T$. The curves in the figure are guides for the eye. The temperatures of the cold block T_C and of the hot block T_H were set such that $(T_H + T_C)/2 = 25$ °C

bunching of the steps and incorporation of impurities on the crystal surface [35], and thus contribute to growing crystals of better quality. Sato et al. also found that the number of crystals deposited at $10\,T$ was much smaller than that in zero magnetic field, as observed for tetragonal lysozyme (Fig. 19.1) and ferritin crystals [6]. Magnetic damping of the convection might also play a role in suppressing secondary nucleation and in decreasing the number of crystals.

In this review of magnetic field effects in protein crystallization, we have shown that a homogeneous magnetic field has significant effects on the number, orientation and habit of the crystals. We have also shown that the degree of the magnetic orientation of the crystals could be controlled by the growth rate and the container geometry in addition to the magnetic field strength. Furthermore, it was found that a magnetic field decreased the growth rate of the crystals and the velocity of the natural convection in an electrolyte solution. It was suggested that the decrease in growth rate in a magnetic field was caused by this magnetic damping of the convection. Such magnetic field effects can also be applied to crystallization of inorganic salts and low molecular weight organic compounds other than proteins. Therefore, a magnetic field can be an effective parameter to advance the state of the art of crystallization from solution.

19.5   Conclusion

The bottleneck in three-dimensional structural analysis of protein molecules is the preparation of large ($> 0.1\,\mathrm{mm}$) single crystals of proteins with suitable high quality. As a possible means for overcoming this difficulty, magnetic field effects have been studied intensively in recent years. Here we outline magnetic field effects on protein crystallization. Crystallization of hen egg-white lysozyme was carried out in static and homogeneous magnetic fields of 0–11 T. It was demonstrated that a magnetic field decreased the number of nuclei that formed, and not only oriented the crystals but changed the crystal habit. The degree of orientation of the crystals depends on the crystal growth rate and container geometry in addition to magnetic field strength. A magnetic field also decreased the growth rate of the crystals. A possible mechanism for this effect is the reduction in protein concentration at the crystal surface resulting from the magnetic damping of convection. Indeed, in situ monitoring of the convection in an aqueous NaCl solution showed that a magnetic field reduced the velocity of buoyant convection.

Acknowledgments

The authors are grateful for partial support by Grants-in-Aid of Scientific Research No. 10555001 and 12750006 (G.S.), No. 10304023 (S.D.) and No. 11305001 (K.N.) of the Japanese Ministry of Education, Science and Culture. We also would like to express our thanks for partial support by funding REIMEI Research Resources of Japan Atomic Energy Research Institute and Mitsubishi Chemical Co. This study was carried out as a part of "Ground Research Announcement for Space Utilization" promoted by NASDA and the Japan Space Forum. We are also grateful for the use of the facilities and assistance of the staff at the High Field Laboratory for Superconducting Materials, Institute for Materials Research, Tohoku University.

References

1. S. Chandrasekhar: *Hydrodynamic and Hydromagnetic Stability* (Oxford University Press, London, 1961)
2. H.A. Chedzey and D.T.J. Hurle: Nature **210**, 933 (1966)
3. H.P. Utech and M.C. Flemings: J. Appl. Phys. **37**, 2021 (1966)
4. K. Watanabe, S. Awaji, J. Sakuraba, K. Watazawa, T. Hasebe, K. Jikihara, Y. Yamada and M. Ishihara: Cryogenics **36**, 1019 (1996)
5. T.M. Rothgeb and E. Oldfield: J. Biol. Chem. **256**, 1432 (1981)
6. G. Sazaki, E. Yoshida, H. Komatsu, T. Nakada, S. Miyashita and K. Watanabe: J. Cryst. Growth **173**, 231 (1997)
7. M. Ataka, E. Katoh and N.I. Wakayama: J. Cryst. Growth **173**, 592 (1997)
8. N.I. Wakayama, M. Ataka and H. Abe: J. Cryst. Growth **178**, 653 (1997)
9. N.I. Wakayama: J. Cryst. Growth **191**, 199 (1998)
10. J.P. Astier, S. Veesler and R. Boistelle: Acta. Crystallogr **D54**, 703 (1998)
11. S. Yanagiya, G. Sazaki, S.D. Durbin, S. Miyashita, T. Nakada, H. Komatsu, K. Watanabe and M. Motokawa: J. Cryst. Growth **196**, 319 (1999)
12. S. Sakurazawa, T. Kubota and M. Ataka: J. Cryst. Growth **196**, 325 (1999)

13. S. Yanagiya, G. Sazaki, S.D. Durbin, S. Miyashita, K. Nakajima, H. Komatsu, K. Watanabe and M. Motokawa: J. Cryst. Growth **208**, 645 (2000)
14. G. Sazaki, S.D. Durbin, S. Miyashita, T. Ujihara, K. Nakajima and M. Motokawa, Jpn. J. Appl. Phys. **38**, L842 (1999)
15. J. Qi, N.I. Wakayama and A. Yabe: J. Cryst. Growth **204**, 408 (1999)
16. T. Sato, G. Sazaki, Y. Katsuya and Y. Matsuura: *Corrected abstracts in XVIIth Int. Union of Crystallorgraphy Congress and General Assembly*, Glasgow, U. K., 4-13 August, 1999: Acta Crystallographica, volume A55, 320 (1999); T. Nakaura, T. Sato, Y. Yamada, S. Saijo, T. Hori, R. Hirose, N. Tanaka, G. Sazaki, K. Nakajima, N. Igarashi, M. Tanaka and Y. Matsuura: in *Proc. Int. Workshop on Chemical, Physical and Biological Processes under High Magnetic Fields*, Omiya, Japan, 24-26 November, 1999: (National Research Laboratory for Magnetic Science, Japan Science and Technology Co., Kawaguchi, 1999)
17. T. Sato, Y. Yamada, S. Saijo, T. Nakaura, T. Hori, R. Hirose, N. Tanaka, G. Sazaki, K. Nakajima, N. Igarashi, M. Tanaka and Y. Matsuura: Acta Crystallogr. **D56**, 1079 (2000)
18. S.-X. Lin, M. Zhou, A. Azzi, G.-J. Xu, N.I. Wakayama and M. Ataka: Biochem. Biophys. Res. Commun. **275**, 274 (2000)
19. A. Katsuki, R. Tokunaga, S. Watanabe and Y. Tanimoto: Chem. Lett. **607** (1996)
20. M. Fujiwara, T. Chidiwa, R. Tokunaga and Y. Tanimoto: J. Phys. Chem. **B102**, 3417 (1998)
21. M. Fujiwara, R. Tokunaga and Y. Tanimoto: J. Phys. Chem. **B102**, 5996 (1998)
22. A. McPherson: *Cryst.lization of Biological Macromolecules* (Cold Spring Harbor Laboratory Press, New York, 1999)
23. G.C. Ford, P.M. Harrison, D.W. Rice, J.M.A. Smith, A. Treffry, J.L. White and J. Yariv: Phil. Trans. R. Soc. Lond. **B304**, 551 (1984)
24. G. Careri, L. De Angelis, E. Gratton and C. Messana: Phys. Lett. **A60**, 490 (1977)
25. D.L. Worcester: Proc. Natl. Acad. Sci. USA **75**, 5475 (1978)
26. L. Pauling: Proc. Natl. Acad. Sci. USA **76**, 2293 (1979)
27. R. Diamond: J. Mol. Biol. **82**, 371 (1974)
28. T. Takeuchi, T. Mizuno, T. Higashi, A. Yamagishi and M. Date: Physica **B201**, 601 (1994)
29. A.J. Malkin, J. Cheung and A. McPherson: J. Cryst. Growth **126**, 544 (1993)
30. G. Sazaki, K. Kurihara, T. Nakada, S. Miyashita and H. Komatsu: J. Cryst. Growth **169**, 355 (1996)
31. N. Ramachandran and K. Mazuruk: J. Jpn. Soc. Microgravity Appl. **15**, 249 (1998)
32. R. Aogaki, K. Fueki and T. Mukaibo: Denki Kagaku **43**, 504 (1975)
33. T.Z. Fahidy: J. Appl. Electrochem. **12**, 553 (1983)
34. N.I. Wakayama: private communication (1999)
35. A.A. Chernov: *Modern Cryst.lography III* (Springer-Verlag, Berlin, 1984) p. 247

20 Magnetoelectrochemistry with a Conducting Polymer

I. Mogi, K. Watanabe and M. Motokawa

The application of additional fields to electrode surfaces provides a new technique for controling electrochemical reactions that might be hard to regulate with conventional electrochemical fields. Irradiation by light or ultrasonic waves and the application of magnetic fields are available for the modification of reactions at electrodes. While light and ultrasonic waves provide energy for electronic and vibrational excitation at the electrode, magnetic fields do not provide such energy but change the direction of ion transport or molecular orientation. Electrolysis in the presence of a magnetic field, which is called magnetoelectrolysis [1], thus allows control of the morphology of electrochemical deposits.

When magnetic fields are applied to the electrodeposition process, convection in the electrolytic solution is induced by the electromagnetic interaction between the Faraday current and the magnetic field. This is well known as the magnetohydrodynamic (MHD) effect [1,2]. On the other hand, in electroless deposition, which results from redox (reduction-oxidation) chemical reactions, macroscopic MHD convection is never caused by magnetic fields because there is no macroscopic current in the solution. The magnetic field effect on the electroless deposition [3] is the Lorentz force acting on the elementary motion of ion diffusion. This effect should be distinguished from the conventional MHD effect and is called "micro-MHD" effect [4,5]. The magnetic field has another influence on the electropolymerization of organic conducting polymers through the diamagnetic orientation. Most organic polymers have such a large anisotropy in diamagnetic susceptibility that they are subject to diamagnetic orientation, resulting in morphological changes, in magnetic fields [6,7].

The early research on magnetoelectrochemistry was focused on the MHD effect in relatively low magnetic fields of no more than 1 T generated by a conventional electromagnet or a permanent magnet [1]. We attempted for the first time electrochemical experiments in high magnetic fields up to 14 T [8] using a Bitter magnet or a superconducting magnet, and found new phenomena concerning the micro-MHD effect [3] and magnetoelectropolymerization (MEP, i.e. electropolymerization in magnetic fields) [9]. Recent development of a cryocooled superconducting magnet [10], which can be easily handled without liquid helium, will provide a breakthrough for electrochemical experiments in high magnetic fields and create new trends and oppotunities in high-field magnetoelectrochemistry.

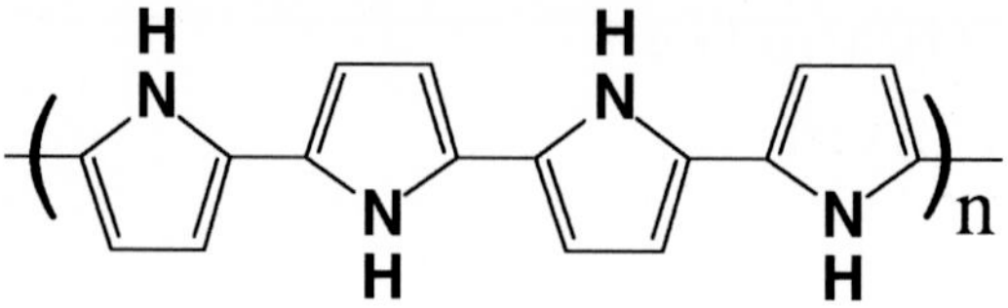

Fig. 20.1. Molecular structure of PPy

In organic electrochemical systems it is an attractive new challenge to explore the effects of diamagnetic orientation by magnetic fields on electropolymerization and polymer electrodes. Organic conducting polymers have attracted much attention as potential materials with many applications to electrode modification and electronics devices. MEP provides a technique for morphology tailoring of organic conducting polymers. The redox behavior depends on the morphology of the conducting polymers containing dopant ions. The morphology tailored by magnetic fields thus allows control of their redox properties.

20.1   Magnetoelectropolymerization

The MEP technique for polymer synthesis is expected to induce oriented polymerization, allowing control of electrochemical properties. We have made an attempt at the MEP of pyrrole and studied the electrochemical properties of the MEP films of polypyrrole (PPy) [9]. PPy is one of the highest potential materials for electronic devices and PPy-modified electrodes have been investigated in connection with electrochemical catalytic properties and biosensing applications. PPy has a linear chain molecular structure (see Fig. 20.1) and is synthesized by oxidative polymerization of pyrrole. The PPy films are prepared by the electropolymerization method on a platinum-disk working electrode at a constant potential of 1.0 V (versus Ag/AgCl) in a 0.1 M pyrrole aqueous solution containing 0.1 M TsONa (sodium p-toluenesulfonate) or Na_2SO_4 as a supporting electrolyte. The oxidative polymerization accompanies anion doping into the film for charge compensation, resulting in an anion-doped PPy film (PPy/TsO or PPy/SO_4). In the MEP process (Fig. 20.2) magnetic fields B are applied parallel to the Faraday currents to eliminate the MHD convection effect. Here the PPy films prepared at 0 and 5 T are called 0T-film and 5T-film, respectively.

Figure 20.3 shows the surface morphology of 0T- and 5T-films of PPy/TsO. Small clusters of PPy as large as ≈ 50 nm aggregate to form the 0T-film. On the other hand, the 5T-film is composed of clusters as large as ≈ 500 nm, each of which is a dense aggregate of the small clusters. This result demonstrates that the MEP results in PPy/TsO film with a denser structure. This morphological change is reflected in the dopant mobility. The ac impedance measurements showed that the dopant mobility in the 5T-film is much smaller than that in the 0T-film.

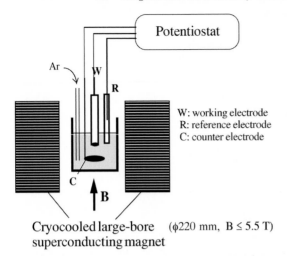

Fig. 20.2. Experimental setup for magnetoelectropolymerization

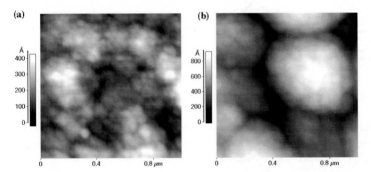

Fig. 20.3. Atomic force microscope images of the surfaces of PPy/TsO: (**a**) 0T-film; (**b**) 5T-film

Figure 20.4 shows the cyclic voltammograms (CVs) of the PPy/TsO films in the 0.1 M TsONa aqueous solution. Cathodic peaks around $-0.5 - -0.7$ V correspond to the reduction of PPy. The CV of the 0T-film has a cathodic peak A at -0.47 V followed by a shoulder B at -0.6 V. The peak A becomes smaller in the CVs of the 0.5T- and 1T-films and diminishes in the 5T-film. On the other hand, the peak B grows with increasing magnetic field.

The reduction process of PPy accompanies the undoping of anions [11], as illustrated in case (a) of Fig. 20.5. When the mobility of the dopant anion within the film is quite small (in case (b) of Fig. 20.5), the reduction process accompanies the cation doping instead of the anion undoping for the charge compensation [12]. The 5T-film of PPy/TsO has the denser structure and a small dopant mobility, and its CV exhibits only the peak B. We thus assumed that the peak A is the PPy reduction accompanying the anion TsO^- undoping and the peak B is the reduction accompanying the Na^+ doping

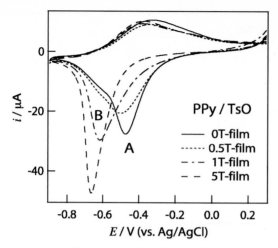

Fig. 20.4. CVs of the MEP films of PPy/TsO

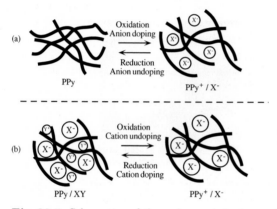

Fig. 20.5. Schematics of the redox and doping-undoping behavior of PPy

from the solution. This assumption was confirmed by measurements of the electrode mass by means of an electrochemical quartz-crystal microbalance. Consequently, the MEP results in a denser structure of the PPy film, and this morphological change causes the change of the doping-undoping behavior of the film. At the reduction process both TsO$^-$ undoping and Na$^+$ doping occur in the 0T-films, while Na$^+$ doping dominantly occurs in the MEP films.

20.2 Magnetoelectropolymerized Film Electrodes

Control of chemical reactions by magnetic fields is one of the current topics in high magnetic field science. Direct control through the Zeeman effect on electron spins is, however, difficult except for radical pair reactions [13] because even at 10 T the Zeeman energy is much smaller than the thermal

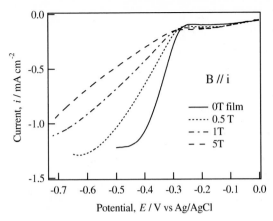

Fig. 20.6. Voltammograms of the proton reduction of the MEP films of PPy/TsO

energy at room temperature. We have proposed a new concept for reaction control by magnetic fields, i.e. the use of a polymer electrode whose properties are controlled by magnetic fields [14]. Several electrochemical reactions were examined on the PPy electrode prepared by MEP. Here we describe two electrochemical reactions that exhibit remarkable magnetic field effects on the electrode activity: the proton reduction on the PPy/TsO film [15] and the p-benzoquinone (Q)/hydroquinone (QH_2) redox reaction on the PPy/SO_4 film [14].

Figure 20.6 shows voltammograms of the proton reduction of the MEP films in a TsOH aqueous solution. It is clearly seen that the electrode activity for the proton reduction decreases with increasing magnetic field during the polymerization. This result means that the proton reduction proceeds easily on the 0T-film but hardly proceeds on the magnetoelectropolymerized films. Such a remarkable change was not observed with the MEP in a magnetic field perpendicular to the Faraday currents.

As for carbon electrodes, the electrode activity depends on the orientation of graphite crystals at the electrode surface [16,17]. We examined proton reduction at basal and edge plane electrodes of pyrolytic graphite in sulfuric acid solution and found that the proton reduction proceeds more easily on the basal plane than on the edge plane. This result gives a hint for the MEP effect.

Mitchell et al. studied the molecular organization of PPy (0T-film) and reported that PPy/TsO has an anisotropic organization with a layered structure of PPy backbone, in which the planes of aromatic rings lie preferentially parallel to the electrode surface [18]. The proton reduction easily proceeds on such a "basal plane" electrode of PPy.

The fact that the magnetic field effect is clearly observed only in fields parallel to the Faraday currents suggests diamagnetic orientation. In this electrode configuration the aromatic ring of PPy is most stable for the orientation

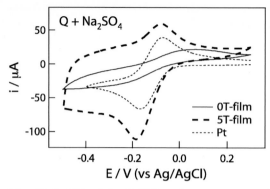

Fig. 20.7. CVs of the Q/QH_2 redox reaction on the 0T- and 5T-films of PPy/SO_4

perpendicular to the electrode surface. Morioka et al. studied the magneto-electropolymerization effects on the molecular orientation of the PPy/TsO film by X-ray scattering and showed that the magnetic fields considerably decrease the degree of aromatic ring orientation parallel to the electrode surface [19]. It is therefore considered that the inactivation for the proton reduction is due to this orientation change of PPy, i.e. the decrease of the "basal plane" part. The inactivation for the proton reduction leads to the advantage of a wide potential window of the electrode in aqueous solutions, and thus MEP allows wider applications of PPy as a modified electrode.

Q and QH_2 undergo a two-electron reaction of

$$Q + 2H^+ + 2e^- \rightleftharpoons QH_2 \tag{20.1}$$

in an aqueous solution. Figure 20.7 shows the CVs of $10\,\mathrm{mM}$ Q in a $0.1\,\mathrm{M}$ Na_2SO_4 aqueous solution on 0T- and 5T-films of PPy/SO_4 and the bare Pt electrode. No clear peak is seen on the 0T-film, while clear redox peaks are seen on the 5T-film and their peak potentials are nearly coincident with those on the Pt electrode. This means that the Q/QH_2 redox reaction hardly proceeds on the 0T-film, but it easily proceeds on the 5T-film.

The PPy/SO_4 film undergoes reduction at about $-0.2\,\mathrm{V}$ and the film is conductive in the more positive potential range. The Q/QH_2 reaction occurs around $-0.1\,\mathrm{V}$, which is close to the reduction potential of the PPy/SO_4 film. It is thus considered that the redox response of the film itself is responsible for the Q/QH_2 electrode reaction. The 0T-film does not exhibit a clear response for the film reduction, but the 5T-film does. This difference is reflected in the redox response for the Q/QH_2 redox reaction.

20.3   Control of the Learning Effect

Organic conducting polymers such as PPy and polythiophene have novel electrochemical properties such as the electrical plasticity [20,21] and the

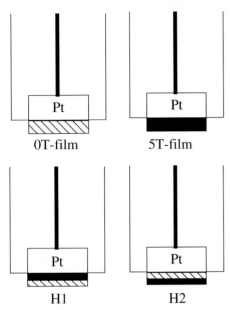

Fig. 20.8. Schematics of 0T-film, 5T-film and hybrid film (H1 and H2) electrodes

learning effect [22], attracting much attention as a material for intelligent molecular devices. Doping and undoping processes in the conducting polymers, involving large morphological and volume changes, proceed gradually by repeating redox cycles. Such gradual changes are considered to be related to the learning of neural elements [22] or the synaptic plasticity [21].

We prepared four kinds of PPy/TsO films with the same total passing charge of $1.0\,C\,cm^{-2}$ ($\sim 2.5\,\mu m$ thickness), as shown schematically in Fig. 20.8 [23]. The hybrid films H1 consist of the inner 5T-film ($0.5\,Ccm^{-2}$) and the outer 0T-film ($0.5\,Ccm^{-2}$), and vice versa for the film H2 .

The learning effect in the dopant-exchange process can be observed by measuring the successive CVs of the film in an electrolytic solution containing a different anion from the dopant [24]. The 30 cycles CVs of the 0T-film were measured in a $LiClO_4$ aqueous solution with a potential sweep range of $-0.6\,V$ to $+0.5\,V$. The reduction current of PPy/TsO around $-0.5\,V$ decreases with repeating potential sweep, while both anodic and cathodic currents around $0\,V$ increase; these correspond to the redox currents of PPy/ClO_4. This indicates that the PPy film undergoes the doping and undoping processes of ClO_4^- instead of TsO^-.

We performed the same experiment for the other films and obtained a "learning curve" for each film, which is the plot of the anodic current at $0\,V$ against the cycle number, as shown in Fig. 20.9. The anodic current of the 0T-film increases gradually up to $101\,mA$ at the last cycle. On the other hand, the 5T-film shows only a small increase up to $36\,mA$ at most. The two

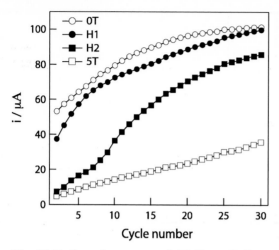

Fig. 20.9. Learning curves of 0T-film, 5T-film and the hybrid films during the dopant-exchange process

hybrid films show quite different learning curves. The learning curve of the H1 film is similar to that of the 0T-film, while it is noted that the curve of the H2 film starts at 7 mA and rises up to 86 mA, showing behavior that is intermediate between that of the 0T- and 5T-films.

The PPy/TsO film has an anisotropic molecular organization, while the PPy/ClO$_4$ film has an isotropic organization [18]. The dopant-exchange process between TsO$^-$ and ClO$_4^-$ thus involves considerable changes in the film morphology, resulting in the gradual progress during repeated potential sweeps. This is observed as the learning effect. In the potential sweep range of -0.6V to 0.5V for the dopant-exchange, the undoping of the TsO$^-$ ion easily proceeds in the H1 and 0T-films, while it hardly proceeds in the H2 and 5T-films, as mentioned above. These features are reflected in the learning curves. There exists a large gap between the learning curves of the 0T- and 5T-films, and the learning curve of the H2 film interpolates between these. These results demonstrate that the learning effect in the dopant-exchange process of the PPy/TsO film can be controlled by the MEP, and that the hybridization of the 0T- and 5T-films allows more precise control of the learning effect. This sould give a hint for the design of intelligent materials.

20.4   Accumulative Effect of the Magnetic Field

Here we describe the magnetic-field-induced (MFI) deactivation of the PPy film [25], which involves a novel effect of magnetic fields, i.e. an accumulative effect. Harada et al. studied the durability of polythiophene films during repeated redox cycles and found the deactivation and the degradation of the films [26]. The former is indicated by a positive shift of the anodic peak in

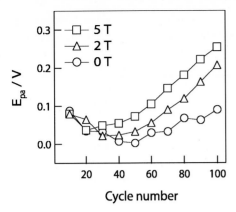

Fig. 20.10. Shifts of the anodic peak potentials in the CVs of the PPy/ClO$_4$ film at 0, 2 T and 5 T

the CV and the latter by a current decrease of the CV. Magnetic fields are expected to affect the morphological changes of polymer films during the redox cycles through diamagnetic orientation.

To examine the durability, the CVs of the PPy/ClO$_4$ films were measured for 100 cycles in a NaClO$_4$ aqueous solution in the potential range from 0.5 V to -0.5 V with and without a magnetic field. The anodic peak potentials E_{pa} of the CVs at 0, 2, and 5 T are plotted as a function of the cycle number in Fig. 20.10. No remarkable difference among these is seen in the first 30 cycles. The magnetic field effect appears around the 40th cycle, and then the peak potential difference with and without the magnetic fields increases with increasing cycle number and magnetic field. This result shows the MFI deactivation of the PPy film.

The fact that the highly positive polarization reactivates the film indicates that the MFI deactivation is caused by the accumulation of cations trapped within the film. It is presumed that the magnetic field changes the alignment of the molecular chains during the doping-undoping processes, and thereby the film morphology becomes different from that in zero magnetic field. Although such a morphological change would be small in one redox cycle, it is accumulated during repeated cycles, affecting the mobility of the dopants. Actually, the MFI deactivation in the PPy films appears clearly after the 40th cycle.

20.5   Future Perspective

We have demonstrated a new method using the MEP technique for the modification of electrochemical reaction media. The control of the morphology and the doping-undoping behavior is just the first step of the application of this method. The MEP-modified electrode will be applied to various elec-

trochemical reactions. The magnetic field response of the redox behavior of
the conducting polymer is a significant basic property for the application
to intelligent molecular devices. A characteristic of MFI deactivation is the
accumulative effect of magnetic fields. Such an effect will serve as a model
experiment for magnetic field effects on biological systems.

The combination of diamagnetic orientation and the MHD effect would
create a novel methed for controling the electrode reaction. The alignment of
polymers on an electrode surface is controlled by the diamagnetic orientation,
and the direction of mass transport is controlled by the MHD convection. Re-
actant species thereby approach the oriented polymer in a certain direction.
This technique could lead to the possibility of an asymmetric synthesis by
magnetoelectrolysis.

Acknowledgment

The experiments with magnetic fields were done in the High Field Laboratory
for Superconducting Materials, Tohoku University.

References

1. T.Z. Fahidy: J. Appl. Electrochem. **13**, 553 (1983)
2. R. Aogaki, K. Fueki and T. Mukaibo: Denki Kagaku **43**, 504 (1975)
3. I. Mogi, S. Okubo and Y. Nakagawa: J. Phys. Soc. Jpn. **60**, 3200 (1991)
4. I. Mogi, S. Okubo and M. Kamiko: Curr. Top. in Crystal Growth Res. **3**, 105 (1997)
5. K. Shinohara and R. Aogaki: Electrochemistry **67**, 126 (1999)
6. G. Maret, K. Dransfeld: In *Strong and Ultrastrong Magnetic Fields and Their Applications*, ed. by F. Herlach (Springer, Berlin, 1985) p. 143
7. G. Maret, N. Boccara and J. Kiepenheuer: *Biophysical Effects of Steady Magnetic Fields* (Springer, Berlin, 1986)
8. I. Mogi, G. Kido and Y. Nakagawa: Bull. Chem. Soc. Jpn. **63**, 1871 (1990)
9. I. Mogi, K. Watanabe and M. Motokawa: Electrochemistry **67**, 1051 (1999)
10. K. Watanabe, S. Awaji, T. Fukase, Y. Yamada, J. Sakuraba, F. Hata, C. K.Chong, T. Hasebe and M. Ishihara: Cryogenics **34**, 639 (1994)
11. R. Qian, J. Qiu and D. Shen: Synth. Met. **18**, 13 (1987)
12. K. Naoi: Denki Kagaku **63**, 109 (1995)
13. H. Hayashi, Y. Sakaguchi and H. Abe: Physica B **164**, 217 (1990)
14. I. Mogi, K. Watanabe and M. Motokawa: Synth. Met. **98**, 41 (1998)
15. I. Mogi, K. Watanabe and M. Motokawa: Mater. Trans. JIM **41**, 966 (2000)
16. R.M. Wightman, E.C. Paik, S. Borman and M.A. Dayton: Anal. Chem., **50**, 1410 (1978)
17. R.M. Wightman, M.R. Deakin, P.M. Kovach, W.G. Kuhr and K.J. Stutts: J. Electrochem. Soc. **131**, 1578 (1984)
18. G.R. Mitchell, F.J. Davis and C.H. Legge: Synth. Met. **26**, 247 (1988)
19. H. Morioka, T. Kimura and R. Aogaki: Trans. Mat. Res. Soc. Jpn. **25**, 65 (2000)

20. M. Iseki, K. Saito, M. Ikematsu, Y. Sugiyama, K. Kuhara and A. Mizukami: J. Electroanal. Chem. **358**, 221 (1993)
21. M. Iseki, K. Saito, M. Ikematsu, Y. Sugiyama, K. Kuhara and A. Mizukami: Mater. Sci. Eng. **C1**, 107 (1994)
22. K. Yoshino, T. Kuwabara and T. Kawai: Jpn. J. Appl. Phys. **29**, L995 (1990)
23. I. Mogi: Microelectron. Eng. **43**, 739 (1998)
24. I. Mogi: Bull. Chem. Soc. Jpn. **69**, 2661 (1996)
25. I. Mogi and K. Watanabe: Bull. Chem. Soc. Jpn. **70**, 2337 (1997)
26. H. Harada, T. Fuchigami and T. Nonaka: J. Electroanal. Chem. **303**, 139 (1991)

21 Highly Oriented Crystal Growth of Bi-Based Oxide High-T_c Superconductors in High Magnetic Fields

H. Maeda, W.P. Chen, K. Watanabe and M. Motokawa

Grain alignment is of great importance for achieving high critical current density in high-T_c superconductors. In recent years intensive research has been focused on obtaining texture in high-T_c superconductors and several techniques have been developed. Among these techniques, magnetic melt processing (MMP), i.e. oriented crystal growth in high magnetic fields, can be applied to both Y-based and Bi-based superconductors. Due to the susceptibility anisotropy of high-T_c superconductors along the crystal axes, some texture can be developed in high-T_c superconductors by melting and solidification in a high magnetic field.

For Bi-based superconductors, precursor composition plays a vital role in the texture development in MMP. Although obvious texture can be obtained by MMP in Bi2212 bulk material with Dy-substitution [1], Pb-substitution [2], MgO-doping and/or Ag-doping [3,4], no texture can be obtained in pure Bi2212 bulk material without any substitution or doping. In this article the effect of precursor composition on the oriented crystal growth of Bi-based superconductors in a high magnetic field is demonstrated in detail in the case of Ag-doping. Ag-doping is widely used in Bi-based superconductors and Ag is also used as a sheath material for various tapes. MMP is also applied to dip-coated monolayer Bi2212/Ag composite tapes with thickness over 60 μm, which shows that MMP can enhance the texture development and therefore increase the critical transport current density J_c in Bi2212 tapes of reasonable thickness.

21.1 Ag-doped Bi2212 Bulk Materials

21.1.1 Experimental Procedure

$Bi_{2.2}Sr_{1.8}CuO_x$ (Bi2201) and $CaCuO_2$ powders were first synthesized by solid state reactions, using high purity chemicals of Bi_2O_3, $SrCO_3$, $CaCO_3$ and CuO. For Bi2201, the synthesis was carried out at 750–800 °C for 30 h in air with two intermediate grindings; for $CaCuO_2$, it was carried out at 800–950 °C for 40 h in air with two intermediate grindings. Bi2201 and $CaCuO_2$ powders were mixed in a ratio of 1:1 and then 10 wt% Ag powder was added to one half of the mixture. After milling in an agate mortar, both of the mixtures with and without Ag-doping were pressed into pellets of 10 mm diameter and 5 mm thickness. Contained in a silver crucible, each pellet was heat-treated

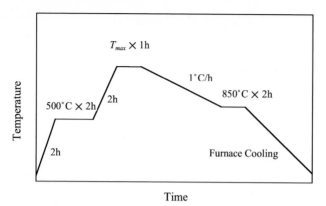

Fig. 21.1. Temperature profile for magnetic melt processing of Bi2212 bulk material

in a vertical tube furnace installed in a helium-free solenoid-type superconducting magnet with a room-temperature clear bore of 52 mm [5]. During the heat-treatment a constant vertical magnetic field H_a of 9 T was applied with pure O_2 gas flowing at a rate of 100 ml/min. The temperature profile for the heat-treatment is shown in Fig. 21.1. The maximum temperature, T_{max}, was changed and investigated in the range of 880–920 °C. The optimum T_{max} is within this temperature range.

For the Ag-doped bulk samples vertical magnetic fields of 0 (without field), 3 and 6 T were also applied through all the heat-treatments, in which the T_{max} was 893 °C.

21.1.2   Crystal Growth and Properties

The bulk samples were cut horizontally at 2 mm in depth from the top surface for X-ray diffraction analysis. For samples that were grown in a H_a of 9 T, two different kinds of diffraction patterns were observed. Ag-doped were samples processed with T_{max} in the range of 890–903 °C exhibit a diffraction pattern with enhanced (00l) reflection lines, while Ag-doped samples processed with other T_{max} values and all pure Bi2212 samples without Ag doping only exhibit a normal diffraction pattern. Two representative patterns are shown in Fig. 21.2; these were recorded for pure Bi2212 (T_{max}: 905 °C) and Ag-doped Bi2212 (T_{max}: 893 °C), respectively. Some reflection lines from the silver crucible are included.

As the enhancement of the (00l) reflection lines represents a grain alignment with the a-b plane parallel to the horizontal cutting surface, it shows that the magnetic field (9 T) initiated oriented crystal growth in those Ag-doped samples processed with T_{max} in the range of 890–903 °C. Bi2212 grains are of plate-like shape with the c-axis in the direction of the thickness. The oriented crystal growth in these Ag-doped samples can be observed. Figure 21.3 is a scanning electron microscope (SEM) fractograph taken from

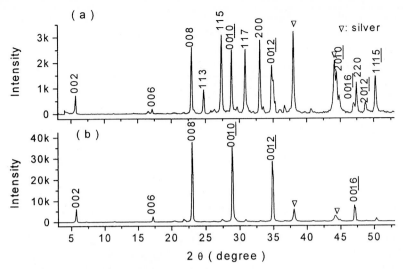

Fig. 21.2. X-ray diffraction pattern of (**a**) pure Bi2212 and (**b**) Ag-doped Bi2212. Both samples were processed in a magnetic field H_a of 9 T

Fig. 21.3. SEM fractograph taken from a vertical fracture surface of an Ag-doped sample prepared in a magnetic field H_a of 9 T (T_{max}: 893 °C)

a vertical fracture surface of the sample whose X-ray diffraction pattern is shown in Fig. 21.3b. In this sample the *a-b* plane of most grains is oriented nearly in the direction perpendicular to the magnetic field H_a applied during its preparation. Bi2212 textured bulk materials prepared by the floating-zone method often have the problem of low density. The samples prepared by MMP, however, show a very dense microstructure.

High-T_c cuprate superconductors are anisotropic in nature and their current carrying capacity is much higher along the a-b plane than along the c-axis. Anisotropy in textured Ag-doped Bi2212 bulk samples was evaluated by a magnetic hysteresis loop measurement. Cubes of $2 \times 2 \times 2\,\mathrm{mm}^3$ were cut from the central parts of the bulk samples and the measurements were carried out on these at $5\,\mathrm{K}$ along two directions: with the measuring field H_m parallel and perpendicular to the magnetic field H_a applied during crystal growth, respectively. It is found that an obvious anisotropy in magnetization hysteresis loops appears in those Ag-doped samples whose X-ray diffraction patterns have shown oriented crystal growth. For the sample shown in Fig. 21.2b the results are given in Fig. 21.4. It is very clear that ΔM, the width of the magnetic hysteresis loop, in the direction $H_m \parallel H_a$ (named $\Delta M_{\parallel}$) is much bigger than that in the direction $H_m \perp H_a$ (named $\Delta M_{\perp}$). The ratio $\Delta M_{\parallel}/\Delta M_{\perp}$ at $0\,\mathrm{T}$ (namely the magnetization anisotropy factor) is over 4. This value is almost the same as those of high-J_c Bi2212 tapes with thin superconducting oxide layers. According to Bean's model, the intragranular J_c value is proportional to ΔM. Thus the bulk has much higher critical current density in the direction perpendicular to H_a.

It should be pointed out that both SEM observation and magnetic hysteresis measurements reveal that no grain alignment is obtained by the application of a H_a of $9\,\mathrm{T}$ in pure Bi2212 and Ag-doped Bi2212 bulk samples with T_{max} outside the range of 890–903 °C, in accordance with the X-ray diffraction result. It is found that in an O_2 atmosphere the melting point of an Ag-doped equimolar mixture of Bi2201 and $CaCuO_2$ is around 890 °C. No melting occurs in the Ag-doped bulk samples prepared with T_{max} in the range of 880–890 °C and it is reasonable that no texture can be developed in these. In Ag-doped samples prepared with T_{max} in the range of 903–920 °C, partial melting occurs, while no oriented crystal growth can be initiated by the magnetic field. This indicates that the crystal growth of Bi-based superconductors in a high magnetic field is sensitive to the maximum temperature T_{max} of the melt processing. Only when the melt is below a certain temperature can obvious texture be developed during solidification. Some irreversible change must occur when the melt is above a certain temperature and this prevents oriented crystal growth initiated by a magnetic field during solidification.

It is well known that Ag-doping considerably decreases the melting point of Bi2212. For the pure Bi2212 samples, inhomogeneous melting occurs when these are processed with T_{max} in the range of 890–903 °C [6]. Figure 21.5 shows a schematic illustration for the inhomogeneous melting in pure Bi2212 samples. The central part of the sample remained solid, while partial melting-solidification occurred in the outer part of the sample. This inhomogeneous melting may have resulted from the decrease in the melting point of the outer part of the sample due to contact with the silver crucible. Pure Bi2212 bulk samples must be processed with T_{max} above 903 °C to obtain homogeneous

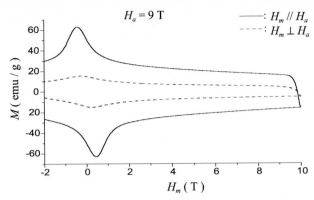

Fig. 21.4. Magnetic hysteresis loops obtained for the Ag-doped sample at 5 K with $H_\mathrm{m} \parallel H_\mathrm{a}$ and $H_\mathrm{m} \perp H_\mathrm{a}$, respectively

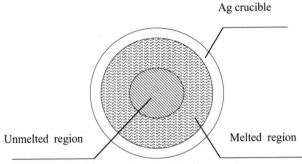

Fig. 21.5. A schematic illustration for inhomogeneous melting observed in pure Bi2212 bulk material

melting in the entire sample, but it is impossible for the texture development according to the result obtained for the Ag-doped samples. Thus some modification in precursor composition is always needed for oriented crystal growth of Bi2212 bulk samples in a magnetic field.

The degree of texture of Bi2212 bulk material can be evaluated quantitatively by the magnetization anisotropy factor and by the ratio of I_{008}/I_{115}, where I_{008} and I_{115} are the intensity of the (008) and (115) reflection lines of the Bi2212 phase, respectively. The results obtained for the Ag-doped bulk samples prepared at 0, 3, 6 and 9 T (T_max: 893 °C) are shown in Fig. 21.6. It is evident that there exists a strong correlation between the degree of texture and the magnetic field H_a up to 9 T. The higher the magnetic field, the better is the texture developed. This suggests that still better texture can be expected with higher magnetic field.

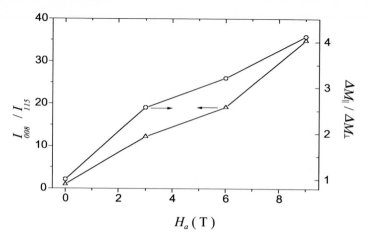

Fig. 21.6. Correlation between field strength H_a and degree of texture evaluated by I_{008}/I_{115} and $\Delta M_\parallel / \Delta M_\perp$ at 0 T, respectively

21.2   Bi2212/Ag Tapes

21.2.1   Experimental Procedure

Monolayer Bi2212/Ag composite tapes were prepared by a dip-coating method, then wrapped with an Ag-foil and finally flattened with the pre-annealing-intermediate-rolling (PAIR) process (here called PAIR tape). The process has been reported in detail in [7]. These Bi2212 layers were 60–80 μm thick. Some of the tapes were already pre-heat-treated by the conventional partial melt-solidification process without magnetic fields (here called H-PAIR tape) in order to examine how the highly textured grain alignment has an effect on the texture growth in high magnetic fields. The H-PAIR tapes have highly textured grains in the thin interface layer between Bi2212 and silver, although in the central part of the Bi2212 oxide core the grain alignment is not so high. Cut into short pieces of 15 mm length, the samples were set horizontally or vertically in a vertical tube furnace installed in the superconducting magnet [5], and heat-treated in a constant vertical magnetic field H_a of 0, 2.5, 5, 7.5 or 10 T. The heat treatment was carried out by a standard partial melt-solidification process with a cooling rate of 2 °C/h (as shown in Fig. 21.7) in a pure O_2 gas flow at a rate of 100 ml/min.

21.2.2   Microstructure and Properties

Because the magnetic field has a tendency to orient Bi2212 grains perpendicular to the field direction, in horizontally set tapes the texture development induced by the vertical magnetic field is in the same direction as that induced by the Bi2212-silver interface, thus the texture development should be enhanced. The critical transport current density J_c of these horizontally

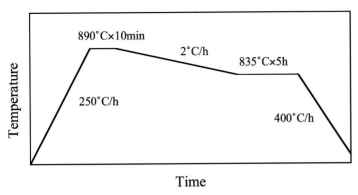

Fig. 21.7. Temperature profile for melt processing of Bi2212/Ag tapes in magnetic fields

set tapes is found to depend strongly on the magnetic field H_a applied in their heat treatment. However, the relationship between J_c and H_a is not monotonous, as seen in Fig. 21.8 [8], which shows the J_c at 4.2 K/10 T versus H_a characteristics for the PAIR and H-PAIR tapes set horizontally and vertically. It is evident that J_c increases greatly with increasing H_a in the range 0–7.5 T. This indicates that J_c for Bi2212/Ag tapes can be effectively increased by MMP through enhancement in texture development. When $H_a = 10$ T, however, J_c decreases dramatically. Liu et al. obtained a similar result in Bi2212/Ag thick films (125 μm in thickness) [9]. In our previous work [4], we have observed a monotonic increase in the texture development in Ag-doped Bi2212 bulk materials with H_a in the range 0–9 T. Up to now we cannot understand the J_c decrease in the tapes processed in a H_a of 10 T. It is also clear that the J_c values of the H-PAIR tapes are slightly higher than those of the PAIR tapes, although the magnetic field dependence of J_c is the same for both tapes. This result is of great interest in showing that the effect of the pre-textured grain alignment remains after repeated heat treatment. A similar idea has been proposed for PAIR tapes by Kitaguchi et al. [10]. On the other hand, the vertically mounted tapes exhibit J_c versus H_a characteristics that are quite different from those of horizontally mounted tapes, as shown in Fig. 21.8. It can be seen that J_c does not change so much with H_a as it does in the horizontal tapes. In the vertical tapes, grain orientation induced by the Bi2212-silver interface is perpendicular to that induced by a vertical magnetic field. The texture induced by the Bi2212-silver interface is strong enough to destroy the texture by MMP. The MMP mainly has an effect on the texture growth in the central part of the tapes. Since transport current flows through the very thin interface grain alignment region in these tapes, J_c does not change so much with H_a.

Figure 21.9 shows two typical SEM micrographs obtained on polished longitudinal sections of two PAIR tapes mounted horizontally and vertically, respectively, in a H_a of 5 T. The thickness of the tapes is about 60 μm. In the

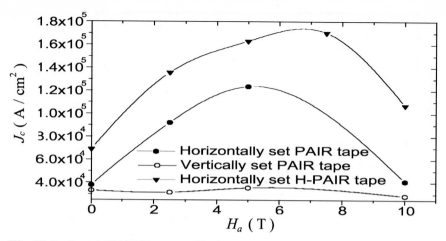

Fig. 21.8. J_c at 4.2 K/10 T versus. H_a characteristics for PAIR and H-PAIR tapes set horizontally and vertically

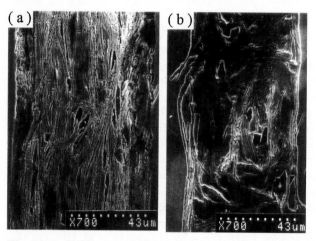

Fig. 21.9. SEM micrograph obtained on longitudinal section of (**a**) a PAIR tape set horizontally and (**b**) a PAIR tape set vertically in H_a of 5 T

horizontal tape, a uniform texture exists in the entire cross-sectional area. It shows that a magnetic field can enhance the texture development in the tapes and high J_c can be obtained in Bi2212/Ag tapes of reasonable thickness. In the vertically set tape, on the other hand, a large portion of the area is non-textured. This results from the interference between the texture development induced by the Bi2212-silver interface and that induced by the magnetic field.

The difference in J_c (ΔJ_c) between tapes mounted horizontally and vertically in the same magnetic field has been estimated for the PAIR and H-PAIR tapes. According to the ΔJ_c versus H_a characteristics, ΔJ_c increases dramatically with H_a up to 5–7.5 T, and the values are almost the same for

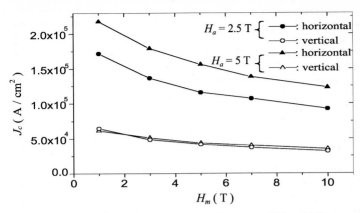

Fig. 21.10. J_c versus. measuring field H_m at 4.2 K for PAIR tapes mounted horizontally and vertically in H_a of 2.5 and 5 T

the PAIR and H-PAIR tapes. This indicates that MMP strongly develops grain alignment mainly in the center part of the Bi2212 core, as described above.

Figure 21.10 shows J_c versus the measuring field H_m at 4.2 K for the PAIR tapes mounted horizontally and vertically in H_a of 2.5 and 5 T. For all the tapes the J_c values gradually decrease with increasing H_m. There is no different behavior in the tapes grown by MMP, suggesting that MMP cannot have an effect on Bi2212 grain formation, except highly oriented grain growth.

21.3 Conclusion

High magnetic fields are effective for initiating highly oriented crystal growth in Bi2212 bulk material and tapes. For Bi2212 bulk material, Ag-doping leads to homogeneous melting at a lower temperature and thus is helpful for texture development. The texture development increases notably with increasing magnetic field in the range of 0–9 T. For monolayer Bi2212/Ag composite tapes with thickness over 60 μm, when the tapes are set perpendicularly to the magnetic field the texture development is obviously enhanced and the transport critical current density J_c increases dramatically with increasing magnetic field strength H_a up to 7.5 T. When H_a increases from 7.5 to 10 T, however, J_c decreases sharply.

Acknowledgement

We would like to thank Prof. P.X. Zhang of the Toyohashi University of Technology for valuable discussions, and Drs. H. Kumakura and H. Kitaguchi of the National Research Institute for Metals for the tape sample preparation. This research is supported by the Japan Science and Technology Corporation.

322   H. Maeda et al.

References

1. S. Stassen, A. Vanderschueren, R. Cloots, A. Rulmont and M. Ausloos: J. Cryst. Growth **166**, 281 (1996)
2. W.P. Chen, H. Maeda, K. Kakimoto, P.X. Zhang, K. Watanabe, M. Motokawa, H. Kumakura and K. Itoh: J. Cryst. Growth **204**, 69 (1999)
3. S. Pavard, C. Villard, R. Bourgault and R. Tournier: Supercond. Sci. Technol. **11**, 1359 (1998)
4. W.P. Chen, H. Maeda, K. Kakimoto, P.X. Zhang, K. Watanabe and M. Motokawa: Physica C **320**, 96 (1999)
5. K. Watanabe, S. Awaji and K. Kimura: Jpn. J. Appl. Phys. **36**, L673 (1997)
6. W.P. Chen, H. Maeda, P.X. Zhang, S. Awaji, K. Watanabe and M. Motokawa: J. Low Temp. Phys. **117**, 771 (1999)
7. H. Miao, H. Kitaguchi, H. Kumakura, K. Togano, T. Hasegawa and T. Koizumi: Physica C **303**, 81 (1998)
8. W.P. Chen, H. Maeda, K. Watanabe, M. Motokawa, H. Kitaguchi and H. Kumakura: Physica C **324**, 172 (1999)
9. H.B. Liu, P.J. Ferreira and J.B. Vander Sande: Physica C **316**, 261 (1999)
10. H. Kitaguchi, H. Miao, H. Kumakura and K. Togano: Physica C **320**, 71(1999)

Index

Printing (Computer to Film): Saladruck Berlin
Binding: Stürtz AG, Würzburg